# Clinical Photomedicine

**CLINICAL DERMATOLOGY**

*Series Editor*

ALAN R. SHALITA, M.D.

*Professor and Chairman*
*Department of Dermatology*
*State University of New York*
*Health Science Center at Brooklyn*
*Brooklyn, New York*

1. Cutaneous Investigation in Health and Disease: Noninvasive Methods and Instrumentation, *edited by Jean-Luc Lévêque*
2. Irritant Contact Dermatitis, *edited by Edward M. Jackson and Ronald Goldner*
3. Fundamentals of Dermatology: A Study Guide, *Franklin S. Glickman and Alan R. Shalita*
4. Aging Skin: Properties and Functional Changes, *edited by Jean-Luc Lévêque and Pierre G. Agache*
5. Retinoids: Progress in Research and Clinical Applications, *edited by Maria A. Livrea and Lester Packer*
6. Clinical Photomedicine, *edited by Henry W. Lim and Nicholas A. Soter*

*ADDITIONAL VOLUMES IN PREPARATION*

Cutaneous Antifungal Agents: Selected Compounds in Clinical Practice and Development, *edited by John W. Rippon and Robert A. Fromtling*

Oxidative Stress in Dermatology, *edited by Jürgen Fuchs and Lester Packer*

Diseases of Connective Tissue in the Skin, *edited by Charles M. Lapiere and Thomas Krieg*

# Clinical Photomedicine

edited by

**Henry W. Lim, M.D.**
**Nicholas A. Soter, M.D.**

*The Ronald O. Perelman*
*Department of Dermatology*
*New York University School of Medicine*
*New York, New York*

**Marcel Dekker, Inc.** **New York • Basel • Hong Kong**

**Library of Congress Cataloging-in-Publication Data**

Clinical photomedicine / edited by Henry W. Lim, Nicholas A. Soter.
p. cm. -- (Clinical dermatology;6)
Includes bibliographical references and index.
ISBN 0-8247-8862-1 (alk. paper)
1. Phototherapy. 2. Photochemotherapy. 3. Photobiology. I. Lim, Henry W. II. Soter, Nicholas A. III. Series.
[DNLM: 1. Photosensitivity disorders. 2. Light--adverse effects. 3. Photochemotherapy. 4. Phototherapy. 5. Skin Diseases--therapy. W1 CL69L v.6 1993 / WB 480 C6405 1993]
RM837.C58 1993
616.5'0631--dc20
DNLM/DLC
for Library of Congress **92-48428**
**CIP**

This book is printed on acid-free paper.



MARCEL DEKKER, INC.
270 Madison Avenue, New York, New York 10016

Current printing (last digit):
10 9 8 7 6 5 4 3 2 1

PRINTED IN THE UNITED STATES OF AMERICA

To Mamie
for her unending patience and support
and Christopher and Kevin
for the joy they bring

—H.W.L.

To my mother

—N.A.S.

# Series Introduction

During the past decade there has been a vast explosion in new information relating to the art and science of dermatology as well as fundamental cutaneous biology. Furthermore, this information is no longer of interest only to the small but growing specialty of dermatology. Scientists from a wide variety of disciplines have come to recognize both the importance of skin in fundamental biological processes and the broad implications of understanding the pathogenesis of skin disease. As a result there is now a multidisciplinary and world-wide interest in the progress of dermatology.

With these factors in mind, we have undertaken a new series of books specifically oriented to dermatology. The series will be purposely broad in focus and will range from pure basic science to practical, applied clinical dermatology. Thus, while there will be something for everyone, all editions in the series should ultimately prove to be valuable additions to the dermatologist's library.

*Clinical Photomedicine*, edited by Drs. Henry W. Lim and Nicholas A. Soter, is an outstanding example of the type of scholarly and practical book we hope will be the hallmark of this series. The editors have assembled a group of authors who are internationally known for their expertise. They have put together a work of both scientific and clinical photomedicine that will be a most valued addition to the libraries of the clinical practitioner as well as of great use to residents, fellows, and other students of photomedicine.

I sincerely hope that you will enjoy reading these books as much as I have enjoyed planning them with the authors and editors of this series.

*Alan R. Shalita*
SUNY Downstate Medical Center
Brooklyn, New York

# Preface

The principles and practice of photomedicine have expanded over the past few years. Observations on the effects of ultraviolet radiation on mediators of inflammation and the immune system have led to better understanding of the sunburn reaction, the photoaging process, and the development of skin cancer. Studies on optics of the skin have led to the new development of laser therapy for various dermatologic conditions, including its use in photodynamic therapy for skin cancers. The clinical application of photochemotherapy has been expanded to include its use as an extracorporeal agent in the treatment of cutaneous T-cell lymphoma and other diseases.

The purpose of this book is to document the expanding field of photomedicine by leading experts in their respective areas. This book is aimed primarily at physicians in the clinical practice of dermatology with an interest in photomedicine. It also will be of value for resident physicians and fellows.

The book is organized into six sections. The introductory section includes the basic principles of photomedicine, including skin optics, light sources, and pharmacology. The next section covers the acute and chronic effects of light on normal skin, including sunburn reaction, photoaging, and skin cancer. Photoimmunology is considered in the next section. Subsequent sections of the book discuss clinical photosensitivity diseases and topical and systemic photoprotection. The last section covers the use of light as a therapeutic agent, which includes phototherapy, photochemotherapy, extracorporeal photopheresis, and the use of lasers in photomedicine and photodynamic therapy.

It is our hope that this book adequately reflects the exciting developments that have occurred in recent years in our understanding of the effects of light on human skin, the resulting extension of knowledge of photosensitivity disorders, and the use of light as a therapeutic agent.

Finally, the authors would like to acknowledge the support from Therakos, Glaxo Inc., and Neutrogena Dermatologics in defraying the cost of color photographs.

*Henry W. Lim*
*Nicholas A. Soter*

# Contents

# Contributors

**R. Rox Anderson, M.D.** Associate Professor, Department of Dermatology, Harvard Medical School, Boston, Massachusetts

**Thomas F. Anderson, M.D.** Associate Professor, Department of Dermatology, University of Michigan Medical School, Ann Arbor, Michigan

**Robin Ashinoff, M.D.** Assistant Professor, Department of Dermatology, New York University School of Medicine, New York, New York

**Paul R. Bergstresser, M.D.** Professor and Chairman, Department of Dermatology, University of Texas Southwestern Medical Center, Dallas, Texas

**Michael H. Berkson, M.D.** Fellow, Department of Dermatology, University of Pennsylvania, Philadelphia, Pennsylvania

**Mary Ruth Buchness, M.D.** Assistant Chief, Dermatology Service, Department of Veterans Affairs Medical Center, and Assistant Professor, The Ronald O. Perelman Department of Dermatology, New York University School of Medicine, New York, New York

**Wai-Kwong Cheong, M.D.** Dermatologist, Department of Dermatology, National Skin Centre, Singapore

**Serge A. Coopman, M.D.** Clinical Consultant, Department of Dermatology, University of Leuven, Leuven, Belgium

**Ponciano D. Cruz, Jr., M.D.** Assistant Professor, Department of Dermatology, University of Texas Southwestern Medical Center, Dallas, Texas

**Ira C. Davis, M.D.** Fellow in Dermatologic Surgery, Department of Dermatology, Bowman Gray School of Medicine, Winston-Salem, North Carolina

**Vincent A. De Leo, M.D.** George Fox Assistant Professor, Department of Dermatology, Columbia University, New York, New York

**Craig A. Elmets, M.D.** Associate Professor, Department of Dermatology, Case Western Reserve University, Cleveland, Ohio

**Roy G. Geronemus, M.D.** Associate Professor, Department of Dermatology, New York University School of Medicine, New York, New York

**Barbara A. Gilchrest, M.D.** Professor and Chairman, Department of Dermatology, Boston University School of Medicine, Boston, Massachusetts

**John Hawk, M.D., F.R.C.P.** Head, Department of Photobiology, St. John's Institute of Dermatology, St. Thomas's Hospital, London, England

**Herbert Hönigsmann, M.D.** Professor of Dermatology, Division of Special and Environmental Dermatology, University of Vienna, Vienna, Austria

**Takeshi Horio, M.D.** Associate Professor, Department of Dermatology, Kansai Medical University, Osaka, Japan

**Irene E. Kochevar, Ph.D.** Associate Professor, Wellman Laboratories of Photomedicine, Harvard Medical School, Boston, Massachusetts

**Henry W. Lim, M.D.** Associate Professor, The Ronald O. Perelman Department of Dermatology, New York University School of Medicine, New York, New York

**Robin Marks, M.B.B.S., M.P.H., F.R.A.C.P., F.A.C.D.** Consultant Dermatologist, Anti-Cancer Council of Victoria, Melbourne, Victoria, Australia

**Thomas Meola, M.D.** Fellow in Photomedicine, The Ronald O. Perelman Department of Dermatology, New York University School of Medicine, New York, New York

**Warwick L. Morison, M.D.** Associate Professor, Department of Dermatology, Johns Hopkins Medical Institutions, Baltimore, Maryland

**Allan R. Oseroff, M.D., Ph.D.** Chairman, Department of Dermatology, Roswell Park Cancer Institute, Buffalo, New York

**Madhu A. Pathak, M.B.B.S. (Tech), M.S., Ph.D.** Senior Associate in Dermatology, Harvard Medical School, Boston, Massachusetts

**Gary S. Rogers, M.D.** Associate Professor, Department of Dermatology, Boston University School of Medicine, Boston, Massachusetts

**Alain H. Rook, M.D.** Associate Professor of Dermatology, University of Pennsylvania, Philadelphia, Pennsylvania

**Shigeru Sassa, M.D., Ph.D.** Associate Professor and Head of Laboratory, Department of Biochemical Hematology, Rockefeller University, New York, New York

**Jerome L. Shupack, M.D.** Professor, Clinical Dermatology, The Ronald O. Perelman Department of Dermatology, New York University School of Medicine, New York, New York

**Arthur J. Sober, M.D.** Associate Professor, Department of Dermatology, Harvard Medical School, Boston, Massachusetts

**Nicholas A. Soter, M.D.** Professor of Dermatology, The Ronald O. Perelman Department of Dermatology, New York University School of Medicine, New York, New York

**Robert S. Stern, M.D.** Associate Professor and Dermatologist, Department of Dermatology, Beth Israel Hospital, Harvard Medical School, Boston, Massachusetts

**Matthew J. Stiller, M.D.** Assistant Professor of Dermatology, The Ronald O. Perelman Department of Dermatology, New York University School of Medicine, New York, New York

**Benjamin R. Vowels, M.D.** Research Assistant Professor, Department of Dermatology, University of Pennsylvania, Philadelphia, Pennsylvania

# I

# Introduction

# 1

# Basic Principles in Photomedicine and Photochemistry

**Irene E. Kochevar**
*Harvard Medical School, Boston, Massachusetts*

Skin is exposed daily to ultraviolet, visible, and infrared radiation from multiple sources including the sun, interior lighting, occupational light sources, and phototherapy systems. Photomedicine is concerned with the responses of tissue to these types of light as well as with the use of light for therapeutic purposes. Responses of skin to light are initiated by the absorption of ultraviolet and visible radiation by molecules in skin. Subsequently, the photoactivated molecules are converted into new compounds (photoproducts) that cause a series of biochemical changes in the tissue such as induction of enzyme activities, secretion of cytokines, and repair of damaged structures. After time, tissue changes can be observed by histologic examination and become apparent as lesions on the skin surface.

Of special importance is the concept that the responses of skin to light are strongly influenced by the color, or wavelength, of light used. Not all wavelengths have the same effects; in fact, specific wavebands are responsible for specific responses. The wavelength of light is important in two ways. First, the wavelength of light controls the depth to which the light energy penetrates into skin, and, second, each of the types of molecules in skin absorb specific wavelengths. Of course, the magnitude of a cutaneous response also reflects the amount of light delivered. However, the magnitude of the response usually does not depend on the rate at which the light energy is delivered.

This chapter focuses on the basic concepts involved in the interaction of ultraviolet and visible radiation with biomolecules. The photochemical reactions of molecules in skin are emphasized. Familiarity with these concepts will lead to a better understanding of the effects of light on skin and of the applications of light in dermatology. The concepts are general and applicable to other light-induced biological processes such as vision and photosynthesis.

## I. WAVELENGTHS AND ENERGY OF LIGHT

Ultraviolet and visible radiation are part of the spectrum of electromagnetic radition (EMR), which also includes very short wavelength radiation such as x-rays and gamma rays and very long wavelength radiation such as microwaves and radiowaves (Table 1). Electromagnetic radiation can be described either as a continuous wave or as a series of packets of energy called photons. The two descriptions are complementary, and each can be best applied to specific aspects of the interaction of light with tissue. In the wave description, EMR is conceived as a continuous wave of regular oscillations of electric and magnetic fields. These fields are perpendicular to each other and to the direction of propagation.

The energy, $E$, of EMR is directly proportional to the frequency, $\nu$, of the oscillations of the waves (Eq. 1)

$$E = h\nu \tag{1}$$

**Table 1** Wavelengths and Energies Associated with Regions of the Electromagnetic Spectrum

| Typical Wavelength (nm) | Spectral Region | Photon Energy (J) |
|---|---|---|
| 0.1 | X-ray | $2.0 \times 10^{-15}$ |
| 10 | Soft x-ray | $2.0 \times 10^{-17}$ |
| 100 | Vacuum UV | $2.0 \times 10^{-18}$ |
| 250 | UVC | $8.0 \times 10^{-19}$ |
| 300 | UVB | $6.6 \times 10^{-19}$ |
| 350 | UVA | $5.6 \times 10^{-19}$ |
| 400 | Violet | $5.0 \times 10^{-19}$ |
| 470 | Blue | $4.2 \times 10^{-19}$ |
| 530 | Green | $3.7 \times 10^{-19}$ |
| 600 | Yellow | $3.2 \times 10^{-19}$ |
| 700 | Red | $2.9 \times 10^{-19}$ |
| 1000 | Near infrared | $1.9 \times 10^{-19}$ |
| 100,000 | Far infrared | $1.9 \times 10^{-21}$ |
| $>10^6$ | Microwaves and radiowaves | $<1.9 \times 10^{-22}$ |

The proportionality constant is Planck's constant, $h$, which equals $6.63 \times 10^{-34}$ J · sec and $\nu$ has units of $sec^{-1}$. Because the product of the frequency, $\nu$, and wavelength, $\lambda$, of EMR is equal to the speed of light in a vacuum, $c$, the energy of EMR is inversely proportional to its wavelength (Eq. 2),

$$E = \frac{hc}{\lambda} \tag{2}$$

where $E$ is the energy of photon (J), $c$ is the speed of light in a vacuum ($3 \times 10^{10}$ cm/sec), and $\lambda$ is the wavelength in cm. Thus, Equation 2 connects the wavelength from the wave description of EMR to the photon of the particle description. The unit of length most often used for wavelengths in the visible and ultraviolet range of EMR is the nanometer (nm, $1 \times 10^{-9}$ m). According to Equation 2, a photon of 300 nm radiation has twice the energy of a photon of 600 nm radiation.

The ultraviolet (UV) and visible portions of the EMR spectrum are of interest in photobiology because photons of light in this wavelength range are sufficiently energetic to cause photochemical reactions in biological molecules. The UV portion of the electromagnetic spectrum that is important in photomedicine is divided into three regions: UVC, 200–290 nm, UVB, 290–320 nm; and UVA, 320–400 nm (Fig. 1). The lower wavelength limit of UVC was designated as

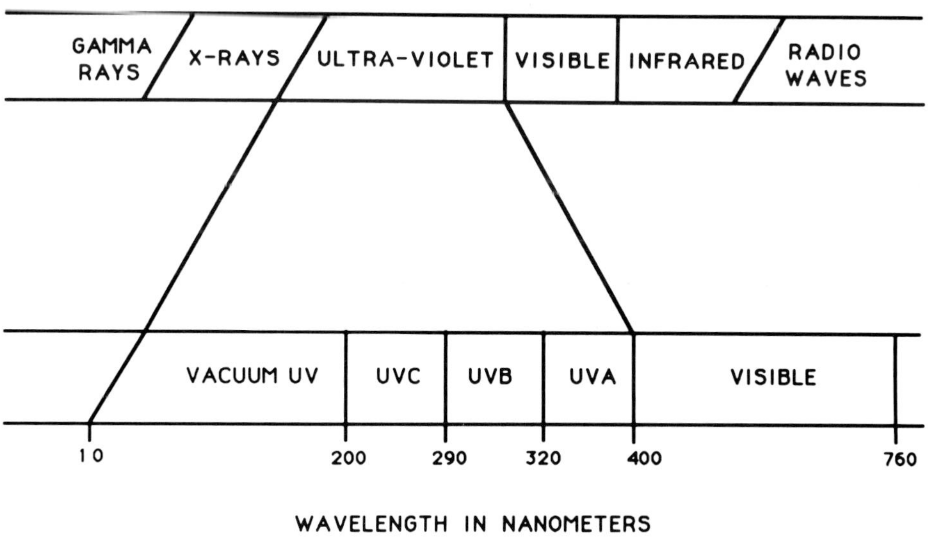


**Figure 1** Wavelength regions of the electromagnetic spectrum.

200 nm because shorter wavelengths cause ionization of many biological molecules and are absorbed by atmospheric oxygen. Radiation biology is the study of the effects of ionizing radiation on tissue; photobiology concerns the biological responses to nonionizing radiation initiated by photochemical reactions. The division between UVC and UVB was established at 290 nm because below this wavelength all radiation from the sun is absorbed by the ozone in the stratosphere. UVC radiation is highly mutagenic and toxic to cells and also is referred to as germicidal or far UV radiation. Germicidal lamps emit mainly at 254 nm.

The UVA portion of the spectrum was divided from UVB at 320 nm because UVA is not strongly absorbed by protein and nucleic acids and does not cause erythema in normal skin at moderate doses in the absence of photosensitizing chemicals. Both UVB and UVC are absorbed by protein and nucleic acids, and cause erythema in skin upon exposure to moderate doses. UVB is also referred to as sunburn radiation and UVA is also called black light. Other nomenclature used by photobiologists include calling the range 200–300 nm "near-UV" and the range 300–400 nm "far UV." It is important to remember that all wavelengths within a particular range (UVA, UVB, UVC) do not have equivalent effects. In fact, in recognition of the greater biological activity of the shorter wavelengths in the UVA range, the wavelengths 320–340 nm are beginning to be called UVA-II and the range 340–400 nm UVA-I.

The visible portion of the EMR spectrum includes wavelengths between 400 and approximately 760 nm. These are the wavelengths that the human retina is exposed to after shorter wavelengths are filtered out by other ocular structures, and to which the retina responds via a photochemical reaction. The colors associated with the visible wavelengths are given in Table 1; note that the photon energy decreases as the wavelength increases.

The energies associated with photons in the UV and visible range are high enough to initiate photochemical reactions. Some reactions require high-energy UV photons such as breaking of carbon halogen bonds in certain photosensitizing drugs. Other reactions, such as isomerization or energy transfer to oxygen, are possible when the chromophore is excited with longer visible wavelengths. Photons of infrared radiation do not have sufficient energy to promote the vast majority of biomolecules to electronically excited states and, therefore, they do not induce photochemical changes. Infrared radiation can be absorbed by molecules to produce molecules with extra vibrational energy and, when the infrared radiation is of sufficiently high intensity, can cause skin damage by thermal effects rather than by photochemical reactions.

In photobiology, the joule (J) is the unit of energy used most frequently to express the amount of light energy. The energy of light delivered to an area of tissue is called the fluence or, less correctly, the dose; it is expressed typically in $J/cm^2$ or $J/m^2$. The rate of delivery of light is expressed in watts (W), which equal J/sec. The rate of delivery of energy per area is called the irradiance or intensity, and the units are $W/cm^2$ or $W/m^2$. Fluence and irradiance are related as shown in Equation 3.

$$\underset{(J/cm^2)}{\text{Fluence}} = \underset{(W/cm^2)}{\text{Irradiance}} \times \underset{(sec)}{\text{Time}} \qquad (3)$$

## II. ABSORPTION OF LIGHT BY MOLECULES IN SKIN

For light to cause a response in skin, the photons incident on the surface must transmit into the tissue, be absorbed by biomolecules, and initiate chemical reactions in the tissue. The light-absorbing molecules are called chromophores. The path of light into and through skin is discussed in Chapter 2. In the absorption process, the electromagnetic energy of the photon is converted into chemical bonding energy of the molecule. The photon is consumed and its energy is now contained in the molecule. After absorbing the photon, the molecule has extra energy and can change chemically to form products (photochemistry), emit light, or generate heat. The most important of these processes in photomedicine is photochemistry because photochemical reactions of biological molecules initiate skin responses to light.

### A. Electronic States of Molecules

The molecular structure of a chemical compound determines which wavelengths of radiation the compound can absorb. The nuclei of the molecule form a framework and electrons are found in defined volumes, called orbitals, in the space around the nuclear framework. According to quantum mechanics, a series of electronic states exists for each compound. Each electronic state corresponds to a specific spatial distribution of electrons around the nuclei. Only certain distributions of electrons are allowed. A specific energy is associated with a molecule in each electronic state (Fig. 2). Thus, each electronic state has a certain molecular energy associated with it, and the series of electronic states is separated by defined energy gaps. Figure 2 shows some of the electronic states of a molecule. The electronic state with the lowest energy is called the ground state or ground electronic state. Electronic states with higher energy are called excited electronic states, or simply excited states, and are designated with higher numbers as their energy increases.

Electronic states of molecules are described as being either singlet states or triplet states. These designations arise because electrons have a property called spin that can have a value of $+\frac{1}{2}$ or $-\frac{1}{2}$. For organic molecules, only the spins on the two electrons with the highest energy in an electronic state are considered. When these electrons have opposite spins, the electronic state is a singlet state. When the electrons have the same spins, the electronic state is a triplet state. Ground state molecules are almost always singlet states. Absorption of energy by

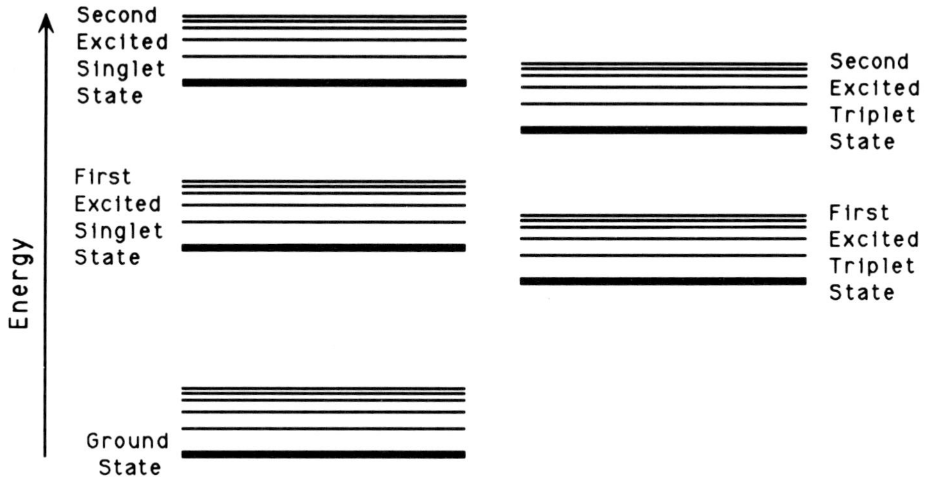


**Figure 2** Energy levels of a molecule. Dark lines represent electronic energy levels and lighter lines are vibrational and rotational levels. At room temperature, molecules are in the ground state and lowest vibrational and rotational levels. Absorption of a photon promotes the molecule to a singlet excited state. Conversion of singlet excited state to triplet excited state results in some loss of energy.

ground state molecules, without changes in the direction of electron spin, results in the generation of excited singlet state. A change in the spin direction of one of the two electrons will result in the generation of excited triplet state, which has a lower energy than the excited singlet state. The photochemical reactions of singlet and triplet states often differ. Figure 2 shows a molecule with two singlet excited states and two triplet excited states.

When a molecule absorbs a photon, the electron distribution in the orbitals is changed and the energy of the photon converts the molecule from its ground state (a singlet state) to an excited singlet state molecule. Not all photons (wavelengths) can be absorbed: only those with an energy that matches the difference between the ground state and the excited singlet states of a molecule. Each chemical compound has a unique set of energy levels determined by the chemical structure of the molecule. Compounds containing many conjugated double bonds (i.e., porphyrins) have small energy gaps between the ground and excited state and absorb visible radiation. Compounds containing fewer rings or conjugated unsaturated bonds have larger energy gaps and absorb higher-energy UV photons (i.e., tryptophan). In reality, the energies of vibrations and rotations in the chemical structure (shown as lighter lines in Fig. 2) combine with the electronic energy to give the total molecular energy. In order to be absorbed, the energy of the photon must match the energy gap between the total energy (electronic, vibrational, rotational) of the ground state and the total energy of an excited singlet state.

## B. Absorption Spectra

The probability that a photon will be absorbed plotted against the wavelength is called the absorption spectrum of a compound. Each compound has a characteristic absorption spectrum. The absorption spectrum of a molecule is composed of numerous lines representing the probability of absorption at each wavelength, that is, at each energy gap. The lines are very close together because of the large number of energy gaps associated with the vibrations and rotations of the molecule. For organic molecules, the lines cannot be differentiated and only the contour of heights is recorded. Absorption spectra show regions of absorption called absorption bands. The wavelengths that have the highest probability of absorption are called the absorption maxima, $\lambda_{max}$, and are identifying characteristics of the compound. In an experimental set-up, the absorption spectrum of molecules in solution is measured with a spectrophotometer. The absorbance, $A$, of a solution containing the chromophore is measured at each wavelength to determine the absorption spectrum.

Skin contains many chromophores absorbing in the visible and UV range. Some, but not all, initiate photobiological responses after absorbing a photon of light. DNA is most important chromophore for photobiological responses in the UVB range (Fig. 3). Aromatic amino acids in proteins, especially tryptophan, also absorb UVB radiation but usually do not initiate biological responses. However, absorption of photons by aromatic amino acids in the stratum corneum is important in photomedicine because this process removes a large amount of the UVB radiation before it reaches the DNA in viable cells. Thus, absorption by the stratum corneum is an important protective factor against UVB effects on skin. Carbohydrates do not absorb above about 240 nm and, therefore, are not chromophores in cutaneous responses to biologically relevant wavelengths. The double bonds in unsaturated fatty acids such as arachidonic acid are not conjugated and do not absorb in the UVB range. 7-Dehydrocholesterol contains conjugated double bonds and, after absorbing UVB radiation, is converted to previtamin $D_3$. Urocanic acid in the epidermis absorbs in the UVB range and is a possible chromophore for certain effects of UV radiation on the immune system. The crosslinking amino acid derivatives, pyridinoline (in collagen) and desmosine (in elastin) in the dermis, also absorb UVB radiation but are not known to initiate photobiological responses. The cofactors nicotinamide adenine dinucleotide (NAD) and flavin adenine dinucleotide (FAD) absorb in the UVA range and may be chromophores for certain responses to UVA.

Chromophores absorbing visible light include hemoglobin, which has a strong absorption maximum in the blue portion (about 410 nm) of visible light and has

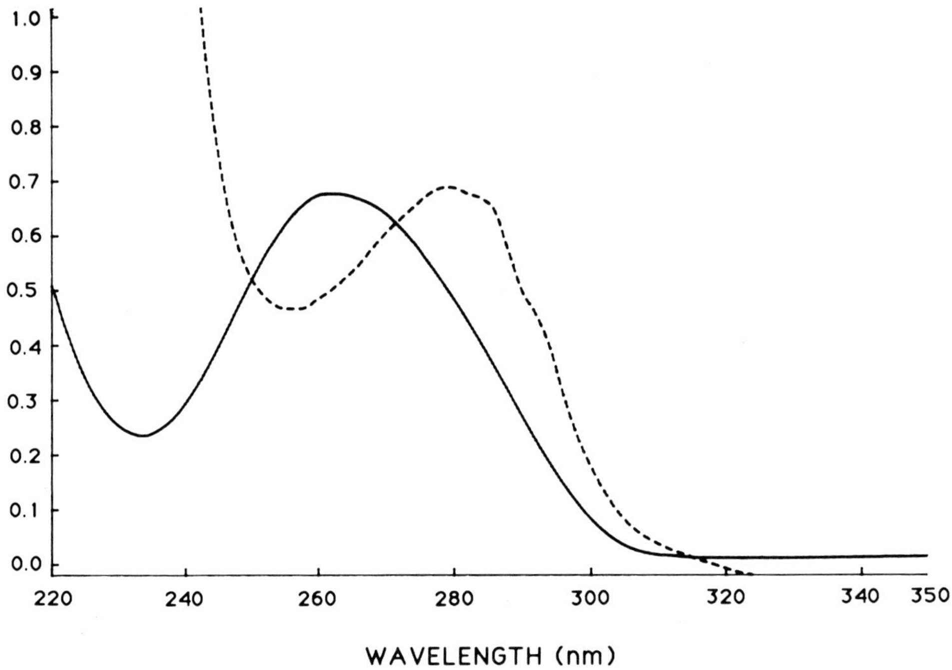


**Figure 3** Absorption spectra of protein (dashed curve) and DNA (solid curve).

lower absorption maxima in the 500–600 nm range. Light absorption by hemoglobin is not associated with photobiological responses in skin, although hemoglobin is the chromophore for certain laser treatments for cutaneous lesions that occur by thermal mechanisms. Protoporphyrin, which has an absorption spectrum similar to that of hemoglobin, is the chromophore for cutaneous photosensitivity in certain porphyrias. Melanin is a very important chromophore in the epidermis because it acts as a filter by absorbing throughout the UVB, UVA, and visible spectrum. The amount of melanin in skin is the major factor in reducing the cutaneous responses, such as sunburn, to UVB radiation. Bilirubin and $\beta$-carotene absorb visible light and are present at low concentrations in normal skin. The light absorption by bilirubin at about 460 nm is the primary step in blue light therapy for neonatal hyperbilirubinemia. $\beta$-Carotene has an absorption maxima in the 450–470 nm range and does not appear to be a chromphore for photobiological responses in skin.

Many cutaneous photosensitivity responses are initiated by the light absorption of chromophores exogenously supplied either topically or systemically. Some exogenous chromophores are used for phototherapy of skin diseases; these include 8-methoxypsoralen (8-MOP), which absorbs in the UVA and UVB region. Cutaneous responses to most photosensitizing drugs such as tetracyclines, chlorpromazine, and amiodarone are elicited with wavelengths in the UVA region. Drug photosensitivity responses to UVB radiation have also been reported.

## III. PHOTOPHYSICAL AND PHOTOCHEMICAL PROCESSES OF MOLECULES IN SKIN

After absorbing the energy of a photon, the excited state molecule has a different distribution of electrons and contains more energy. Thus, the excited state molecule can undergo chemical reactions and physical processes not possible for ground state molecules. The processes available to molecules in excited states are shown in Figure 4. Not all processes are equally likely to occur for each type of molecule. The probability that a process will occur depends on the structure of the molecule. Each process will be discussed for molecules in general and related to specific biomolecules.

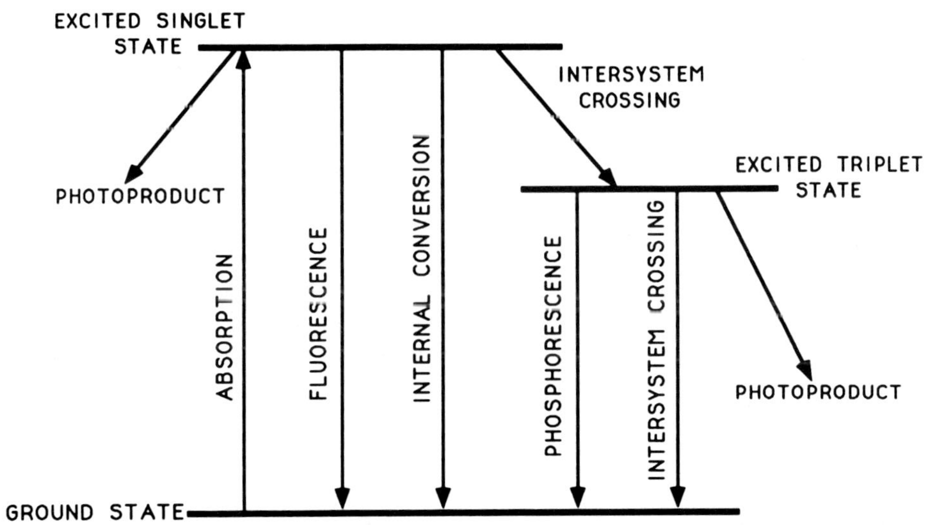


**Figure 4** Processes of excited states. Absorption of a photon by a molecule creates the excited singlet state, which can emit fluorescence, lose energy as heat (internal conversion), undergo a photochemical reaction to result in formation of photoproduct, or transform into a triplet excited state. The triplet excited state can emit phosphorescence, photochemically react, or return to the ground state.

## A. Excited Singlet States

The excited singlet state molecule formed upon absorption of a photon typically has a lifetime of less than 50 nsec ($10^{-9}$ sec) and sometimes as short as a few psec ($10^{-12}$ sec). Emission of fluorescence and internal conversion are photophysical processes that occur very rapidly and return the excited singlet state molecule to the ground state without producing a chemical change. Fluorescence is the emission of a photon of light by the excited singlet state. The wavelengths of fluorescence emission of a molecule are at longer wavelengths than the absorption wavelength of the molecule. Fluorescence has many applications in dermatology. For example, a Wood's lamp, which emits mainly in the UVA range, is used to detect fluorescence associated with certain cutaneous changes such as porphyrins in the urine of patients with porphyria. Fluorescence from dyes such as fluorescein and rhodamine is used for fluorescence microscopy. The fluorescence of the crosslinking amino acid derivatives desmosine and pyridinoline is used to detect these compounds in elastin and collagen, respectively.

Internal conversion is a photophysical process that returns the excited singlet state molecule to the ground state without a chemical change. The energy of the excited state is dissipated as heat, and internal conversion is described as a "radiationless decay" process. Usually the amount of heat produced is not detectable because it is very small and dissipates quickly in the tissue. When a high-intensity light source is used that excites many molecules in a small volume of tissue, a temperature rise may result if the molecules return to the ground state by internal conversion. A laser is almost always required to produce a large number of excited state molecules in a small tissue volume. The thermal energy produced upon rapid internal conversion of hemoglobin molecules after excitation by a pulsed dye laser has been used in laser treatment of port wine stains. The heat deposited is believed to cause coagulation of blood and denaturation of proteins.

When the excited singlet state molecule undergoes a photochemical reaction it is converted into a different compound, the photoproduct. Because the excited singlet state exists for a very short time, it does not usually have time to encounter and react with other molecules. Thus, almost all of the photochemical reactions of excited singlet state molecules are unimolecular, such as the isomerization of urocanic acid (described below).

Another photophysical process of excited singlet states is intersystem crossing, in which the excited singlet state is converted to an excited triplet state molecule. This is often the slowest photophysical process because conversion between singlets and triplets is not allowed by quantum mechanics due to the need to flip the direction of the spin on the electron. The triplet state has a lower energy than the excited singlet state and a different distribution of electrons and, thus, different chemical reactivity. The probability of intersystem crossing varies with the structure of the compound from only about 1% for fluorescein to more than 50% for protoporphyrin.

## B. Excited Triplet States

The triplet excited state exists for a longer time than the singlet state. The lifetime of an excited triplet state molecule in biological systems is limited mainly by the interaction of the triplet state with oxygen. During this interaction, the energy of the triplet state molecule is transferred to a ground state oxygen molecule. This process is called energy transfer and is described later. The products of energy transfer are the ground state of the molecule and an excited state of oxygen called singlet oxygen, $^1O_2{}^*$. Typical lifetimes for a triplet state molecule in oxygenated tissue are a few $\mu$sec ($10^{-6}$ sec), whereas in the absence of oxygen the lifetimes may be as long as a few msec ($10^{-3}$ sec). Triplet state molecules formed in solids at low temperatures can exist for several seconds. Because of their longer lifetimes, it is possible for excited triplet state molecules to encounter other molecules and react with them. The longer lifetime of triplet state molecules compared to singlet state molecules is the reason that most photochemical reactions of chromophores with biological molecules involve the triplet state.

The excited triplet state molecule can emit a photon of light called phosphorescence. Phosphorescence is observed at even longer wavelengths than the fluorescence of the same molecule because the energy gap between the triplet state and the ground state is smaller than that between the excited singlet state and the ground state (Fig. 2). Phosphorescence emission is usually not strong enough to be seen in tissue at room temperature. The triplet excited state may convert into the singlet ground state by intersystem crossing. This process is usually slower than photochemical reactions of triplet states in tissue.

## C. Quantum Yield

For any excited state molecule, singlet or triplet, the probability that each of the processes described above will occur depends on the molecular structure of the chromophore and the availability of other molecules with which to react. The term quantum yield is used to describe the likelihood of these processes.

$$\text{Quantum yield} = \frac{\text{Number of molecular events of interest}}{\text{Number of photons absorbed}}$$

For example, the quantum yield for fluorescence is the number of photons emitted divided by the number of photons absorbed. The quantum yields for many of the processes of biological chromophores have been measured in solution but not in tissue. The quantum yield for each process of an excited state has a value between 0 and 1.

The quantum yield for a photochemical reaction that initiates a biological response is only one of the factors that determines the magnitude of the response. Other factors such as repair processes and thresholds influence the translation of a photochemical event into a skin lesion.

## IV. PHOTOCHEMISTRY IN SKIN

Several different types of photochemical reactions occur after chromophores in skin absorb photons. Some of the photochemical reactions involve endogenous chromophores such as DNA and urocanic acid. Other photochemical reactions involve chromophores that are added to skin (exogenous chromophores). These molecules are called photosensitizers because they render skin abnormally sensitive to wavelengths of light absorbed by the chromophore. Photosensitizers include many drugs and chemicals encountered in the environment or in work places. The photosensitizer absorbs UV or visible radiation and an excited state, usually the triplet, initiates chemical reactions with normal cellular molecules. Some photosensitizers are endogenous chromophores found at higher than normal levels in photosensitive skin, such as protoporphyrin in patients with certain porphyrias. Several of the more frequently observed photoreactions of endogenous chromophores and photosensitizing compounds are described here.

Photoaddition reactions usually involve interaction of an excited state molecule with a ground state molecule. In the photoproduct, the two molecules are covalently linked. The excited state molecule may react with a ground state molecule of the same type. For example, dimers are formed between adjacent thymines on a strand of DNA when one of the thymines is photoactivated (Eq. 4). These photoproducts have been measured in human skin exposed to UV radiation. UVB radiation is most effective but dimers were detected after exposure of skin to wavelengths as long as 375 nm (1). Cyclobutyl pyrimidine dimers and another type of dimer involving the carbon–nitrogen exocyclic double bond of cytosine (called the 6–4 photoproduct) are believed to be the major products responsible for mutagenic and toxic effects of UV radiation (2).

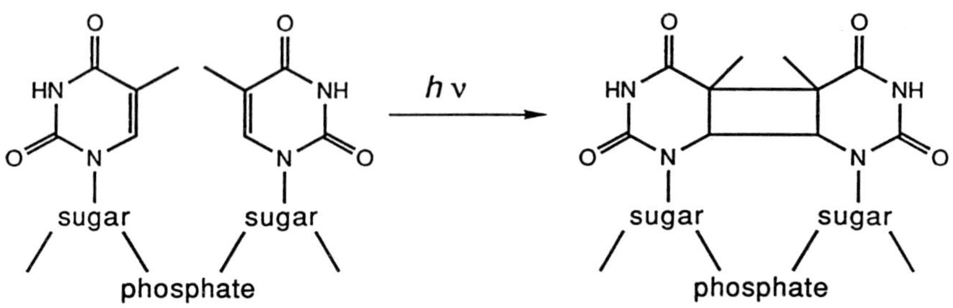

(4)

Photoaddition reactions often involve addition of the excited state molecule to a different type of molecule in its ground state. For example, irradiation of skin treated with the photosensitizing drug 8-MOP with wavelengths absorbed by the 8-MOP results in formation of covalent linkages with pyrimidine bases in DNA to form cycloadducts (Eq. 5). The cycloadduct absorbs UV radiation subsequently and may covalently link to another pyrimidine in the opposite strand of DNA to form a crosslink between the two DNA chains. Adducts of 8-MOP and DNA have been detected in human skin after topical application of 8-MOP and exposure to UVA radiation (3). These photoproducts have been shown to be involved in the phototoxicity of 8-MOP in human skin and in the cytotoxic and mutagenic properties of psoralens in cell culture (4).

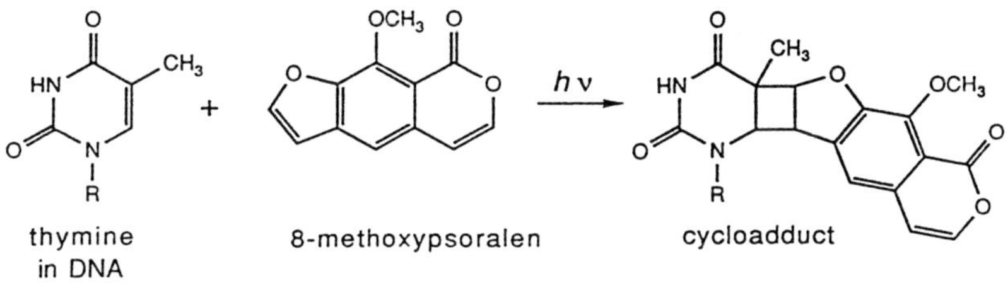


(5)

*Cis-trans* photoisomerization reactions of cutaneous chromophores appear to be involved in several cutaneous responses to light. Photoisomerization reactions occur in molecules containing at least one carbon–carbon double bond that is not in a ring structure. The *cis* and *trans* forms of the molecule interconvert as shown for *trans*-urocanic acid in Equation 6. Photoisomerization of bilirubin in skin is another example. The visual process is also initiated by a photoisomerization reaction; in this case, the chromophore is retinal rhodopsin, which isomerizes after absorbing visible light. The change in molecular shape that occurs when a molecule changes between *cis* and *trans* forms alters the solubility of the molecule or its ability to bind to sites in tissue. Formation of *cis*-urocanic acid by UVB irradiation of skin is believed to be one of the events for certain photoimmunologic responses (5).

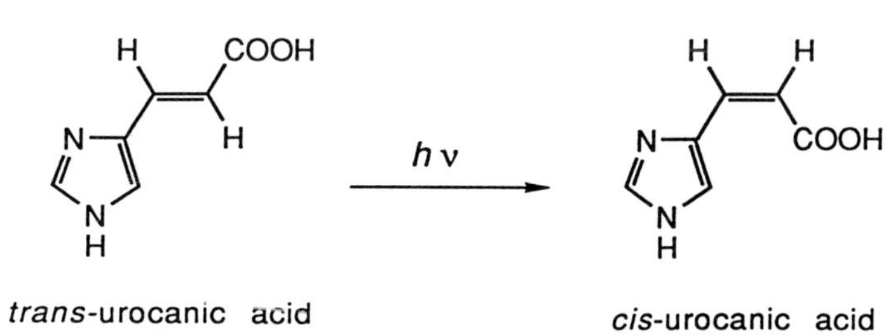


(6)

The energy transfer from excited triplet state molecules to ground state oxygen in tissue forming singlet oxygen, $^1O_2^*$ (Eq. 7), is an example of a general reaction in which energy is transfered from the excited triplet state to a ground state molecule. Oxygen is exceptional in that its ground state is a triplet rather than a singlet. Formation of singlet oxygen ($^1O_2^*$) is frequently cited as the first step in the photosensitivity associated with porphyrias and certain drugs. Singlet oxygen is highly reactive toward many biological molecules. For this reason, the lifetime of $^1O_2^*$ in tissue is very short; it has been estimated to be about 50 nsec in cell membranes (6). Singlet oxygen causes oxidation of amino acids, nucleic acids, and unsaturated lipids. Histidine, tryptophan, and cysteine are the amino acids most readily oxidized. The products of $^1O_2^*$ with unsaturated fatty acids include hydroperoxides (Eq. 8) that can subsequently decompose to free radicals that initiate lipid and protein oxidation by a chain reaction. The oxidation products formed by the interaction of $^1O_2^*$ with cell constituents can lead to cell toxicity.

$$^3DYE^* + O_2 \longrightarrow DYE + {}^1O_2^* \qquad (7)$$

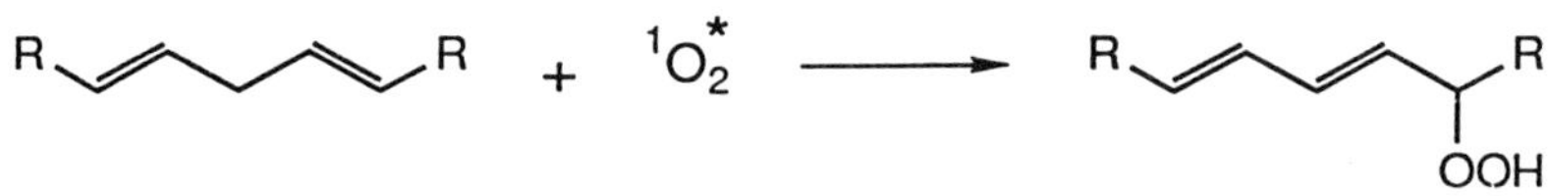


(8)

Electron transfer from excited state chromophores to molecules in the tissue, especially to oxygen, can also result in cell damage. The product of electron transfer to oxygen is superoxide anion, $O_2^{\cdot-}$, and a cation radical of the dye (Eq. 9). Superoxide anion is not very reactive toward biomolecules and can be removed by the enzyme superoxide dismutase contained in many cells and interstitial fluids. Superoxide anion can be converted into a highly reactive hydroxyl radical, $OH^\cdot$, with the aid of a transition metal catalyst. Hydroxyl radical is highly reactive and causes damage almost indiscriminately to all biomolecules. Thus, its lifetime is very short and it can only react with molecules very close to the site where it is generated. Hydroxyl radical is an important toxic species for the effects of ionizing radiation on cells.

$$^3DYE^* + O_2 \longrightarrow DYE^{\cdot+} + O_2^{\cdot-} \qquad (9)$$

Excited state molecules undergo bond cleavage to form radicals that subsequently react with biomolecules. The energy required for this reaction is higher than for most other photochemical reactions, so that bond cleavage reactions usually only occur upon UV excitation of a chromophore. Certain phototoxic drugs, such as chlorpromazine (Eq. 10) and amiodarone, contain halogen substituents, and the radicals formed by cleavage of carbon–halogen bonds are believed to underlie the mechanism for their phototoxicity.

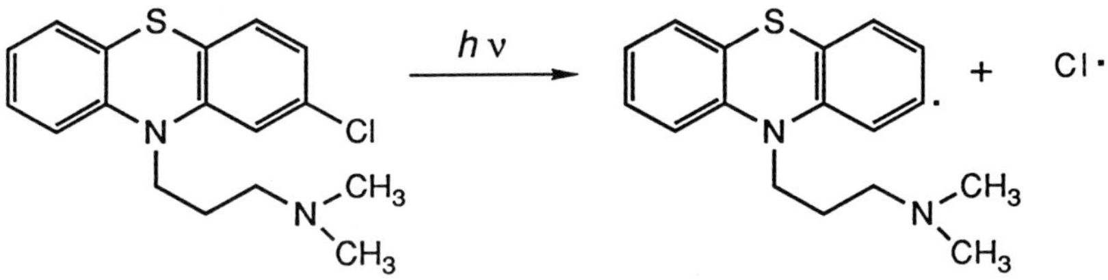


(10)

## V. PHOTOBIOLOGICAL RESPONSES IN SKIN

Many processes intervene between absorption of a photon by a cutaneous chromophore and the final result, a cutaneous photosensitivity response. A goal of photobiology and photomedicine is to learn which chromophores are involved in a specific biological response to light and which biochemical processes translate the photochemistry of the chromophore into the biological response.

### A. Action Spectrum

The action spectrum is an important tool for photobiologists in identifying the chromophore in skin likely to be responsible for a particular response. An action spectrum describes the efficiency of each wavelength at producing a particular response. When the photons in a certain wavelength range are particularly effective, fewer photons are needed in that range than in other wavelength ranges. Peaks in the action spectrum correspond to the most effective wavelengths. Ideally, the action spectrum for a response corresponds to the absorption spectrum of the chromophore. The action spectra for several cutaneous responses have been measured, for example, erythema, melanogenesis, and altered immune response. Action spectra for photosensitivity caused by exogenous agents have also been reported. An action spectrum for erythema is given as an example in Figure 5. The action spectrum may differ from the absorption spectrum of a compound in solution because of filtering by other components of the tissue or because the in vivo absorption spectrum is altered due to its cellular environment. This is the case for the action spectrum for erythema. The chromophore is believed to be DNA, which has an absorption maximum at 260 nm. However, the action spectrum maximizes at about 250 and 295 nm because proteins in the upper layers of the epidermis filter out light around 280 nm.

To construct an action spectrum, skin is exposed to varying doses of

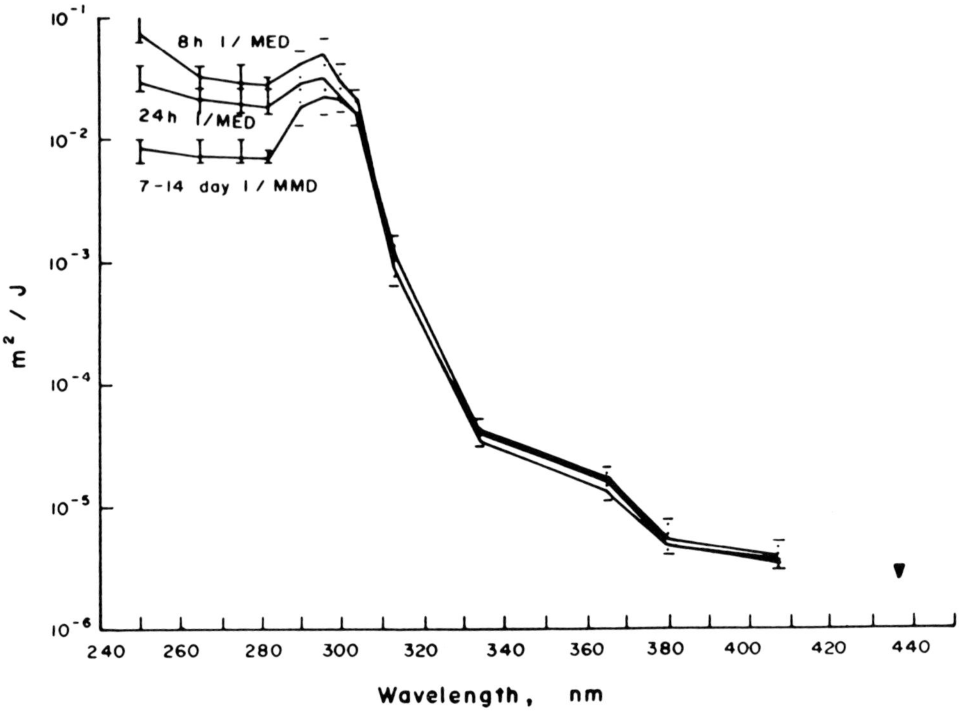


**Figure 5** Action spectrum for delayed erythema in white human skin. The reciprocal of the minimal erythema dose ($m^2/J$) is plotted against wavelength. (From Parrish et al. Photochem Photobiol 1982; 36:187–191.)

radiation of several wavelengths. These studies require use of a powerful light source such as a 2500 W xenon arc lamp in combination with a monochromator, which isolates narrow wavelength bands for the irradiation. Either a dose–response curve is obtained or, for many cutaneous responses, the minimal effective dose is determined at each wavelength. The slope of the dose–response curve or the reciprocal of the minimal effective dose is plotted against wavelength. Most action spectra have used energy units to express the amount of light delivered per area of skin (i.e., $J/cm^2$). The dose should ideally be expressed in the number of photons delivered (rather than the energy) to an area because the response is initiated by absorption of photons and the energy per photon varies with the wavelength of radiation. A simple way to calculate the number of photons at a given wavelength is the following relationship:

$$\text{Number of photons in 1 J} = 5 \times 10^{15} \times \text{wavelength (nm)}$$

For practical objectives, such as determining which wavelengths should be blocked to prevent a certain photosensitivity reaction, it may only be necessary to determine the broad bands (UVB, UVA, or portions thereof) that cause the response. The range of wavelengths causing a photosensitivity response can be elucidated without determining the exact action spectrum. This type of measurement can be made with a 150 W xenon arc lamp and the appropriate filters.

### B. Reciprocity between Time and Intensity

For most cutaneous photosensitivity responses the magnitude of the response is only dependent on the number of photons absorbed and not on the rate at which the photons are absorbed. This concept is called reciprocity between time and intensity. It derives from the paradigm in which a photochemical reaction is the primary event and the biological response is proportional to the number of molecules of a specific photoproduct produced in tissue. The number of photoproduct molecules is proportional to the number of photons delivered to the tissue. When the rate of photon delivery is doubled, the time required to deliver the same number of photons is halved. Time and intensity reciprocity has been supported for several cutaneous responses. For example, the same number of photons produced the same erythema response when the photons were delivered over several hours as when they were delivered in a few seconds (7). Reciprocity of time and intensity is expected to be violated when repair processes occur during long delivery times and for other reasons. For example, during chronic exposure of mouse skin to UVB in experimental photoaging studies, a greater response was observed when the light was delivered over a longer period (8).

## VI. SUMMARY

Responses of skin to light are initiated by the absorption of photons by chromophores in the tissue. The wavelengths of light absorbed by each chromophore are determined by its chemical structure. After absorbing a photon of optical radiation, the chromophore is in an excited state from which it can react to form photoproducts. The presence of photoproducts in cells initiates biochemical processes that eventually lead to observable lesions. These basic concepts can be applied to all photobiological processes. The identity of the chromophores and the biochemical mechanisms for light-induced responses in skin are only partially known and are topics of active research.

## REFERENCES

1. Freeman SE, Hacham H, Gange RW, Maytum DJ, Sutherland M. Wavelength dependence of pyrimidine dimer formation in DNA of human skin irradiated in situ with ultraviolet light. Proc Natl Acad Sci USA 1989; 86:5605–5609.

2. Ananthaswamy HN, Pierceall WE. Molecular mechanisms of ultraviolet radiation carcinogenesis. Photochem Photobiol 1990; 52:1119–1136.
3. Yang XY, Gasparro FP, DeLeo VA, Santella RM. 8-Methoxypsoralen–DNA adducts in patients treated with 8-methoxypsoralen and ultraviolet A light. J Invest Dermatol 1989; 92:59–63.
4. Papadopoulo D, Averbeck D. Genotoxic effects and DNA photoadducts induced in Chinese hamster V79 cells by 5-methoxypsoralen and 8-methoxypsoralen. Mutat Res 1985; 151:281–291.
5. DeFabo EC, Noonan FP. Mechanism of immune suppression by UV irradiation in vivo. I. Evidence for the existence of a unique photoreceptor in skin and its role in photoimmunology. J Exp Med 1983; 157:84–98.
6. Kanofsky JR. Quenching of singlet oxygen by human red cell ghosts. Photochem Photobiol 1991; 53:93–99.
7. Park YK, Gange RW, Levins PC, Parrish JA. Low and moderate irradiances of UVB and UVC irradiation are equally erythemogenic in human skin. Photochem Photobiol 1984; 40:667–669.
8. Kiss I, Chen S, Tramposch KM. The effect of high and low ultraviolet B dose exposure on the degree of hairless mouse wrinkling. Photochem Photobiol 1991; 53:109–112.

# 2

# Optics of the Skin

**R. Rox Anderson**
*Harvard Medical School, Boston, Massachusetts*

The skin is a highly dynamic, essentially protective organ with specialized functions. Specific mechanisms for photoprotection, coloration, and photobiological responses have evolved, all of which are influenced heavily by the optical properties of skin. Cutaneous optical properties are dynamic, varying with each heart beat (blood volume changes), within minutes of sun exposure (immediate pigment darkening), hours of sun exposure (sunburn), days of sun exposure (tanning), months and years of sun exposure (photoaging). There is a huge variation in melanin content among human races, accounting for colors from pale pink to jet black. A wonderful range of skin hues is produced by natural, pathologic, or exogenous pigments. The entire profession of dermatology relies in essence on skin optics, since all light returning from the skin used for clinical diagnosis depends on this. Photomedicine in particular relies on tissue optics, since the deposition of energy that begins all photobiological responses depends directly on the distribution of light and chromophores (absorbing molecules) within the tissue. The development of selective laser surgery techniques in dermatology came directly from studies of skin optics, and these form an interesting subset of photobiological responses.

The goals of this chapter are to 1. present a useful, approximate model of skin optics; 2. elucidate different mechanisms of skin photoprotective response for different parts of the optical spectrum; 3. present data on skin optical properties; 4. discuss ways of altering skin optics that are useful in phototherapy; and 5. present the basis for laser therapy that relies on heterogeneous optical absorption.

## I. BASIC OPTICS

Two basic processes govern the optics of skin: absorption and scattering. Absorption is best understood by considering that light consists of discrete packets (called photons) of energy. Absorption is the loss of a photon when its energy is invested in some transition within an atom or molecule. In the ultraviolet (UV) and visible spectrum, absorption leads to specific electron transitions in the absorbing molecule (chromophore). The energy of the excitation transition corresponds to the energy of the photon, which is inversely related to wavelength by Planck's law. Therefore, UV and visible absorption tends to occur in discrete wavelength bands, and molecules have characteristic or unique absorption spectra. With a few rare exceptions, all effects of light on matter begin with absorption. The absorbed photon energy is typically dissipated as heat, but may also lead to photochemical reactions from the electronically excited state(s) produced, or to re-emission of light in the form of fluorescence or phosphorescence, as the excited molecule returns to its ground state.

Scattering is defined broadly as any process in which the direction of propagation of light is altered, and occurs due to fluctuations or boundaries of refractive index as light encounters matter. Without scattering, the skin would return no light at all and would be invisible. Scattering greatly increases the path length that light must take in penetrating into the skin, and essentially combines with absorption, therefore, to limit the depth of light penetration. Understanding scattering requires consideration of the wave nature of light, since the character of scattering depends greatly upon the relative size of the scattering object and the wavelength of light.

Scattering by very small particles (defined as less than about 1⁄10 the wavelength, that is, less than about 50 nm for visible light scattering) was first described in detail by Lord Rayleigh, and hence bears his name. Rayleigh, or molecular, scattering is weak, almost isotropic (equal in all directions), and strongly dependent on wavelength, varying inversely with the fourth power of wavelength. Thus, shorter wavelengths are much more strongly scattered by small particles or molecules. Scattering by particles about equal to the wavelength (i.e., between about 1⁄10 and 10 times the wavelength) is called Mie scattering, after the physicist who first produced a solid theory describing it. Mie scattering is much stronger than Rayleigh scattering, is much less dependent on wavelength, and tends to be increasingly forward-directed for larger particles. Figure 1 shows the broad range of scattering behavior.

A cloud viewed against the blue sky provides an excellent example of Mie vs. Rayleigh scattering, respectively. The sky is blue because of the strong inverse dependence of Rayleigh scattering on wavelength. Thus, the shorter (blue) visible wavelengths are scattered more intensely by the sky, whereas the longer (red) wavelengths tend to penetrate through the atmosphere. In contrast, clouds, which are composed of aerosol water droplets, are white or grey because of the relative independence of Mie scattering with wavelength. Although a cloud is much smaller and contains much less scattering material than the sky, Mie scattering is intense, so that the cloud appears bright, and casts a shadow. It is no surprise that the dominant scattering in skin is from particles, fibers, and

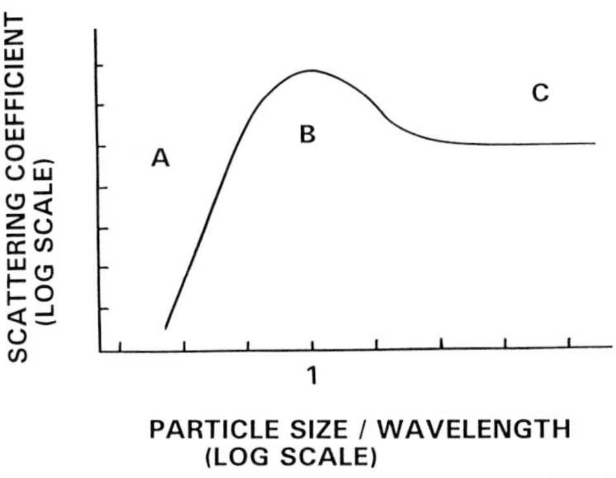


**Figure 1** Scattering of electromagnetic radiation varies greatly in strength and direction, depending on the size of scattering particles relative to the wavelength. A. Rayleigh (molecular) scattering is nearly isotropic and varies inversely with the fourth power of wavelength. B. Scattering is maximum with particles about the size of the wavelength, and becomes forward-directed. C. Scattering by very large particles is highly foward-directed, and nearly independent of wavelength. In skin, all three types occur, but scattering is dominated by large particles (see text).

structures about equal to or larger than the wavelength of light. Scattering in the dermis by collagen fibers varies inversely with wavelength (1,2).

When the scattering object becomes very large in relation to wavelength, Mie's theory reduces to familiar optical principles such as the reflection and refraction of light by mirrors, windows, lenses, and other objects, which are governed by the refractive index of the material. Refractive index is defined as the ratio of the speed of light in a vacuum compared to that in the material, and is usually designated as $n$.

Absorption and scattering combine in a complex fashion to determine the optics of tissue. The majority of light returning ("remitted") from within the skin is composed of photons that have been scattered many times, taking a random path that happens to lead outside the skin again. Recently, significant progress has been made using mathematical models of radiation transport for analysis of tissue optical properties, although none has yet been proven truly appropriate to living skin. These models can be divided into those based on phenomenologic definitions (Kubelka-Munk model; see reference 1), diffusion theory and numerical approximations (2,3), and probability or Monte Carlo models that simulate the "random walk" of photons in a scattering medium (4).

## II. SKIN OPTICS

It is helpful to present a broader picture before giving detail. In essence, the stratum corneum and epidermis are a compact, thin, predominantly light-absorbing layer, well-endowed with specific pigments for UV photoprotection by absorption. The dermis, in contrast, is a thick, relatively acellular layer in which light scattering plays a much more important role, with its absorption over much of the optical range, including UVA, localized largely in blood vessels. With this in mind, skin optics will now be discussed layer by layer, going into the skin. For each layer, the discussion will also flow from shorter (UV) to longer (visible, then infrared) wavelengths.

## A. Skin Surface (Skin–Air Interface)

In skin, refraction and reflection appear to play a significant role only at the skin surface, where there is a sudden and major change in refractive index in going from air (n = 1.0) to stratum corneum (n = 1.55) (5). This change in refractive index causes about 5% of normally incident light to be reflected from skin (1), with the remaining 95% entering the skin. This quantitatively minor reflection accounts for our ability to see the skin surface at all. When the skin surface reflection is eliminated from view using polarized light techniques, it is impossible to differentiate surface contour, yet much easier to view deeper structures such as vessels, infiltrates, or pigmentation (6). Simple polarized light techniques are useful in enhancing the view of either surfaces or subsurface skin structure (6).

In exfoliative disorders such as psoriasis, there may be multiple air–skin interfaces due to multiple layers of scale. Because each skin–air interface reflects about 5% of normally incident light, and more for oblique angles of incidence, psoriatic plaques can appear white or “micaceous” (mica looks this way for the same reason). Therefore, much less of the incident light even enters the underlying skin in scaly lesions. Application of transparent lubricants, which fill the air spaces between scales with a refractive index much closer to that of the skin, allow for greater light delivery into plaques yet have relatively little effect on normal skin. This is well known clinically, when mineral oil is applied to “view through the scale.” It is an important maneuver in optimizing phototherapy or photochemotherapy, because UV-transparent lubricants preferentially increase the UV delivered into scaly plaques (7,8). There is very little wavelength dependence to this effect, and, therefore, lubrication of psoriatic skin is useful for both the UVB and UVA regions of the spectrum.

Another major influence of reflection at the skin–air interface has been only recently recognized, and can produce much greater light intensity within the skin than that of the incident light itself striking the skin surface (9). This apparent paradox occurs because of the strong internal reflection of back-scattered light when it encounters the skin–air interface. Over half of the back-scattered light is reflected back into the skin at each encounter with the skin surface (10); thus, what we see returning from within the skin is only a fraction of that actually back-scattered. Near the surface (i.e., within the epidermis and probably upper papillary dermis) the light intensity can become many times that of the incident light itself. In more precise terms, the ratio of the internal light intensity to that of a wide, collimated beam such as sunlight or many sources is given by:

$$Ii/Io = 1 + 6R$$

where $R$ is the diffuse reflectance of the skin and the coefficient value 6 is determined by the refractive index (see reference 9 for a complete derivation). For fair white skin, typical values of $R$ and of the increased intensity within the epidermis resulting from this effect are estimated in Table 1. Note that the influence is maximal in the red part of the spectrum, but significant in the UV region as well.

## B. Stratum Corneum and Epidermis

Many authors have measured the transmission of UV radiation through isolated epidermis, and have shown that both thickness and melanin pigmentation play important roles (1,11–14). Table 2 lists typical transmission values for full-thickness epidermis of fair white and black skin, for comparison.

In the UVC and UVB regions, there is a strong, broad absorption band in stratum corneum centered near 275 nm due to the aromatic amino acids tryptophan, tyrosine, and phenylalanine (1). In a study by Sandritter (15), microspectrophotometry of fresh epidermal sections showed the UV absorption maximum to be near 260 nm in basal keratinocytes, but the absorption maximum in upper epidermis and stratum corneum near 275 nm. This is consistent with the loss of DNA (absorption maximum at 260 nm) and increase in protein (absorp-

**Table 1** Approximate Increase in Light Intensity Within Fair White Epidermis Due to Internal Reflection of Back-Scattered Light from a Nearly Collimated Incident Source

| Wavelength (nm) | Diffuse reflectance $(R)$ | Internal/external light intensity |
|---|---|---|
| 300 | 0.04 | 1.2 |
| 350 | 0.10 | 1.6 |
| 400 | 0.15 | 1.9 |
| 500 | 0.30 | 2.8 |
| 600 | 0.50 | 4.0 |
| 700 | 0.60 | 4.6 |

**Table 2** Approximate Typical In Vitro Abdominal Human Stratum Corneum and Full-Thickness Epidermal Transmission of UV, Visible, and Near-Infrared Radiation for Fair White and Black Skin

| | Transmittance (total) | | | |
|---|---|---|---|---|
| | Fair white | | Black | |
| Wavelength (nm) | stratum corneum | epidermis | stratum corneum | epidermis |
| 250 (UVC) | 0.01 | 0.003 | 0.001 | 0.0001 (?) |
| 260 | 0.005 | 0.001 | 0.0005 | 0.00005 |
| 270 | 0.003 | 0.0005 | 0.0003 | 0.00003 |
| 280 (UVB) | 0.005 | 0.001 | 0.0007 | 0.00007 |
| 290 | 0.04 | 0.01 | 0.007 | 0.001 |
| 300 | 0.15 | 0.08 | 0.03 | 0.005 |
| 310 | 0.3 | 0.15 | 0.06 | 0.02 |
| 320 (UVA) | 0.4 | 0.25 | 0.1 | 0.03 |
| 330 | 0.55 | 0.35 | 0.12 | 0.05 |
| 340 | 0.65 | 0.4 | 0.15 | 0.06 |
| 350 | 0.7 | 0.45 | 0.2 | 0.08 |
| 400 (blue) | 0.75 | 0.5 | 0.3 | 0.15 |
| 500 (green) | 0.8 | 0.6 | 0.45 | 0.3 |
| 600 (orange) | 0.85 | 0.7 | 0.6 | 0.45 |
| 700 (red) | 0.9 | 0.8 | 0.7 | 0.60 |
| 1200 (near infrared) | 0.95 | 0.9 | 0.95 | 0.9 |

Source: Refs. 1, 13.

tion maximum near 275 nm) during keratinocyte maturation and upward migration. Urocanic acid, which is a deamination product of histidine and appears to be selectively retained in human epidermis (16), absorbs in a band peaked at 265–270 nm (Fig. 2).

All of these major epidermal UVB chromophores have little absorption above 320 nm. Melanin is the only substantial epidermal chromophore with absorption extending into the UVA, visible, and near-infrared part of the spectrum. Amino acids and DNA are obligatory constituents of keratinocytes, and absorb well in the UVC and UVB spectral regions. Thus, even vitiliginous or albino skin can mount a UVB photoprotective response via epidermal hyperplasia, an epidermal response that occurs best after UVB (vs. UVA or UVC) irradiation (17). One would strongly expect vitiliginous or albino skin to mount a poor or nonexistent photoprotective response to UVA or psoralen/UVA (PUVA) photochemotherapy, although this author is unaware of studies aimed at this hypothesis.

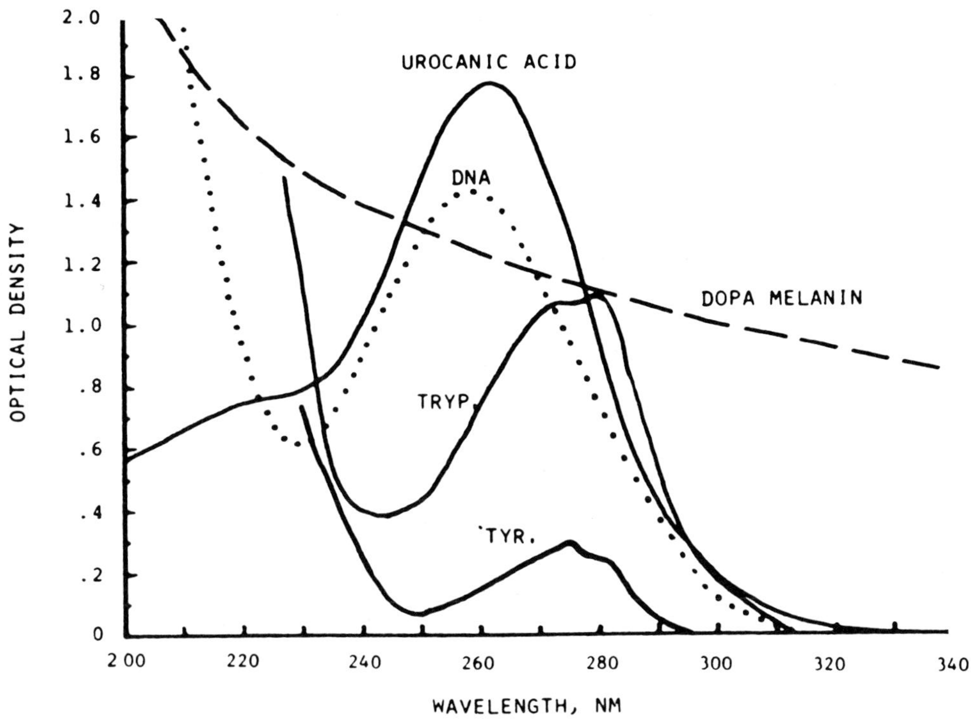


**Figure 2** Ultraviolet absorption spectra of the major epidermal chromophores: dihydroxyphenylalanine (DOPA)-melanin (1.5 mg/100 ml aqueous solution), urocanic acid ($10^{-4}$ M), calf thymus DNA (10 mg/100 ml aqueous, pH 4.5), tryptophan ($2 \times 10^{-4}$ M, pH 7), tyrosine ($2 \times 10^{-4}$ M, pH 7). (From Ref. 1.)

Urocanic acid and melanin both play a role as "endogenous sunscreens," among other possible functions. Urocanic acid is present in eccrine sweat, but it is likely that most of the urocanic acid present in stratum corneum is derived from filaggrin catabolism. Urocanic acid is readily water extractable, and its removal affects skin response to UVB. Bathing or extraction into water of clean white skin sites for 30 min lowers the UVB minimal erythema dose by approximately 35% (R. R. Anderson and R. W. Gange, unpublished observations). Changes in the spectral transmittance of UVB and UVC radiation through in vitro human epidermis after topical applications of water are accounted for spectrally and quantitatively by the amount of urocanic acid extracted into the external water bath. It is interesting to speculate that eccrine sweating, which frequently occurs due to heating during prolonged sun exposure, may play some role in immediate photoprotection. The water present in sweat evaporates, leaving a urocanic-acid rich layer on the skin surface. It is unknown whether urocanic acid synthesis per se is influenced by UV exposure. Urocanic acid has been hypothesized to initiate photobiological responses, notably photocarcinogenesis (18), although the mechanisms involved remain speculative.

Melanin clearly plays a major role in photoprotection of the skin, although its primary function in many species, and perhaps in humans, is probably for camouflage (19). Black persons are much less prone to skin cancer and photoaging than white persons. There is no established pathway in human skin for degradation of melanin, but Negroid melanosomes, which occur singly within keratinocytes (20), are more stable as organelles, resulting in intact melanosomes within black stratum corneum, in contrast to the "melanin dust" of white stratum corneum. It is unknown how this difference in physical cross-section affects the optical absorption cross-section in black vs. white stratum corneum, but one would expect melanin in the "dust" form to be more photoprotective due to a decrease in the optical sieve effect. The induction of increased melanogenesis by light does not necessarily require a preceding inflammatory reaction, occurs in melanocytes isolated by tissue culture, and appears to be cumulative over many days of suberythemogenic exposures (21). However, the action spectrum for tanning in white subjects after single exposures closely parallels that for delayed erythema (22), suggesting similar primary photochemistry leading to each response. Action spectra for multiple exposures typical of both environmental exposure and phototherapy are unknown.

Epidermal hyperplasia following UVA is less than that after UVB at erythemogenically equivalent exposures (17), suggesting that UVB may induce more transport of newly formed pigment to the upper epidermis. The anecdotal clinical observation that UVA-induced tanning lasts much longer than UVB-induced tanning is consistent with this hypothesis. The major factor differentiating skin types and races is a genetically determined capability for constitutive and facultative tanning. However, for chronic exposure in any given individual, the amount and distribution of epidermal melanin involve a negative feedback loop, because the stimulus for tanning is UV exposure to viable epidermis (or, more likely, to melanocytes themselves), which in turn varies inversely with the amount of melanin present in the upper epidermis and stratum corneum.

The absorption spectrum of human eumelanin is unknown, mainly because melanin has not been extracted in soluble form without potential alteration of its chromophoric structure. Synthetic dihydroxyphenylalanine (DOPA)-melanin appears to have a 300–1200 nm absorption spectrum very similar to that of eumelanins (1), and is also very similar to epidermal melanin absorbance, derived by analysis of diffuse reflectance spectra from skin of varying pigmentation. However, there is considerable uncertainty not only over melanin absorption spectra but also over why this insoluble heteropolymer has such a broad, featureless UV-visible absorption spectrum. Melanin has many electronic properties of solid state semiconductors, and its photoelectronic spectra suggest many overlapping conduction bands; the absorption spectrum has also been suggested to be influenced by strong particle scattering (23). Melanin absorption extends to approximately 1200 nm; beyond this near-infrared wavelength there is essentially no difference in the optics of black vs. fair white skin (1). However, the major effect of 1064 nm laser pulses in pigmented skin is the selective destruction of pigmented epidermal and follicular cells (24), because even the low absorption by melanin at this wavelength dominates over other pigments.

Humans are unusual among mammals in their retention of carotenoids. When present in high concentration systemically, lipophilic carotenoids such as beta-carotene partition sufficiently to stratum corneum to account for the yellow palms and skin of carotenemic individuals. Beta-carotene's role in protecting against phototoxicity in patients with porphyrias cannot be explained solely on the basis of its absorption of visible light (i.e., a sunscreen-like protection). Quenching of excited triplet states and photochemical intermediates is likely. Indeed, carotenoids present in high concentration in photosynthetic plants appear to function in this manner.

## C. Dermis

Optical scattering by human dermis is due in large part to collagen fibers, and varies inversely with wavelength (1), to the power of approximately 3⁄2, in the UVA and visible wavelengths. This is far less than the inverse fourth-power dependence of Rayleigh scattering, and implies that the dominant scattering centers are on the order of 0.2–0.5 $\mu$m. Dermal scattering, at least in the visible spectrum, is largely forward-directed, with the mean scattering angle per scattering event being about 20 degrees at the 632 nm helium–neon (HeNe) laser wavelength (25). Bloodless full-thickness dermis appears glistening white. However, the decreased dermal scattering and anisotropy of longer (red) wavelengths results in their significantly deeper penetration into the dermis. Whenever there is but a thin layer of dermis overlying a cavity, or a black pigment embedded within the dermis, blue colors result (26,27). This is because of the greater relative loss of back-scattered longer (red) wavelengths, which suffer a greater total path length due to reduced scattering at longer wavelengths, and are more apt to be absorbed or lost. Known as the Tindall effect, this accounts for the blue color of nevus of Ota, blue nevi, carbon tattoos, the blue ocular sclera of osteogenesis imperfecta, a "five o'clock shadow," and the normal blue-eyed iris, none of which contain blue pigments. Blue colors throughout the animal kingdom tend to result from a light-scattering layer on top of an absorbing layer, rather than specific blue pigments. Even the blue jay has no blue pigment in its feathers.

The penetration of UV radiation into human dermis in vivo is poorly known, despite its obvious importance to human photobiology and phototherapy. Major dermal proteins contain few aromatic amino acids in comparison to epidermis. There is normally no melanin in dermis, and little or no urocanic acid. Estimates of dermal UV penetration based on in vitro measurements of scattering (1) suggest that UVB radiation reaching the dermis is attenuated by half in approximately 30 $\mu$m, whereas UVA radiation may be attenuated by half in 50–100 $\mu$m of dermis, but these are crude estimates. The absorption of light by oxyhemoglobin in cutaneous blood in vivo can be estimated by analysis of reflectance spectra (28,29), which suggests that on the order of 10% of incident UVA radiation may be absorbed within blood vessels in fair white skin. The successful use of blue light for treatment of neonatal hyperbilirubinemia indicates that significant 460 nm radiation reaches papillary dermal vessels, since the unconjugated bilirubin present is mainly bound to serum albumin. It is unclear whether changes in peripheral circulating T cells observed after PUVA photochemotherapy or UVB radiation are due to radiation absorbed by circulating lymphocytes per se.

Despite the lack of good optical data for dermis in the UV, it is likely that dermal optics account for the depth to which acute and chronic responses to different wavelengths are noted. Longer wavelengths penetrate deeper into dermis. From 254 nm through 365 nm, the progressively steeper dose-response curve for inflammatory response to longer wavelengths suggests that deeper vessels are involved in response to longer wavelengths. Chronic photobiological responses ("photoaging") of skin also appear to present a correlation between the primary site for response and depth of optical penetration. Animal studies implicate UVA predominantly in dermal changes, such as skin sagging. Very little is known about the primary photochemistry, molecular mechanisms, or action spectra involved in chronic effects on dermis.

In the visible portion of the spectrum, dermal optics are somewhat better defined, due mainly to interest in developing laser treatments for portwine stains (2,3,30,31). Absorption by hemoglobins dominates dermal optical absorption for wavelengths less than 610 nm (1) (Fig. 3), such that the selective deposition of laser energy occurs in dermal blood vessels (30).

The superficial venous plexus juxtaposed to the papillary dermis accounts for most of the dermal absorption of wavelengths less than 600 nm. The absorption coefficient of venous blood is many times that of the surrounding dermis at all visible wavelengths less than about 600 nm. Ten years ago, Anderson and Parrish (30) developed a theory based on this strong preferential absorption by blood in the 577 nm region to devise a microscopically selective laser treatment for vascular malformations (e.g., portwine stains and telangiectases). The process, now called selective photothermolysis (31), involves preferential absorption of a laser pulse shorter than the time associated with passive cooling of the "target" vessels by thermal conduction. Heat is therefore produced in dermal microvessels at a rate faster than it is removed, and the vessels transiently

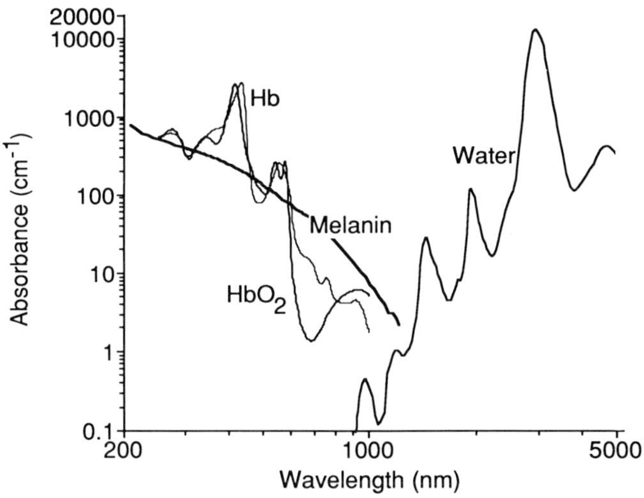


**Figure 3** Absorption spectra of major skin chromophores. The values shown are absorption coefficients for pure water, human blood at 11 g/100 ml hemoglobin concentration (oxygenated and deoxygenated), and DOPA-melanin, which has a 300–1200 nm spectrum that is very similar to pigmented epidermis, at 15 mg/100ml concentration in water. The DOPA-melanin concentration shown is approximately equivalent to heavily pigmented (dark brown) human epidermis.

become extremely hot, probably about 120°C, which is at the threshold for injury (32). The resulting injury causes thrombosis and a necrotizing vasculitis (33,34), leading to vessel destruction without widespread dermal injury. This process is presently accomplished with pulsed tunable dye lasers producing 300–600 $\mu$sec pulses delivering typically 6–8 J/cm$^2$ at the skin surface per pulse. The use of wavelengths somewhat longer than the absorption maximum of oxyhemoglobin at 577 nm produces deeper, still selective, vascular injury (35) due to better penetration into the dermis. The actions prectrum for selective photothermolysis in cutaneous blood vessels as measured by either immediate or delayed purpura formation, however, closely tracks the absorption spectrum of venous blood (36).

Beyond 600 nm, there is a sudden decrease in absorption by hemoglobins (except methemoglobin, which is normally present at less than 0.5% of total). For example, two-thirds of 580 nm light encountering a 30 $\mu$m diameter venule is absorbed in a single pass, whereas only 10% is absorbed at 600 nm, and less than 1% at 700 nm. This, together with a decrease in dermal scattering, and decreased absorption by melanin at longer wavelengths (1), causes a dramatic increase in the penetration of light for wavelengths longer than 600 nm (red and near-infrared light). In the near-infrared region, light penetrates well into the dermis and subcutaneous tissues (37,38). The weak absorption bands of oxyhemoglobin (an electronic excitation) and of water (a vibrational excitation) near 950 nm combine to produce slightly less penetration at this wavelength. The most penetrating optical wavelength is approximately 1150 nm, at which about a quarter of the incident radiation traverses the entire dermis. Beyond this wavelength region, absorption by water becomes significant (Figure 3).

Thus, the 600–1200 nm region has been dubbed a "window" region, and is the wavelength region of choice for photodynamic therapy of solid tumors (39). This application requires deep penetration and relatively uniform light distribution. The majority of photodynamic therapy photosensitizers under development are phototoxic in this wavelength region, and include porphyrin derivatives and precursors, phthalocyanines, chlorins, and purpurins. In general, these cause phototoxicity by transfer of energy between an excited triplet state of the dye and molecular oxygen. The singlet oxygen produced is then capable of oxidative reactions with macromolecules, while the dye is returned to its ground state and may absorb another photon again. The transition required from ground state to singlet oxygen requires approximately 1 eV, which must come from the photon energy, and corresponds to a wavelength of approximately 1200 nm. Because some energy is usually lost in the dye excitation process, drug phototoxicity is probably fundamentally limited to wavelengths less than about 1000 nm. Photobiological reactions related to laser heating do not intrinsically depend on the photon quantum energy.

Within the highly penetrating 600–1100 nm region, the major cutaneous chromophore is melanin, which is normally present only in epidermis and hair follicles. Selective photothermolysis of melanin-containing cells occurs with submicrosecond laser pulses spanning the entire absorption spectrum of melanin from UV through infrared wavelengths (24,31,40).

In guinea pig skin, red and near-infrared laser pulses lead to permanent depigmentation of hair follicles, inducing leukotrichia, whereas UVA laser pulses lead only to transient epidermal depigmentation (24). This is a clear example of different biological responses due to different light penetration depth within the dermis, and shows that red and near-infrared laser pulses are capable of killing melanized cells deep in the dermis, without producing scarring. A clinical study using Q-switched ruby laser pulses (694 nm, deep red light) to treat nevus of Ota has shown very promising clinical results (41).

Abnormal endogenous or exogenous pigments abound in skin and can drastically change optics of the skin. Melanin abnormally present in the dermis causes a blue color, as described above. Purpura, erythema, and cyanosis result from changes in blood content or oxygenation. Jaundice results from the 460 nm absorption band of bilirubin. Hemochromatosis arises from deposition of ferrihemoglobin pigments. Tattoos and other dermal deposits from systemic metal or drug ingestion occur in a wide variety of colors. Tattoos, and perhaps other optically distinct abnormal dermal pigments, can be removed with minimal risk of scarring using selective photothermolysis (42).

At still longer wavelengths in the far-infrared region, water and other molecules have characteristic, strong absorption bands due to vibrational excitations (Fig. 3). Far-infrared radiation from the surgical $CO_2$ laser, which emits at 10,600 nm, is mainly absorbed in the first 20 $\mu$m of wet tissue. This laser is useful as a vaporizing and cutting instrument, but is wholly incapable of producing photochemical reactions due to the low photon energy at this long wavelength.

## III. OPTICAL SPECTROSCOPY AS A DIAGNOSTIC TOOL

Structure and composition of matter can be investigated by analyses of the absorption, reflection, fluorescence, phosphorescence, or nondestructive effects of light. While highly developed for astronomy, physics, and analytical biochemistry, spectroscopy of living skin has received relatively little attention. This is probably because of the impressive diagnostic power of another unique optical instrument: a pair of eyes connected to the rest of the human brain. However, the ability to quantify or delineate cutaneous response is of considerable interest in photomedicine.

Reflectance spectroscopy can be used to measure pigmentation or inflammatory responses of human skin reproducibly (43). Almost every intensive care unit now uses small "pulse oximeters" to monitor peripheral arterial oxygen saturation. These devices measure the fluctuating decrease in light reflectance caused with each heartbeat's delivery of arterial blood to the skin, at wavelengths capable of differentiating oxygenated from reduced hemoglobin. Studies using lasers for time-dependent (radar-like) reflection measurements, and other specialized techniques, may eventually have medical utility (44).

The fluorescence of skin is very poorly characterized. Fluorescence is used clinically with Wood's lamp (a UVA source with little or no visible light output), to detect pigmentation abnormalities (45) or fluorescent substances that occur secondary to infections. Fluorescence spectra of human skin appear to change with chronic sun exposure, perhaps due to alterations in dermal elastin (46,47). Tissue autofluorescence appears to derive mainly from constituents of elastin (fluorophore unknown), collagen (pyridinoline crosslinks), aromatic amino acids (predominantly tryptophan and its oxidation products), nicotinamide adenine dinucleotide (NAD), and perhaps precursors or products of melanin (48). The recent development of photodynamic therapy of cancer has emphasized a longstanding clinical need to quantify tissue concentrations of cytotoxic drugs, which in this case can probably be done well optically. It is likely that specific, noninvasive, and useful optical diagnostics will be developed over the next decade. The possibility of using specific fluorescent probes for in vivo diagnosis remains essentially unexplored.

## IV. SUMMARY

The optics of skin are highly dynamic, vary widely with pigmentation and other factors, and constitute part of the skin's functions of photoprotection and appearance. Despite the complexity with which light absorption and scattering interplay to create skin optics, a useful and approximately accurate model can be described. In the UVC and UVB regions, strong absorption by protein, DNA, urocanic acid, and melanin in stratum corneum and epidermis forms a remarkably effective barrier against actinic damage. Urocanic acid is extractable from skin into water, which increases sensitivity to UVC and UVB radiation. Melanin is the only major epidermal pigment capable of UVA photoprotection. Although strongly influenced by race and heredity, melanogenesis is induced by multiple suberythemogenic exposures, especially in the UVA region. Epidermal hyperplasia and melanogenesis are major photoprotective responses of the skin. The optics of the dermis are dominated by light scattering, which varies inversely with wavelength. Very little is known about the UV optical properties of dermis. The absorption of light by hemoglobins in dermal vessels is strong, except for red and near-infrared wavelengths, which penetrate deeply into the skin and beyond. Selective vascular absorption of yellow–orange laser pulses has been developed as a tool for the treatment of microvascular malformations with minimal scarring, which relies directly on differential optical absorption within the dermis. Some benign pigmented lesions and tattoos can likewise be removed with selectively absorbed laser pulses. In the highly penetrating red and near-infrared regions, deep phototoxicity is possible with photodynamic therapy, using oxygen-dependent photosensitizers. In the far infrared region, strong absorption by tissue water allows uniform, efficient tissue removal by surgical lasers such as the $CO_2$ laser. Noninvasive optical diagnostic techniques are still in an early stage of development.

## REFERENCES

1. Anderson RR, Parrish JA. The optics of human skin. J Invest Dermatol 1981; 77:13–19.
2. Van Gemert MJC, Jacques SL, Sterenborg HJCM, Star WM. Skin optics. IEEE Trans Biomed Eng 1989; 36:1146–1153.
3. Welch AJ, Yoon G, van Gemert MJC. Practical models for light distribution in laser-irradiated tissue. Lasers Surg Med 1987; 6:499–493.
4. Wilson BC. A Monte Carlo model for the absorption and flux distributions of light in tissue. Med Phys 1983; 10:824–827.
5. Scheuplein RJ. A survey of some fundamental aspects of the absorption and reflection of light by tissue. J Soc Cosmet Chem 1964; 15:111–112.
6. Anderson RR. Polarized light examination and photography of the skin. Arch Dermatol 1991; 127:1000–1005.

7. Parrish JA. Phototherapy and photochemotherapy of skin diseases. J Invest Dermatol 1981; 77:167–171.
8. Anderson RR, Parrish JA. Optical properties of human skin. In: Regan JD, Parrish JA, eds. The Science of Photomedicine. New York: Plenum Press, 1982: 147–164.
9. Anderson RR, Beck H, Bruggemann U, Farinelli W, Jacques SL, Parrish JA. Pulsed photothermal radiometry in turbid media: internal reflection of backscattered radiation strongly influences optical dosimetry. Appl Optics 1989; 28:2256–2262.
10. Judd DB. Fresnel reflection of diffusely incident light. J Res Natl Bur Stand 1942; 29:329–332.
11. Bruls WAG. Optical Properties of Human Epidermis. PhD thesis, University of Utrecht, Utrecht, The Netherlands, 1984.
12. Bruls WAG, van der Leun JC. Forward scattering properties of human epidermal layers. Photochem Photobiol 1984; 40:231–242.
13. Everett MA, Yeargers E, Sayre RM, Olson RL. Penetration of epidermis by ultraviolet rays. Photochem Photobiol 1966; 5:533–542.
14. Lucas NS. The permeability of human epidermis to ultraviolet radiation. Biochem J 1930; 25:57–70.
15. Sandritter W. Z Pathol 1953; 64:520–530.
16. Baden HP, Pathak MA. The metabolism and function of urocanic acid in skin. J Invest Dermatol 1967; 48:11–17.
17. Rosario R, Mark GJ, Parrish JA, Mihm MC Jr. Histological changes produced in skin by equally erythemogenic doses of UV-A, UV-B, UV-C and UV-A with psoralens. Br J Dermatol 1979; 101:299–308.
18. Reeve VE, Greenoak GE, Canfield PJ, Boehm-Wilcox C. Topical urocanic acid enhances UV-induced tumor yield and malignancy in the hairless mouse. Photochem Photobiol 1989; 49:459–464.
19. Morison WL. What is the function of melanin? Arch Dermatol 1985; 121:1160–1163.
20. Szabo G, Gerald AB, Pathak MA, Fitzpatrick TB. Racial differences in the fate of malanosomes in human epidermis. Nature 1969; 222:1081–1084.
21. Parrish JA, Zaynoun S, Anderson RR. Cumulative effects of repeated subthreshold doses of ultraviolet radiation. J Invest Dermatol 1981; 76:356–359.
22. Parrish JA, Jaenicke KF, Anderson RR. Erythema and melanogenesis action spectra of normal human skin. Photochem Photobiol 1982; 36:187–191.
23. Wolbarsht ML, Walsh AW, George G. Melanin: a unique biological polymer. Appl Optics 1981; 20:2184–2188.
24. Anderson RR, Margolis RJ, Wantenabe S, Flotte T, Hruza GJ, Dover JS. Selective photothermolysis of cutaneous pigmentation by Q-switched Nd:YAG laser pulses at 1064, 532, and 355 nm. J Invest Dermatol 1989; 93:28–32.
25. Jacques SL, Alter CA, Prahl SA. Angular dependence of HeNe laser light scattering by human dermis. Laser Life Sci 1987; 1:309–333.
26. Findlay GH. Blue skin. Br J Dermatol 1970; 83:127–134.
27. Anderson RR. The physical basis of brown skin colors (melanoderma). In: Fitzpatric TB, Wick M, Toda K, eds. Brown Melanoderma. Tokyo: University of Tokyo Press, 1986: 3–7.

28. Anderson RR, Hu J, Parrish JA. Optical radiation transfer in the human skin and applications in in vivo remittance spectroscopy. In: Marks R, Payne PA, eds. Proceedings of the Symposium on Bioengineering and the Skin. Lancaster, UK: MTP Press Ltd, 1981: 245–254.
29. Wan S, Anderson RR, Parrish JA. Analytical modeling for the optical properties of the skin with in vitro and in vivo applications. Photochem Photobiol 1981; 34:493–499.
30. Anderson RR, Parrish JA. Microvasculature can be selectively damaged using dye lasers: a basic theory and experimental evidence in human skin. Lasers Surg Med 1981; 1:263–266.
31. Anderson RR, Parrish JA. Selective photothermolysis: precise microsurgery by selective absorption of pulsed radiation. Science 1983; 220:524–527.
32. Paul BS, Anderson RR, Jarve J, Parrish JA. The effect of temperature and other factors on selective microvascular damage caused by pulsed dye laser. J Invest Dermatol 1983; 81:333–335.
33. Gay-Grosier F, Polla LL, Tschopp J, Schifferli JA. Complement activation by pulsed tunable dye laser in normal skin and hemangioma. J Invest Dermatol 1990; 94:426–431.
34. Greenwald J, Rosen S, Anderson RR, Harrist TJ, MacFarland F, Noe J, Parrish JA. Comparative histologic studies of the tunable dye (at 577 nm) laser and argon laser: the specific vascular effects of the dye laser. J Invest Dermatol 1981; 77:305–310.
35. Tan OT, Murray S, Kurban AK. Action spectrum of vascular-specific injury using pulsed irradiation. J Invest Dermatol 1989; 92:868–871.
36. Levins PC, Grevelink JM, Anderson RR. Action spectrum of immediate and delayed purpura in human skin using a tunable pulsed dye laser. J Invest Dermatol 1991; 96:588.
37. Wan S, Parrish JA, Anderson RR, Madden M. Transmittance of nonionizing radiation in human tissues. Photochem Photobiol 1981; 34:679–681.
38. Hardy JD, Hammel GT, Murgatroyd D. Spectral transmittance and reflectance of excised human skin. J Appl Physiol 1956; 9:257–264.
39. Dougherty TJ. Photodynamic therapy: status and potential. Oncology 1989; 3:67–73.
40. Dover JS, Margolis RJ, Polla LL, Watanabe S, Hruza GJ, Parrish JA, Anderson RR. Pigmented guinea pig skin irradiated with Q-switched ruby laser pulses: morphologic and histologic findings. Arch Dermatol 1989; 125:43–49.
41. Taylor CR, Flotte T, Michaud N, Jimbow K, Anderson RR. Q-switched ruby laser (QSRL) irradiation of benign pigmented lesions: dermal vs. epidermal. Lasers Surg Med Suppl 1991; 3:65 (abstr).
42. Taylor CR, Gange RW, Dover JS, Flotte TJ, Gonzalez E, Michaud N, Anderson RR. Treatment of tattoos by Q-switched ruby laser. Arch Dermatol 1990; 126:893–899.
43. Kollias N, Baqer AH. Quantitative assessment of UV-induced pigmentation and erythema. Photodermatology 1988; 5:53–60.
44. Wilson BC, Anderson RR. Laser reflectance spectroscopy of tissue. In: Pratesi R, ed. Lasers in Medicine and Biology, NATO Proceedings. In press.

45. Gilchrest BA, Fitzpatrick TB, Anderson RR, Parrish JA. Localization of melanin pigmentation in the skin with Wood's lamp. Br J Dermatol 1978; 96:245–248.
46. Leffell DJ, Stetz ML, Milstone LM, Deckelbaum LI. In vivo fluorescence of human skin, a potential marker of photoaging. Arch Dermatol 1988; 124:1514–1518.
47. Anderson RR. In vivo fluorescence of human skin (letter). Arch Dermatol 1989; 125:999–1000.
48. Fellner MJ, Chen AS, Mont M, McCabe J, Baden M. Patterns and intensity of autofluorescence and its relation to melanin in human epidermis and hair. Int J Dermatol 1979; 18:722–730.

3

# Light Sources in Photomedicine

**Thomas F. Anderson**
*University of Michigan Medical School*
*Ann Arbor, Michigan*

The first light source used in photomedicine was natural sunlight. Although this continues as a common source, due to variations in intensity, temperature, weather, and the constraints of modesty, alternative artificial sources of radiation have been developed over the years. The history of photomedicine parallels the development of these light sources. In this chapter, the artificial light sources commonly used in photomedicine will be discussed (Table 1) with the exception of those used in extracorporeal photopheresis, photodynamic therapy and laser photomedicine, which are discussed in separate chapters.

## I. THE CARBON ARC LAMP

The carbon arc lamp used by Neils Finsen in his Nobel Prize-winning work in the treatment of cutaneous tuberculosis was the first artificial source used in photomedicine (1). Electromagnetic radiation is released when electric current is passed between electrodes forming an arc or spark gap. As the carbon in the electrodes vaporizes and is consumed, a continuous spectrum of radiation between 200 and 1500 nm is emitted. The emission spectrum can be altered with the addition of specific metals, such as cerium in the Sunshine carbon electrode. Although used in industry, the carbon arc lamp is unsuitable for medical office practice since it produces a variable radiant output, results in considerable smoke and odor, and is expensive and cumbersome to use (1,2).

## II. CLOSED VAPOR ARC LAMPS

The open carbon arc led to the development of closed vapor arc lamps (2,3). By passing an electrical arc through a silica glass-jacketed envelope containing a gas

**Table 1** Summary of Light Sources Used in Photomedicine

| Light Source | Emission Spectrum | Erythema Dosage (UVB) | Photopatch Test Dosage (UVA) | Advantages | Disadvantages |
|---|---|---|---|---|---|
| Sunlight | Continuous 286–4000 nm | 20 min, latitude 40 on June midday | Glass or Mylar filter 60 min | Economical, large field | Variable unpredictable output |
| Carbon arc | Continuous 200–1500 nm, peak 800–1000 nm (dependent on trace metals) | 10–30 mJ/cm$^2$, 5–30 sec at 25 cm | 0.5–1.5 J/cm$^2$, 3 min at 25 cm | Historical value: Finsen Nobel Prize research | Fumes and variable irradiance as electrodes are destroyed |
| Gas discharge arcs | | | | | |
| Low-pressure mercury “cold quartz” lamps | Discontinuous, predominantly 253.7 nm | 30 sec at 25 cm | Not suitable | Cannot produce 2+ erythema; germicidal | Limited uses |
| Medium-pressure mercury “hot quartz” lamp | Discontinuous with peaks at 254, 263, 297, 303, 365, 405, 440 nm | 10–30 mJ/cm$^2$, 30–60 sec at 45 cm | 0.5–4.0 J/cm$^2$, 10–60 min at 45 cm using window glass filters | Relatively inexpensive, intense UVB | 3 min warm-up; source for both therapy and diagnosis; medium field size |
| Medium-pressure “metal halide” | Continuous spectrum 295–420 nm | 10–40 secs at 10 cm (40 mJ/cm$^2$) | 30–90 sec at 40 cm using special filter | High-intensity, UVA/SUP | 3 min warm-up; medium field size; bulb replacement at 1,000 hr; heat production; moderately expensive |

| | | | | | |
|---|---|---|---|---|---|
| Wood's lamp–above with nickel oxide filter | 340–450 nm | Not suitable | 5–10 min at 15 cm | Fluorescence test | Not suitable for phototherapy |
| Xenon arc lamp (150–1000 Watt) | Continuous 290–410 nm | 10–30 mJ/cm$^2$ = 12 sec at 55 cm | 1–5 J/cm$^2$ with Schott and IR filters (90–120 sec) | With appropriate filters, can be used in research | Excessive IR technical, problems; small field size |
| Fluorescent Lamps | | | | | |
| UVB: "Sunlamp" | Continuous, 270–390 nm, peak 313 nm | 10–30 mJ/cm$^2$ = 60–120 sec at 25 cm | Not suitable | Large field size when mounted in banks | Lamp output degrades with use, 70% at 500 hr |
| Intermediate UVA/UVB | Continuous 300–400 nm | 30 min at 25 cm | — | Immediate warm-up time | High costs for cabinets and bulb replacement |
| UVA: "black light" | Continuous 310–450 nm, peak 365 nm | — | 5–30 min at 25 cm | Excellent source for phototesting | Energy per unit area less than mercury lamps |
| Visible: "special blue day light, etc." | Continuous 400–700 nm | — | — | Range of spectral distribution, variety of purposes | Long treatment times; multiple ballasts and cabinet construction result in heat production |
| Incandescent lamps | Visible IR | — | — | "Heat lamp" | Poor UV producers |

Abbreviations: IR, infrared; SUP, selective ultraviolet phototherapy vapor heated to a high temperature and pressure, a variety of emission spectra can be produced depending on the gas vaporized and the pressure used.

## A. Xenon Arc Lamps

Xenon vapor compressed and heated to 20–40 atms produces a relatively continuous spectrum from 240 to 11,000 nm, similar to sunlight with the exception of an increased content of ultraviolet radiation. Other gases or mixtures can be used to alter the spectral output, such as zirconium, argon, and mercury. With the use of a Corning or Schott (WG system) filtering system, the xenon arc lamp can be incorporated into a device called a solar simulator, designed for a variety of phototesting needs (4,5). Excessive production of visible and infrared radiation can be reduced with the use of a quartz-jacketed water filter or dichroic reflectors (which transmit infrared and visible light while relecting ultraviolet radiation). A xenon arc solar simulator is capable of producing very intense radiation in the range of 10–30 $mJ/cm^2$ in 5–10 sec at 10 inches. However, they are limited by a very small field size of 3–10 cm. In addition, they are cumbersome to use, expensive, and are usually reserved for experimental investigation or sunscreen evaluation. Extremely high intensity radiation often necessary for industrial uses can be produced by intermittent pulse discharges of high-pressure xenon lamps. Whether these pulse lamps will prove to be useful for phototherapy or phototesting requires further evaluation.

Because of the relatively high intensity output of the solar simulator, the addition of spectral grating or a series of narrow-band pass filters combined with appropriate radiometry can be used as a monochrometer. This is an instrument capable of generating an action spectrum by comparing the biological effectiveness of the series of narrow wave bands of relatively equal energies.

## B. Mercury Vapor Arc Lamp

The mass production of the quartz-jacketed mercury vapor arc lamp led to an increase in ultraviolet phototherapy from the 1920s to the present (2,6–9). By passing an electrical arc through mercury vapor sealed in a silica glass (quartz) envelope, a discontinuous spectrum of characteristic peaks and valleys is produced (Fig. 1). At low temperature and pressure, 90–95% of the radiation produced is a peak at 254 nm. Other peaks produced include 185, 248, and 263 nm within the ultraviolet C (UVC) spectrum; 297, 303, and 313 nm in the UVB range; and 365, 405, 440, 554, and 578 nm in the UVA and visible spectra. Since 185 nm photons are capable of ionizing oxygen to ozone, all mercury vapor lamps must be used in a well-ventilated area. As the temperature and pressure within the lamp are raised, the relative intensities of the peaks shift, with less UVC and more UVB radiation produced. Operating at a low temperature and pressure, the "cold quartz" lamp has been used for its germicidal

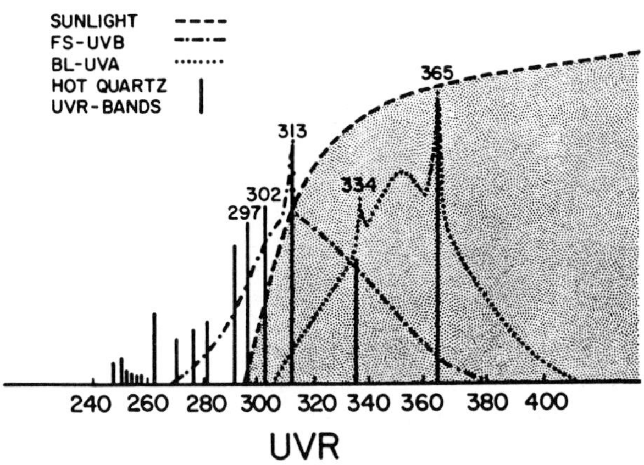


**Figure 1** By plotting relative intensity vs. wavelength differences between the various fluorescent lamps, one can visualize sunlight and hot quartz sources. Note the relative continuous spectrum of sunlight, and fluorescent lamps compared to the discontinuous bands seen in the hot quartz lamp. Because certain wavelengths of ultraviolet light are more efficient in producing specific photobiological events (action spectrum), care must be taken to match the proper light source with the desired result. (From Ref. 14.)

properties and for acne phototherapy for many years. It is obviously not useful for UVB or UVA phototesting or phototherapy.

The "hot quartz" (Fig. 2) or alpine sunlamp, which operates at higher temperature and pressure, produces a significant proportion of its output in the UVB range and is commonly used in dermatology. A typical minimal erythema dose of 10–30 mJ/cm$^2$ results from exposure for 30–60 sec at a distance of 45 cm from the lamp to the skin. With the use of a glass filter to block UVB, these lamps can be used for phototesting (7). However, because of their discontinuous spectrum, they are not ideal. Because these lamps produce only limited radiation between 320 and 360, they are not appropriate for psoralen photochemotherapy.

Hot quartz lamps have a medium exposure field size: approximately ½ to ⅓ of a human adult body can be exposed at 45 cm, thus necessitating shielding and fractionated exposure when total body irradiation is needed. Hot quartz lamps can only be used in well-ventilated areas, produce excessive infrared and heat, and require a warm-up time of 3–15 min to stabilize their spectral distribution and intensity. Despite these drawbacks, they are commonly used in phototherapy to treat localized areas. In addition, their intensity can be increased by shortening the lamp to skin distance, limited only by the infrared effects on the skin. The intensity of radiation is approximated by the inverse of the distance squared. Thus by cutting the distance in half from 45 to 22 cm, the intensity is increased approximately four fold (1,9). A flexible fiberoptic or liquid light-conducting guide can be adapted to a mercury lamp for high-intensity, small field size work (10).

**Figure 2** An adjustable, portable hot quartz lamp can be used for sites needing extra treatment and for difficult-to-reach areas. Since intensity of ultraviolet light approximately varies inversely with the square of the distance to the exposed target and is most intense when delivered perpendicular to the target, care must be taken to position the lamp the same way with each exposure.

### C. Metal Halide Lamps

An improvement over the hot quartz sun lamp is the metal halide source (11,12). This is a medium to high-pressure mercury vapor lamp to which metal halides have been added to produce other radiation peaks, filling the mercury spectrum into a relatively continuous spectral distribution. With the addition of filters to block UVC and UVB, these lamps produce more continuous and intense UVA, suitable for light testing or psoralen photochemotherapy. With the removal of a filter, they are excellent UVB sources as well.

## III. FLUORESCENT LAMPS

With the development of fluorescent tube technology in the 1940s, and the popularization of the use of the fluorescent "Zimmerman" cabinet in the 1960s, followed by the development of photochemotherapy and UVB treatment protocols in the 1970s and 1980s, the practitioner now has a wide variety of light sources from which to choose (1–3,13,14). Fluorescent lamps consist of a low-pressure mercury discharge source, housed in a thin glass tube whose inner surface has been coated with a special phosphor-filter. These phosphors absorb 254 nm radiation and subsequently readmit a continuous spectrum of longer-wavelength radiation. The fluorescent spectral distribution is dependent on the phosphor mixture utilized and can be designed to deliver a wide range of emission spectrums including visible light, UVA, and UVB with little to no infrared radiation. The intensity can be improved by increasing the length of the tube and with the addition of internal or external reflectors. Fluorescent lamps are typically available in 2, 4, and 6 foot long tubes (in addition to other lengths), and can be mounted in banks or in cabinets designed specifically for phototherapy or phototesting. To stabilize current and voltage necessary for a stable spectral distribution, fluorescent lamps need to be electrically balanced with appropriate "ballasts." Although fluorescent lamps do not produce a significant amount of heat themselves, unfortunately, the traditional ballasts do produce heat and therefore fluorescent cabinets need adequate cooling and ventilation. Newer "electronic" ballasts can operate with less heat production and improved ultraviolet intensity; however, they are much more susceptible to variations in current and line voltage. A bank of fluorescent tubes has the advantage of a larger field size at a relatively less cost and shorter "warmup time" than arc lamps. Unfortunately, the fluorescent intensity and spectral distribution of these lamps degrade with time, with only 70–80% of the original output after 500 hr of use. For filtering purposes or protection against lamp breakage, fluorescent lamps can be covered with a UV-transparent Teflon or UVB-filtering Mylar plastic sheathing. However, the use of these filtering sheaths can theoretically result in some spectral instability and excess loss of intensity over time due to the absorption of radiation altering the filtering effects in a process called "solarization" (15).

Because fluorescent tube nomenclature is not completely standardized, when switching from one manufacturer to another, care must be taken to measure the spectral power distribution or obtain a copy of this from the manufacturer (16–18). In general, lamps are designated by length (in inches, for example, FS40, F72, F74) and spectral emission (for example, CW for cool white visible, BL for black light, or S for sunlamp). Other designations include operating current (HO for high output, VHO for very high output) and endfitting designation (bipin, double contact, or "slim line" single contact). Each fluorescent lamp has its specific and appropriate ballast, therefore, lamps are not interchangeable even if endfittings are the same. A typical Westinghouse UVB sunlamp would be designated FS20, FS40, or FS60 depending on the length in inches. Examples of UVA lamps would be the Sylvania (GTE) FR74T12-PUVA (where R stands for internal reflector); Westinghouse F72T12/BL-S/HO or F72T12/BL-S/SHO. These lamps would correspond to the Philips F72T12BLHO or BLVHO UVA lamps.

A typical UVB fluorescent "sunlamp" containing, for example, a thallium-activated calcium–phosphate–phosphor mixture produces a continuous spectrum from 270 to 390 nm peaking at 313 nm. For example, a Westinghouse FS sunlamp produces approximately 60% of its total ultraviolet radiation as UVB (18). By altering the phosphor-filtering system, the Philips Company has developed a narrow–band fluorescent UVB lamp peaking at about 310–313 nm (19). Although these lamps require a higher dosage and longer treatment time, they have been proposed to be more effective and safer with less erythema than traditional broad-band UVB lamps.

Utilizing a cerium-activated calcium–phosphate–phosphor mixture, for example, two major UVA spectral distribution fluorescent lamps are available commercially. Lamps designed for psoralen photochemotherapy produce less then 3% UVB with approximately 20% UVA radiation in the 320–340 nm range necessary to activate psoralens maximally. Black lights designated BL for nonmedical use, such as in tanning parlors, may produce as little as 1% UVB but 99% of their UVA may be greater than 340 nm. These fluorescent lamps producing "long" UVA are not appropriate for UVB phototherapy and psoralen/UVA (PUVA) (16,18,20). Care must be taken when replacing lamps in phototherapy panels and cabinets not to use these nonmedical UVA lamps just because a wholesale distributor can underbid a competitor for their "UVA" lamp. The typical minimum phototoxic dosage of a bank of UVA lamps is 0.5–4 J/cm$^2$. This may be delivered in 2–10 min for a PUVA lamp or in 15–16 min for a BL lamp.

A fluorescent lamp with a peak rating midway between PUVA fluorescent lamps and sunlamps has been produced by the Klafs Sunlight Corporation. The Metec Helarium (Wolff-B1-12-100 watt) fluorescent tube produces around 30% of its UVA in the 320–340 nm short UVA range and around 8% UVB (18). This lamp has been promoted as a combination UVA/UVB selective ultraviolet light phototherapy (SUP) source, but in theory could also be useful for psoralen photochemotherapy (21). Visible light-producing fluorescent lamps used for general illumination emit negligible amounts of UVA and UVB. Therefore, office workers undergoing psoralen photochemotherapy need not worry about exposure to these lamps, although patients with marked photosensitivity (e.g., those with actinic reticuloid) may be affected by this radiation (22). Concern has been expressed that all fluorescent lamps produce more short-wavelength UVC radiation than is found in natural outdoor sunlight exposure (due to the filtering effects of the ozone layer) (23). Since UVC is a potent mutagen and can cause severe eye irritation, it has been proposed that all fluorescent lamps be filtered with plastic diffusers or other filters that block out UVC. Fluorescent visible phototherapy has two major applications: the therapy of neonatal hyperbilirubinemia and the treatment of winter depression.

Based on the principle that the absorption of blue visible light (between 420 and 490 nm) produced the isomerization and subsequent excretion of bilirubin, neonates with hyperbilirubinemia are commonly treated with so-called bili lights (1,24). The light sources commonly used include daylight F20T12/D, blue F20T12/B, and special blue F20T12/BB. Increasing the spectral output between 420 and 490 nm appears to be associated with increased efficacy, resulting theoretically because the blue emission spectrum closely approximates the absorption maximum of bilirubin. Banks of hemicylindrical units of these fluorescent lamps are available to be used with a variety of neonatal isolette units. Guidelines for the use of these lamps have been published.

An exciting new use for visible phototherapy is the treatment of winter depression or seasonal affective disorder (25). This involves daily exposure to full-spectrum fluorescent light from a panel of lamps providing direct illumination at the patient's eyelevel with about 2500 lux (Vita-Lite, Duro-Test Corporation) for 2–6 hours or 10,000 lux (Color-Gard, Duro-Test Corporation) for 30 min. (Lux is defined as the metric unit of illumination: 1 lumin/m$^2$, also called the meter candle.) Unlike all other forms of phototherapy, this treatment depends on direct or indirect exposure of the patient's eyes. In addition to bright light therapy for winter depression, treatment of subsyndromal "winter doldrums" as well as exploratory applications for nonseasonal circadium rhythm adjustment (jet lag, sleep timing, etc.) is being evaluated. Although ophthalmologic examinations of winter depressives treated with high-intensity bright light have thus far revealed no treatment-related abnormalities, concurrent follow-up ophthalmologic examinations are recommended with this treatment.

## VI. INCANDESCENT LAMPS

Incandescent lamps are designed with a coiled tungsten filament contained in an evacuated or inert gas-filled glass envelope (3). Electrical resistance produces operating temperatures around 3000 Kelvin emitting mainly infrared and visible radiation. Evaporation from the filament causes dark deposits to form on the glass envelope, which results in reduced transmission and ultimately breakage of the filament with time. The incorporation of halogen gases and the use of silica glass for the envelope increase ultraviolet production. These lamps are not efficient producers of ultraviolet light and have little place in clinical practice.

## V. COMPARISON OF PHOTOTHERAPY EQUIPMENT

Practicing dermatologists utilize three types of ultraviolet light sources in the treatment of psoriasis and other conditions and in phototesting (Table 1). The hot quartz medium-pressure mercury vapor lamp has been the most commonly utilized. However, in recent times fluorescent panels and cabinets and, to a lesser extent, the metal halide lamp systems have become the most common radiant light sources. Hot quartz lamps are usually used to treat limited psoriatic disease such as that occurring on the palms, soles, elbows, and knees. This is due in part to their moderate to small field size and fact that the intensity of radiation varies approximately with the inverse square of the distance from the light source to the target. Because of the portability of these lamps, they can also be used for bedridden patients with generalized psoriasis by treating the patient in four to eight segments. Care must be taken to shield previously treated areas to avoid double exposure. Although the hot quartz lamp is a relatively inexpensive source of intensive UVB radiation at short distances, it is losing popularity to fluorescent lamp systems. This is due in part to the fact that they produce ozone and heat and require adequate ventilation, in addition to their need for a minimum of 3 min lamp warm-up time and required cooling-off period between treatments if the lamp is extinguished. Because these lamps produce excess UVC radiation compared with fluorescent lamps, they produce erythema at lower intensities or minimum erythema doses than fluorescent lamps (13,26). Because UVC radiation is not as effective as UVB in the treatment of psoriasis, these lamps are only effective if patients are pushed to mild or moderate erythema. Although protocols have been developed for using hot quartz lamps for phototesting and/or photopatch testing (7), theoretically, they are not the best light sources due to their discontinuous spectrum.

Metal halide lamps, which were designed to produce a more continuous ultraviolet spectrum than mercury vapor lamps, would be theoretically better suited for localized photopatch testing. These units are designed with reflective backing to increase intensity and filterage designed to block UVC and maximize UVB or, alternatively, UVA irradiation. With a system of three vertical banks of five metal halide lamps per bank placed in the corners of an isosceles triangle, the field size of this metal halide system is comparable to a fluorescent cabinet and an entire patient may be irradiated (10). The metal halide lamp systems share with the hot quartz lamps the problem of heat and ozone production and the need for ventilation. Also with the combination of the metal halide lamps, the expense of this system is great. Although, in general, mercury vapor and metal halide light sources degrade more slowly than fluorescent lamps, the filtered ultraviolet radiation system may require a longer warm-up period for irradiant stability and may decrease in intensity more quickly than an unfiltered light source due to "solarization" or ultraviolet degradation of the filters (15).

The most common ultraviolet radiation source used for the phototherapy and photochemotherapy of psoriasis and other disorders is a system of fluorescent tubes arranged in banks and cabinets to maximize field size (Fig. 3). Depending

**Figure 3** Fluorescent cabinets can be designed to deliver both UVA and/or UVB with individual dosimetry controls. A time totalizer for lamp use can help track when the fluorescent tubes need replacement.

on the manufacturing process and phosphor-filtering system used, a range of spectral distributions can be produced with the intensities of the panels and cabinets dependent on the length, wattage, presence of an internal or external reflective coating, and current and voltage stability related to the selection of the proper ballasts. Morrison and Pike have published careful spectral distribution measurement data on the most commonly used fluorescent ultraviolet light sources (18). The Federal Bureau of Radiation Control has set standards for the appropriate spectral distributions to be used in UVB or PUVA medical radiation equipment (20). Black lights or UVA lights used in the tanning industry are not appropriate for medical phototherapy. A fluorescent lamp (Wolff-Helarium-B1-12-100W) distributed by the Klafs Sunlight Corporation produces an ultraviolet spectral distribution midway between typical UVB and PUVA lamps. A system of these lamps has been promoted as useful for combination UVA/UVB selective ultraviolet light phototherapy (SUP) popular in Europe (21,27). Although treatment times are much longer, this system can improve psoriasis and eczematous conditions such as atopic dermatitis with little or no irritating erythema. A comparison of the relative spectral distribution of erythemogenic UVB (+ UVC) versus short UVA (used in PUVA) and long UVA (used for tanning) is found in Table 2.

A cabinet or bank of fluorescent lamps has the advantage over the hot quartz system of a relatively large field size, a more immediate warm-up time, and relatively less heat production. Although the lamps themselves do not become significantly warm, the ballasts required to support multiple lamps in an enclosed cabinet system often result in excessive heat production, which could theoretically become a cardiovascular stress when treatment times become very long (28). Therefore, some fluorescent cabinet systems also require individual air conditioning or adequate venilation to remove excess heat. In addition to fluorescent cabinets, smaller units have been designed for localized treatment, such as of the palms and soles, scalp, or other localized areas. The major disadvantage of fluorescent lamp systems is that the output and spectral distribution of fluorescent lamps degrade with time and usually require replacement after 1,000 hr of use.

A number of factors affect the choice of phototherapy apparatus to be purchased by a given medical practice. The estimated frequency and type of skin disease to be treated must be compared to the proposed budget to determine whether a less costly, lower-intensity unit would be satisfactory or a very expensive, high-intensity unit capable of delivering either UVA or UVB is needed. Other factors that affect choice include space, ventilation, and electrical requirements of a potential location for equipment, and amount of training and background of the personnel utilizing the equipment. An apparatus recommended for home use would certainly differ in cost, size, and output from that to be used for a busy hospital service (29). A list of manufacturers who distribute phototherapy products can be found in Table 3.

**Table 2** Spectral Power Distribution of Erythemogenic UVB (+ UVC) and Short or Long UVA Expressed as Percentages of Total Emission in the UV Wave Band

| | WAVELENGTH (nm) | | |
|---|---|---|---|
| UV Sources | 250–320 % UVB (UVC) | 320–360 % Short UVA | 360–400 % Long UVA |
| Houva-Lite | 2.5 | 73.5 | 24.0 |
| Sylvania BL | 3.1 | 67.5 | 29.4 |
| Sylvania PUVA | 2.3 | 63.7 | 34.0 |
| General Electric BLB | 1.5 | 64.3 | 34.3 |
| General Electric BL | 3.8 | 38.1 | 58.1 |
| Sunlight (est) | 3.0 | 34.0 | 62.0 |
| Westinghouse BL | 0.4 | 10.9 | 88.7 |
| Metec Helarium | 7.7 | 65.3 | 27.0 |
| Sunlamp (FS) (20%<300) | 61.7 | 33.3 | 5.0 |
| Alpine UV (70%<300) | 60.4 | 5.5 | 34.1 |
| UVA Sun 5000 | — | 12.5 | 87.5 |
| Dermalight Metal Halide (est) | 1.0 | 36.0 | 63.0 |

An ideal light source for UVB phototherapy of psoriasis would maximize UV betweeen 300 and 320 nm while minimizing UVC. The best PUVA source would theoretically maximize 320–360 nm while minimizing UVB.

Lamp Sources:

Dermalight Systems, Studio City, CA (Dermalight 2005-3-metal halide)
General Electric Co., Baltimore, MD (F40BLB blacklight; F40BL blacklight)
Hanovia, Newark, NJ (Alpine UVB "Hot Quartz")
Klafs Sunlight Corp., Arlington Heights, IL (Bl-12-100W Metec)
Mutzhas Corp, Munich, Germany (UVA Sun 5000-metal halide)
National Biol. Corp., Cleveland, OH (F24T12-BL-HO Houva-Lite)
Sylvania, Danvers, MA (FR40T12 PUVA Lifeline; F40/350 BL blacklight)
Westinghouse Elect. Corp., Bloomfield, NJ (F40BL blacklight; FS40 "Sunlamp")

Sources: Refs. 2,18,29.

Although certain safety features of phototherapy equipment are specified by the Bureau of Radiation Health mandated by government regulation (30,31), there are a number of features that are important to compare beyond cost. Engineering design, including the size, type, number, and configuration of the light sources utilized have been evaluated; it has been demonstrated that the

**Table 3** Representative Manufacturers of Phototherapy Equipment (Not Inclusive)

Airan Ltd.
Lampputtie 4 Ltd.
Helsinki, Finland

Atlantic Ultraviolet (UVB, UVA, home)
250 North Fehr Way
Bay Shore, NY 11706
(516) 586–5900

Canterbury Instruments
124 Millstrood Road
Whitstable, Kent, U.K.

Coast Air Ltd.
Long Spring, Porters Wood
Valley Road Industrial Estate
St. Albans, Herts A13 6 NG, U.K.

Conrad-Hanovia
100 Chestnut Street
Newark, NJ 07105

Cooper-Hewitt Corp. (UVB, home)
20 Kenton Lands Road
Erlanger, KY 41018
1 (800) 544-3757

Daavlin Company (UVB, UVA)
P.O. Box 626
Bryan, OH 43506
(419) 636-6304

Derma-Control, Inc. (UVB, UVA, home)
9416 Gulf Stream Rd.
Frankfurt, IL 60423
(815)469-8027

Derma Light Systems (UVB, UVA, metal halide)
13135 Ventura Blvd. Suite 306
Studio City, CA 91604
1 (800) 882-8321

Dixwell
17-19 Rue Mazagran
69007 Lyon, France

Duffield Medical Equipment
Units 1 & 2, Goods Wharf
Good Road, Belper
Derbyshire DE5 1 UU, U.K.

Environmental Growth (UVB, UVA)
P.O. Box 390
Chagrin Falls, OH 44022
(216) 247-5100

Fuller Ultraviolet Corp
14806 S. Drexel Ave.
Dolton, IL 60419
(312) 841-4240

General Theraphysical (Hanau, UVB)
1520 Washington Ave.
St. Louis, MO 63103
(314) 231-9643

Klafs Sunlight Corp. (UVB, UVA, home)
210 Campus Drive
Arlington Heights, IL 60004

National Biologic Corp. (UVB, UVA, home)
1532 Enterprise Parkway
Twinsburg, OH 44087
1 (800) 338-5045

North Penn Mfg. Co. (UVB, home)
1100 Adams Avenue
Philadelphia, PA 19124
(215) 288-0805

Psoralite Corporation
1401 Pineview Dr.
Columbia, SC 29209
1 (800) 331-3534

RA Fischer (UVC, Wood's)
517 Commercial Street
Glendale, CA 91203
(818) 241-1178

| | |
|---|---|
| Richmond Light Co. (UVB, home)<br>2301 Falkirk Drive<br>Richmond, VA 23236<br>(804) 276-0559 | UltraDerm Systems (UVB, UVA, home)<br>Professional Arts Bldg.<br>2201 S. Michigan Ave.<br>Saginaw, MI 48602<br>(517) 792-6100 |
| Sallman-SUP-Lampen<br>Werrestrasse 94<br>D-4900 Herford<br>05221/24862 Germany | Ultralite Enterprises, Inc. (UVB, UVA)<br>390 Farmer Ct., P.O. Box 1895<br>Lawrenceville, GA 30246<br>(404) 936-0594 |
| Spectronics (UVB, Wood's)<br>956 Brush Hollow Rd.<br>Westbury, NY 11590<br>(516) 333-4840 | Waldmann-Elektromedizinische (UVB, UVA)<br>Gerate<br>Postfach 1240<br>D-7220 VS-Schwenningen, Germany |

design of a number of lamps equally spaced totally surrounding a patient is superior to the original Zimmerman cabinet concept of two fluorescent tubes in each corner of a box (32–35). Since the intensity of all ultraviolet light falls off each end of a fluorescent tube, it is important that the floor of the cabinet be raised so that a patient's feet are elevated a few inches above the beginning of the fluorescent tube. Both the horizontal and vertical variability of the ultraviolet intensity must be taken into consideration when one is purchasing an ultraviolet light cabinet. A stacked bank of mercury halide lamps may have a more uniform vertical intensity than a bank of fluorescent lamps (11). However, a bank of closely spaced reflected fluorescent lamps surrounding a patient may produce a more stable horizontal variability than a grouping of three stacked panels of metal halide lamps.

There have been very few published *direct* comparisons of phototherapy equipment in clinical practice (27,36–38). Fluorescent systems may have an advantage over mercury arc equipment, producing a higher efficacy to erythema ratio (13). As might be expected, a larger number of fluorescent lamps is associated with increased UV intensity and shorter treatment times. An increased number of light sources and the use of reflecting material undoubtedly result in a greater uniformity of ultraviolet flux as well (34).

Another technical feature of great importance is the dosimetry control unit. This engineering feature can vary from a simple timer and on/off switch to a computerized panel that can indicate dosage in both time and energy per unit area of specific amounts of UVA and UVB. Folding cabinets and panels can be built or purchased to conform to small space requirements (29). Comfort and safety features that are important to compare include the mechanism for controlling temperature and ventilation, mechanism for shielding ultraviolet from staff, the type and position of hand rails, protection from bulb breakage, method for patient visualization (Fig. 4), and dosimetry control fail-safe features, including

**Figure 4** A visible translucent, ultraviolet opaque window, located above the control panel, for direct visualization of patients is an excellent safety feature. As an alternative, some manufacturers utilize an overhead mirror for patient visualization. (From Ref. 14.)

back-up timers and automatic shut-off systems (31). Acrylic sheaths or metal barriers or screening have been used to prevent accidental contact of patients with lamps and protect from bulb breakage (Fig. 5). The ease with which these protective systems can be removed is an important feature that determines down time when lamp replacement is necessary.

**Figure 5** Safety standards direct that patients must not be allowed to come into direct contact with the lamps. A thin-wire mesh screen can be utilized or ultraviolet translucent plastic panels or sheathing can protect patients. However, the ease with which lamps can be changed is an important factor in the maintenance of phototherapy equipment.

## VI. EQUIPMENT ACCESSORIES

Eye protection is an absolute necessity for patients and staff during treatment (39,40). In the case of psoralen photochemotherapy, patients must also be protected from direct or window glass sun exposure until blood levels of psoralens are neglible. Typical office lighting from visible fluorescent lamps is not a significant risk to these patients. Wraparound UV-opaque sunglasses, goggles, and ultraviolet filtering-coated lenses may be obtained from ultraviolet light equipment manufacturers, or directly from the companies listed in Table 4.

To be clinically meaningful, measurement of ultraviolet output from light sources using radiometers must be properly performed (41, 42). The spectral response of the radiometer must be correlated with the spectral distribution of the

**Table 4** Photoprotective Eye Apparatus

| | | |
|---|---|---|
| **Goggles Worn During Treatment** | | |
| NOIRettes | Recreational Innovations Co.<br>P.O. Box 159<br>South Lyon, MI 48178<br>(313) 769-5565 | |
| Super Sunnies | Lucas Products Corp<br>Toledo, OH 43612 | |
| **Wraparound Glasses (UVA and UVB)** | | |
| Blak-Ray | UVP, Inc. (formerly:<br>Ultraviolet Products Inc.)<br>5100 Wallnut Grove Avenue<br>San Grabiel, CA 91778<br>(213) 285-3123 | Model: Contrast Control<br>Spectacles UVC-303; yellow,<br>tinted, or clear, worn over<br>prescription frames.<br>Cost: $6.00 |
| Jelly Bean | American Optical Safety Products<br>327 N. Main St<br>Bluffton, OH 45817<br>(800) 722-9921 | UV Safety glasses<br>Variety of colors<br>Cost: $2.30 + shippping |
| NOIR | Recreational Innovations Co.,<br>P.O. Box 159<br>South Lyon, MI 48178<br>(313) 769-5565 | Models 101 or 501: amber<br>Models 102 or 502: grey–green<br>Worn over prescription frames.<br>Cost: $13.50 + $2.10 shipping<br>& handling |
| Retna-Spec | Eye Communications<br>870 South Myrtle Ave.<br>Monrovia, CA 91016 | Amber, Green, or Red<br>Worn over prescription frames<br>Cost: $15.00 + shipping |
| Silver-Shield goggles | Dioptics Medical Products<br>15550 Rickfield Blvd., Suite C<br>Irvine, CA 92718<br>(714) 859-7111 | Tinted brown or gray<br>Cost: $5.00 + shipping |
| **Clear UVA-Blocking Lens** | | |
| UV 400 (Orcolite) | Dioptics Medical Products,<br>order thru local optician<br>(patient brochures available) | Clear or tinted<br>Cost: approx. $10.00 more<br>than ordinary prescription<br>lenses; determined by optician |
| UV 400-clip-ons | Dioptics Medical Products,<br>order through above address | Amber or gray tint<br>Worn over prescription frames<br>Cost: $18.00/pair through<br>dermatologist |

ultraviolet light sources. For example, a broad-band UVA detector might give a high "UVA" reading from a blacklight used in a tanning parlor, although, as discussed previously, this light source is not appropriate for psoralen photochemotherapy (14,16). Radiometer (and solar simulator) manufacturers or distributers are listed in Table 5.

**Table 5** Solar Simulators/Monochromometers and Radiometers

| |
|---|
| EG & G Electro-optics<br>35 Congress Street<br>Salem, MA 01970<br>(508) 745–3200 |
| Eppley<br>Sheffield Avenue<br>Newport, RI 02840<br>(401) 847-1020 |
| International Light<br>Dexter Industrial Green<br>Newburyport, MA 01950<br>(617) 465-5923 |
| Spectral Energy<br>67 Woodland Avenue<br>Westwood, NJ 07675<br>(201) 664–0876 |
| Optronics Lab<br>Orlando, FL 32811<br>(407) 422–3171 |
| Oriel Corporation<br>250 Long Beach Boulevard<br>Stratford, CT 06497<br>(203) 357-1600 |
| Solar Light Company<br>6655 Lawnton Avenue<br>Philadelphia, PA 19126 |
| Xenon Corporation<br>66 Commerce Way<br>Woburn, MA 01801<br>(617) 938-3594 |

## REFERENCES


1. Harber LC, Bickers, DR. Photosensitivity Diseases: Principles of Diagnosis and Treatment. Philadelphia: W.B. Saunders, 1989.
2. Harber LC, Bickers DR, Epstein JH, Pathak MA, Urbach F. Report on ultraviolet light sources. Report by the Task Force on Photobiology of the National Program for Dermatology. Arch Dermatol 1974; 109:833–834.
3. Phillips R. Sources and Application of Ultraviolet Radiation. London: Academic Press, 1983.
4. Berger DS. Specification and design of solar ultraviolet simulators. J Invest Dermatol 1969; 53:192–199.
5. Wulf HC, Brodthagen H. Solar simulators: modifications for testing with visible light. Acta Derm Venereol (Stockh) 1982; 63:240–242.
6. Beckett RH. Modern Actinotherapy. London: William Heinemann Medical Books, 1955.
7. Epstein S. Simplified photopatch testing. Arch Dermatol 1966; 92:216–220.
8. Fitzpatrick TB, ed. Sunlight and Man: Normal and Abnormal Photobiologic Responses. Tokyo: University of Tokyo Press, 1974.
9. Magnus IA. Dermatological Photobiology. London: Blackwell Scientific Publications, 1976.
10. Plewig G, Hofmann C, Braun-Falco O, Nath G, Kreitmair A. A new apparatus for the delivery of high intensity UVA and UVA + UVB irradiation, and some dermatological applications. Br J Dermatol 1978; 98:15–23.
11. Chue B, Borok M, Lowe NJ. Phototherapy units: comparison of fluorescent ultraviolet B and ultraviolet A units with a high-pressure mercury system. J Am Acad Dermatol 1988; 18:641–645.
12. Mutzhas MF, Hölzle E, Hofmann C, Plewig G. A new apparatus with high radiation energy between 320–460 nm: physical description and dermatological applications. J Invest Dermatol 1981; 76:42–47.
13. Diffey BL, Farr PM. An appraisal of ultraviolet lamps used for the phototherapy of psoriasis. Br J Dermatol 1987; 117:49–56.
14. Anderson TF. Phototherapy equipment: selection of apparatus. In: Lowe NJ, ed. Practical Psoriasis Therapy. Chicago: Yearbook Medical Publishers, 1985:52–67.
15. Diffey BL. The stability of light sources: implications for photobiological studies. Photochem Photobiol 1988; 47:317–320.
16. Fleming MG, Brody N. Biologic effectiveness of black light fluorescent lamps for PUVA. Letter to the editor. J Am Acad Dermatol 1985; 12:894.
17. Cole CA, Forbe PD, Devies RD. Different biologic effectiveness of blacklight fluorescent lamps available for therapy with psoralens plus ultraviolet A. J Am Acad Dermatol 1984; 11:599–606.
18. Morison WL, Pike RA. Spectral power distributions of radiation sources used in phototherapy and photochemotherapy. J Am Acad Dermatol 1984; 10:64–68.
19. Van Weelden H, De La Faille H, Young E, Van Der Leun JC. A new development in UVB phototherapy of psoriasis. Br J Dermatol 1988; 119:11–19.

20. Devices used as a radiation source (in combination with psoralen drug) in the photochemotherapy for psoriasis. Fed Reg February, 1960.
21. Paul BS, Stern RS, Parrish JA, Arndt KA. Low-intensity selective UV phototherapy. Arch Dermatol 1983; 119:122–124.
22. Morison WL, Strickland PT. Environmental UVA radiation and eye protection during PUVA therapy. J Am Acad Dermatol 1983; 9:522–525.
23. Elwood JM, Maxwell KJ. UV radiation from fluorescent lights. Lancet 1983; 2:579.
24. Sisson TRC. Visible light therapy of neonatal hyperbilirubinemia. In: Smith CK, ed. Photochemical and Photobiological Reviews, Vol. 1. New York: Plenum Press, 1976:241–268.
25. Termon M, Remé CE, Rafferty B, Gallin PF, Termon JS. Bright light therapy for winter depression: potential ocular effects and theoretical implications. Photochem Photobiol 1990; 51:781–792.
26. Parrish JA, Jaenicke KF. Action spectrum for the therapy of psoriasis. J Invest Dermatol 1981; 76:359–364.
27. Jekler J, Larkö O. Combined UVA-UVB versus UVB phototherapy for atopic dermatitis: a paired-comparison study. J Am Acad Dermatol 1990; 22:49–53.
28. Ciafone RA, Rhodes AR, Audley M, Freedberg IM, Abelmann WH. The cardiovascular stress of photochemotherapy (PUVA). J Am Acad Dermatol 1980; 3:499–505.
29. Jordan WP, Clarke AM, Hale RK. Long-term modified Goeckerman regimen for psoriasis using an ultraviolet B light source in the home. J Am Acad Dermatol 1981; 4:584–91.
30. Environmental Health Criteria 14. Ultraviolet radiation. Geneva: World Health Organization, 1979. Finland: United Nations Environmental Health Organization, World Health Organization, and International Radiation Protection Association, 1979.
31. Bickford ED, Berger DS, Corth R, Crouch CL, Daniels F, Griffith JW, Guth SR, Ham WT, Hughes PC, Job DD, Lewis AL, Matelsky I, Neer RM, Parr WH, Parrish JA, Sausville JW, Sliney DH, Thorington L. Risks associated with use of UV-A irradiators being used in treating psoriasis and other conditions. Lighting Design Appl 1979.56–58.
32. Nachtwey DS, Rundel RD. A photobiological evaluation of tanning booths. Science 1981; 211:405–407.
33. Androphy EJ, Eaglestein WH. Letter to the editor. UV-B-energy distribution in phototherapy equipment. Arch Dermatol 1984; 120:17.
34. Horwitz NS, Frost P. A phototherapy cabinet for ultraviolet radiation therapy. Arch Dermatol 1981; 117:469–473.
35. Zimmerman MC. Ultraviolet light therapy. Arch Dermatol 1958; 78:646–652.
36. Roenigk HH Jr. Comparison of phototherapy systems for photochemotherapy. Cutis 1977; 20:485–489.
37. Roenigk HH Jr, Martin JS. Photochemotherapy for psoriasis. Arch Dermatol 1977; 113:1667–1670.

38. Farr PM, Diffey BL, Marks JM. Phototherapy and dithranol treatment of psoriasis: new lamps for old. Br Med J 1987; 294:205–207.
39. Davey JB, Diffey BL, Miller JA. Eye protection in psoralen photochemotherapy. Br J Dermatol 1981; 104:295–300.
40. Davies JE. Sore eyes due to accidental exposure to ultraviolet B. Lancet 1984; 2:983.
41. Mackenzie LA. UV radiometry in dermatology. Photodermatology 1985; 2:86–94.
42. Diffey BL. A comparison of dosimeters used for solar ultraviolet radiometry. Photochem Photobiol 1987; 46:55–60.

# 4

# Pharmacology of Therapeutic Agents in Photomedicine

**Ira C. Davis**
*Bowman Gray School of Medicine, Winston-Salem, North Carolina*
**Matthew J. Stiller and Jerome L. Shupack**
*New York University School of Medicine, New York, New York*

Agents used in photomedicine include anthralin, tar, and psoralens, which interact with ultraviolet light in a therapeutic manner. Other drugs used in photomedicine are antimalarials, azathioprine, cyclosporine, and retinoids. The above agents will be discussed in this chapter. Therapeutic agents, such as glucocorticosteroids, antihistamines, and methotrexate, with no unique application in photomedicine will not be discussed.

## I. ANTHRALIN

Anthralin (Fig. 1) is a synthetic compound structurally related to chrysarobine, a Goa powder derivative, which was found to be an efficacious antipsoriatic compound. Its pharmacology and mechanisms of action have been extensively reviewed (1–3). Oxygen radicals generated by anthralin are probably responsible for many of its therapeutic and irritant properties (1). The oxidation products of these free radicals are responsible for the staining seen with these preparations.

Anthalin can inhibit synthesis of nuclear and mitochrondrial DNA as well as proliferation of psoriatic keratinocytes in vitro. There is a decrease in polyamine synthesis and cGMP levels in anthralin-treated psoriatic skin. A decrease in arachidonic acid levels unaccompanied by a significant decrease in leukotriene $B_4$ or 12-hydroxyeicosatetraenoic acid concentrations was noted as well (4). Furthermore, anthralin acts as an electron donor to the mitochondrial electron transport chain, resulting in inhibition of adenosine triphosphate (ATP) synthesis (5). On ultrastructural analysis, mitrochondrial swelling is noted after single

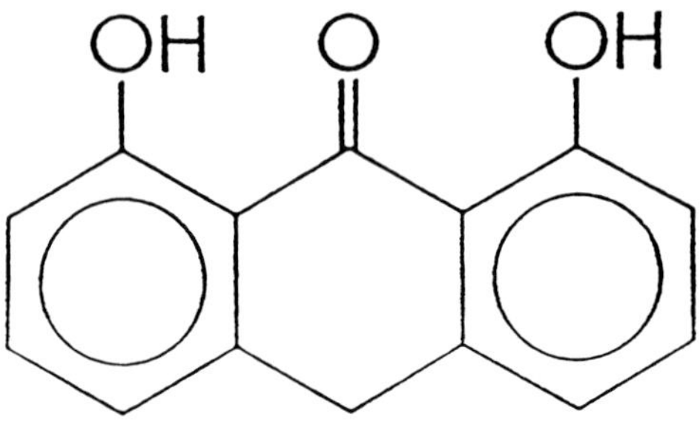

**Figure 1** Structure of anthralin.

applications of anthralin to psoriatic lesions (6). With repeated use of the medication, a reduction in the number of mitochondria is noted (7). Finally, anthralin inhibits human polymorphonuclear leukocyte chemotaxis and mitogen-stimulated lymphocyte proliferation in vitro. A reduction in the number of neutrophils was noted in anthralin-treated psoriatic plaques after 2 weeks; T cells and monocyte–macrophage (CD14) numbers were decreased after 8 weeks (3). Despite these molecular, metabolic, and cellular effects of anthralin, a clear understanding of the primary factor or factors responsible for anthalin's efficacy remains to be elucidated.

Anthralin has been used in conjunction with ultraviolet (UV) B and coal tar baths (Ingram method), psoralen and UVA (PUVA), and etretinate. Maximal inflammatory response is noted at 48–72 hr. Major side effects include erythema, staining, and pigmentation. Contact dermatitis has also been reported. The incidence of erythema increases with higher concentrations of anthralin. Pretreatment with UVB or topical PUVA reduces the erythema. Short contact therapy has been advocated also in order to decrease irritation and staining. Application of 10% triethanolamine after removal of anthralin has resulted in a reduction in anthralin-induced inflammation without affecting therapeutic efficacy (8). Hair will stain purple under alkaline conditions, but has been reported to stain yellow under acidic conditions (9).

## II. TARS

Tar products used in dermatology are derived from the distillation of wood, or from the condensation of the gases produced when coal is heated to form coke (10). Crude coal tar and its extract liquor carbonis detergens (LCD) are used more commonly than the wood tar derivatives. Coal tar has been used topically in photomedicine since the 1920s, when Goeckerman introduced it in the treatment of psoriasis in conjunction with ultraviolet light. Because coal tar is a complex mixture containing over 10,000 different compounds, there is no means to standardize the product to compare the potency of various tar preparations.

Tar is a phototoxic substance with an absorption spectrum in the UVA range. In fact, Parrish and colleagues (11) have shown that photosensitizing dosages of UVA are as effective as UVB in clearing psoriasis in patients treated with a 5% tar gel. The use of UVA, however, is limited by the long exposure times necessary to deliver the photosensitizing dose and by an acute burning pain experienced by the patient when treated with a tar–UVA regimen. Because, in practice, tar is commonly used with UVB, a synergistic effect between tar and UVB has been proposed. Tar induces a decrease in epidermal thickness in healthy volunteers, and suppression of DNA synthesis in a hairless mouse model (10,12). Adverse effects include irritant or allergic contact dermatitis, folliculitis, and phototoxic reactions.

In the last decade, controversy has arisen regarding the efficacy of coal tar products in phototherapy. Lowe and colleagues (12) noted that 1% and 5% crude coal tar in petrolatum as well as a 5% tar extract in an oil base were clinically superior to their respective vehicles when they were combined with suberythemogenic dosages of UVB. However, in a bilateral comparison study, Stern et al (13) noted no statistically significant differences between suberythemogenic dosages of UVB given three times a week with either tar oil or its vehicle, each applied to one-half of the body.

The clinical efficacy of coal and wood tars must be underscored by the potential mutagenicity of the polycyclic aromatic hydrocarbon constituents. DNA adducts in the skin of psoriatic patients treated with a 5-day course of coal or juniper tar have been reported (14). Similar findings were reported in mice and human skin explants treated with these pharmaceutical tar products. However, a review of the literature revealed no increased incidence in skin cancer in patients with psoriasis treated with coal tar (15). Retrospective studies of over 500 patients with psoriasis or atopic dermatitis treated with the Goeckerman regimen and followed up 25 years after the initial therapy noted no increase in skin cancer compared to annual age specific incidence rates (16,17). The cooperative PUVA group is the only clinical study to note an increase in skin carcinoma in patients with a previous history of high exposure to tar and/or UVB (18). Further studies are needed to clarify the role of tar in phototherapy with regard to both efficacy and carcinogenicity.

## III. PSORALEN

Psoralen is the general term used to describe chemical compounds called furocoumarins, which are found in four to five major plant families. The major psoralens used in dermatology are 8-methoxypsoralen (8-MOP), 5-methoxypsoralen (5-MOP), and trimethylpsoralen (TMP) (Fig. 2).

The use of 8-MOP in the United States began in 1951 for vitiligo and in 1962 for psoriasis. In 1974, Parrish's group reported their success in the treatment of psoriasis using oral 8-MOP and ultraviolet A (UVA).

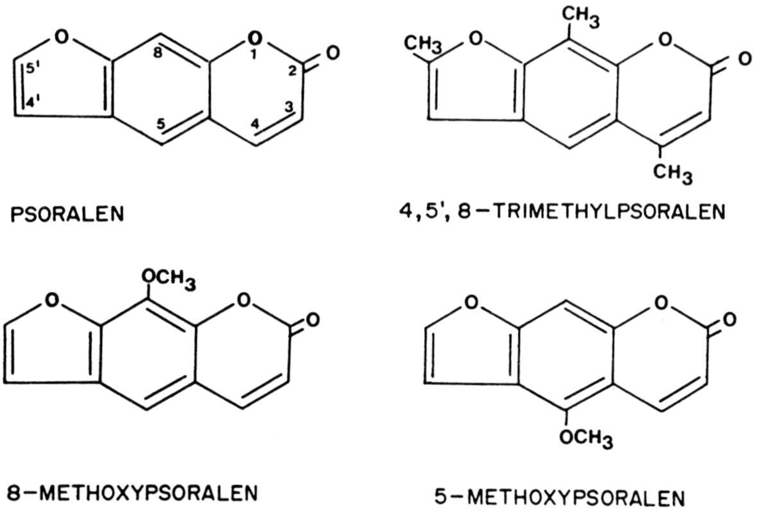


**Figure 2** Structure of the psoralens.

The entry of psoralens into the cell is not well understood. Diffusion, active transport, endocytosis, and phagocytosis may play a role. Although psoralens preferentially bind to DNA, psoralens can be detected in the cytoplasm of epidermal cells in all layers and are associated with lipoproteins and keratins. The pharmacologic action of psoralen is initiated by UVA photostimulation, which can be classified into direct (type I) or indirect (type II) reactions (19).

Type I reactions occur when the psoralen excited by UVA reacts with a target other than oxygen, such as DNA, cell membranes, or proteins. This reaction is mediated by the transfer of an electron or hydrogen atom to the particular cellular component. Initially, psoralens intercalate with DNA to form a molecular complex held together by van der Waals' forces and hydrogen bonding. 5-MOP is able to intercalate at 3 sites every 100 nucleotides, while 8-MOP intercalates at twice as many sites. This occurs preferentially where the base sequence alternates. After UVA irradiation, psoralens form covalent bonds with DNA pyrimidine bases, resulting in the formation of monofunctional and bifunctional adducts. Bifunctional adducts differ from monofunctional adducts by exhibiting interstrand crosslinkage. Monofunctional adducts induce mutagenicity and inhibit DNA synthesis, while bifunctional adducts cause cell death (19).

Type II reactions are energy transfer reactions. The excited psoralen reacts with molecular oxygen to form singlet oxygen ($^1O_2^*$), superoxide anions ($O_2^{\cdot -}$), and hydroxy radicals ($OH^{\cdot}$) (19). Singlet oxygen and superoxide anions can produce DNA strand breaks. These free radicals also cause cellular damage, which can contribute to the development of the erythema and vesiculation seen after photochemotherapy (19).

Psoralens are used in a systemic or topical fashion. Systemic psoralen preparations include 8-MOP, 5-MOP, and TMP, while 8-MOP and TMP are available as topical preparations. The rates of absorption of the psoralens vary depending upon the formulation (20,21). The ingestion of a heavy meal prior to psoralen ingestion may result in a decreased peak concentration as well as a longer time interval to reach this peak level (22). Seventy-five to 90% of 8-MOP is bound to plasma proteins (22). Protein binding rates for 5-MOP and TMP are 93–99% and 98%, respectively. The liver metabolizes psoralens and is also largely responsible for the first-pass effect, which results in decreased drug bioavailability. The elimination half-life for 8-MOP and 5-MOP is approximately 1 hr. After oral ingestion of 40 mg of $^{14}$carbon-labeled 8-MOP, 75% and 14% of the radioactivity was recovered in the urine and feces, respectively (23). Only trace amounts of the parent compound are found in these biological specimens. More than 90% of 8-MOP and TMP is excreted in the urine after 12 hr as their respective metabolites (24).

The conventional crystalline form of 8-MOP was the most commonly utilized psoralen until the introduction of 8-MOP liquid in a soft gelatin capsule. The liquid formulation has multiple advantages over the previous crystalline form (20). Peak serum levels are reached in 1–2 hr, in contrast to a mean of 4 hr for the older preparation. The new preparation is more rapidly eliminated, and peak serum concentrations are three times higher with the liquid form. As a consequence, a 25–40% decrease in the minimal phototoxic dose is noted. A 40–50% decrease in total cumulative dose is observed in patients treated with the liquid formulation. There is less intraindividual variation in serum levels with the liquid–gelatin formulation, and diet does not affect absorption.

5-Methoxypsoralen (5-MOP) has been used as an alternative to 8-MOP in Europe (25). A 1.2 mg/kg 5-MOP dose is almost as effective as the 0.6 mg/kg of liquid form of methoxsalen, and pruritus and severe erythema were less frequently noted with 5-MOP. The total UVA dose needed for clearance was the same as in 8-MOP PUVA therapy. However, 5-MOP-treated patients required an increased number and longer duration of treatment.

Oral TMP has been used to treat vitiligo. High dosages of the medication (100–400 mg) are required to achieve measurable serum levels (20).

Adverse effects of PUVA therapy may be classified into acute and chronic effects. Acute side effects include erythema, vesicles, pruritus, and nausea. Less commonly, headache and dizziness can occur. Drug fever due to PUVA therapy and to 8-MOP alone has been noted.

The chronic side effects of major concern are the risk of malignancy and the development of cataracts in patients on long-term PUVA therapy. The members of the American 16-center PUVA follow-up study (26) have found an increased risk for squamous-cell carcinoma in patients exposed to high dosages of PUVA, defined as at least 160 PUVA treatments (approximately 2000 $J/cm^2$). An increase in genital tumors has been noted in men treated with PUVA in this study (27). Swedish investigators (28) also noted a dose-dependent increase in the incidence of squamous cell carcinoma in patients treated with 8-MOP and UVA. In constrast, no increased incidence of squamous cell carcinoma was noted in the topical or systemic TMP-treated patients who constituted approximately 20% of the study population. Finally, increased incidences of respiratory carcinoma in both sexes, pancreatic cancer in males and, kidney and colon cancer in females were found. However, other European investigators have failed to show an increase in the development of cutaneous carcinoma in patients without a history of previous exposure to carcinogens such as arsenic, ionizing radiation, and methotrexate (29).

Animal studies report the development of cataracts after exposure to 8-MOP and UVA (30). Psoralens are capable of photobinding to the lens upon exposure to UVA light. This effect lasts as long as psoralens are present in the lens, which is from 12 to 24 hr after ingestion of drug. No statistically significant correlation between symptomatic cataracts and PUVA exposure has been noted to date in humans. However, in patients receiving more than 100 PUVA treatments, a statistically significant increase in the development of nuclear sclerosis and posterior supcapsular opacities was noted (31). Therefore, protective eyewear should be worn for at least 24 hr after psoralen ingestion.

Other chronic side effects seen with PUVA include xerosis, hyperpigmentation, pigmented macules, actinic damage, chronic phototoxicity, hypopigmentation, nail pigmentation, skin aging, and worsening of psoriasis or other underlying skin diseases.

Khellin, a furanochrome that is structurally similar to 8-MOP, has been used with UVA in the treatment of vitiligo (32). This compound is less phototoxic than 8-MOP and forms monofunctional adducts with DNA.

## IV. ANTIMALARIALS

Hydroxychloroquine, chloroquine, and quinacrine (Fig. 3) are the antimalarials used in photomedicine to treat solar urticaria, chronic actinic dermatitis, and porphyria cutanea tarda. The first two are substituted 4-aminoquinolones. Quinacrine has an additional benzene ring and is an acridine derivative.

The aminoquinolones are rapidly and completely absorbed from the gastrointestinal tract. Peak plasma concentrations occur 1–6 hr after oral ingestion (33). Fifty percent of the drug is bound by plasma proteins, and the half-life is 50 hr. In contrast, 80–90% of quinacrine is bound to plasma proteins and peak plasma levels are reached in 8–12 hr. The antimalarials reach a steady-state level in 4–6 weeks (34). These agents are widely distributed throughout the body.

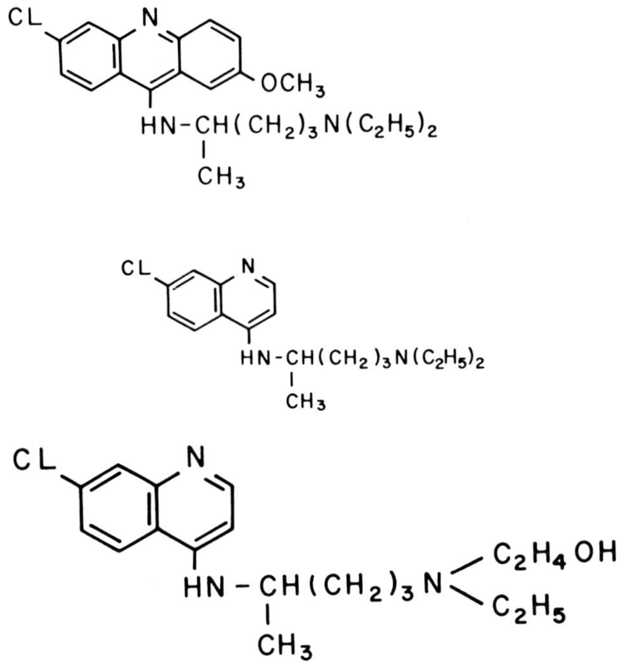

**Figure 3** Structure of antimalarials: quinacrine (top), chloroquine (middle), and hydroxychloroquine (bottom), the antimalarial agent most widely used in photodermatology.

Drug tissue concentrations can be up to 20,000 times higher than the plasma concentration. Approximately 70% of the drug is excreted in an unchanged form (35). Biotransformation produces metabolites that retain some antimalarial activity. The site of biotransformation has not been clearly delineated but may possibly occur in the liver. Single-dose studies show that three to four times more hydroxychloroquine is excreted in the feces than in the urine (35). This ratio is reversed for chloroquine. Detectable amounts of chloroquine have been found in the urine and plasma up to 5 years after drug discontinuation.

Multiple mechanisms of action have been postulated for the antimalarial medications (34, 36). The medications may have a weak sunscreen effect. In addition, they have multiple anti-inflammatory properties. They affect lysosomal functions, inhibit hydrolytic enzymes such as hyaluronidase and collagenase, and inhibit chemotaxis of neutrophils, eosinophils, and macrophages. Antimalarials also have antiprostaglandin ($PGF_2$, in particular), antiserotonin, and antihistaminic effects. Antimalarials inhibit immune complex formation, antibody production, and interleukin-1 secretion by monocytes. Another effect is the intercalation of antimalarials with DNA resulting in inhibition of DNA and RNA polymerases. Chloroquine, used in the treatment of porphyria cutanea tarda, forms a water-soluble complex with porphyrins; it also reduces iron stores in the liver (37,38).

Gastrointestinal and dermatologic side effects are most commonly noted. Gastrointestinal effects include nausea, vomiting, epigastric pain, weight loss, diarrhea, anorexia, and abdominal bloating. The most common cutaneous side effect is pigmentary alteration (34). Blue–gray pigmentation is seen with all the antimalarials in 10–30% of patients, affecting the pretibial surface, nail bed, face, and hard palate. Hemosiderin, melanin, and sulfur particles in macrophages and the extracellular matrix have been implicated as the source of pigmentation. After discontinuing use of the medication, the discoloration decreases or disappears. In one-third of cases, quinacrine can cause a yellow coloration that will fade over the course of months. Antimalarials may rarely exacerbate psoriasis. Morbilliform, urticarial, lichenoid, and exfoliative dermatitides may occur. In rare cases, bleaching of hair in light-haired persons is noted. Central nervous system and hematologic abnormalities are uncommon.

The major concern with the use of chloroquine and hydroxychloroquine is the possibility of irreversible retinopathy. The incidence of retinopathy, which is related to the daily dose, is low (approximately 300 reported cases worldwide) (39). Severe retinopathy may progress even after discontinuation of medication. Amsler grid testing every 2 weeks offers the best screening test to detect early retinopathy and visual field defects (40). Although corneal deposition and a loss of accommodation due to ciliary body dysfunction may occur, these complications are rarely symptomatic and do not require drug discontinuation (41). Corneal deposits and edema have been noted with quinacrine (42). However, only one case of quinacrine-associated retinopathy has been reported. The patient took the medication, which she had already taken for 12 years, for an additional 4 years without any progression of her retinopathy (43).

Hydroxychloroquine and chloroquine should not be used concomitantly, since the risk of retinopathy is likely to be additive. To minimize ocular toxicity, Mackenzie recommended that the dosage of chloroquine and hydroxychloroquine should not exceed 3.5–4.0 and 6.0–6.5 mg/kg lean body weight/day, respectively (44). Dosage should be adjusted for patients with renal or hepatic insufficiency and hypoalbuminemia.

## V. AZATHIOPRINE

Azathioprine (Fig. 4), a purine analogue, is well absorbed after oral intake, and is metabolized to 6-mercaptopurine, which is not cytotoxic, and to 6-thioinosinic acid and 6-thioguanylic acid which are cytotoxic (45). Conversion primarily by erythrocytes and the liver to inactive metabolites occurs within 10 hr. These

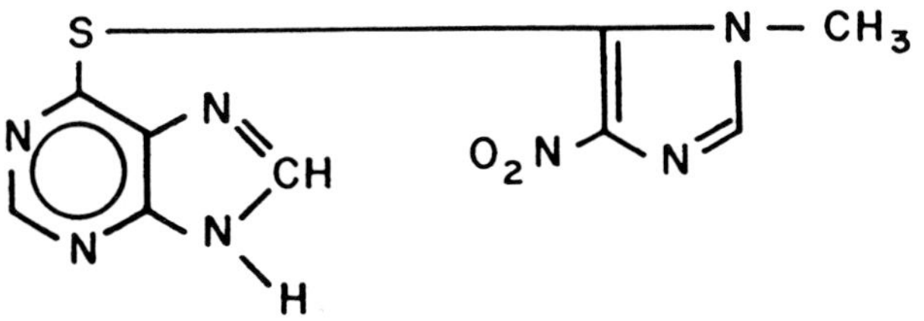


**Figure 4** Structure of the purine analogue azathioprine.

metabolites are excreted in the urine. Azathioprine is metabolized by xanthine oxidase. Therefore concomitant allopurinol administration will increase azathioprine levels. Azathioprine inhibits DNA and RNA synthesis and mitoses by incorporating into nucleic acids. Both B- and T-lymphocyte immune responses are affected (43).

Major adverse effects involve the hematologic, hepatic, and gastrointestinal systems. In patients with rheumatoid arthritis, nausea, vomiting, diarrhea, mucosal ulcers, unspecified skin eruptions, and leukopenia were commonly noted (46). Hepatotoxicity rarely occurs, and is reversible if therapy is promptly discontinued (45). In patients with rheumatoid arthritis treated with azathioprine, there is no increase in the incidence of internal neoplasms (46). There is conflicting evidence regarding higher rates of skin cancer reported with azothioprine. An increase in sister chromatid exchange has been demonstrated in cells harvested from cultures of phytohemagglutinin-stimulated whole blood from patients treated with this drug (47). A synergistic effect was noted when these cells were exposed to an ultraviolet light source that emitted wavelengths ranging from UVC (250 nm) through visible light (450 nm), with a peak wavelength in the UVA spectrum (365 nm). Other adverse effects include infection (bacterial, viral, fungal, mycobacterial), pancytopenia, thrombocytopenia, anemia, pancreatitis, cholestasis, reversible periportal fibrosis, and hypersensitivity reactions.

## VI. CYCLOSPORINE

The immunosuppressive drug cyclosporine (Fig. 5), an 11 amino acid polypeptide, was isolated from the fungi *Trichoderma polysporum* and *Cylindrocarpon lucidium*, in 1972. Its most common dermatologic use is in the treatment of psoriasis; in photomedicine it has been used in the treatment of chronic actinic dermatitis (48).

Considerable interpatient variation in cyclosporine absorption is observed (49). Peak blood levels occur 4 hr after oral ingestion. Milk and juice do not affect blood cyclosporine levels, which are increased by food. Approximately 33% of the drug is absorbed and up to 27% is eliminated after its first passage through the liver.

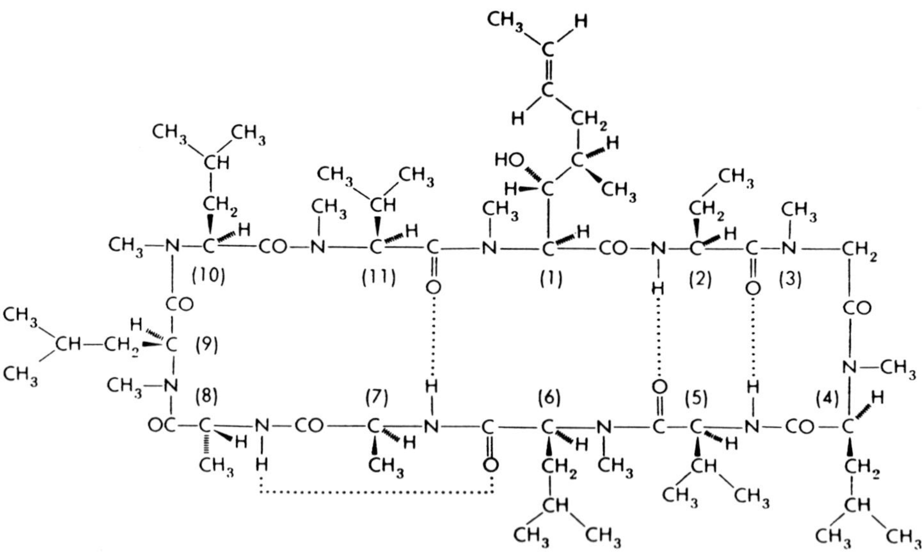

**Figure 5** Structure of the cyclic peptide cyclosporine A.

Cyclosporine is metabolized by the cytochrome P-450 IIIA family. Other drugs metabolized by this cytochrome system will increase cyclosporine levels. Concomitant use of cyclosporine will likewise increase the concentration of these drugs, which include erythromycin, ketoconazole, diltiazem, verapamil, nicardipine, corticosteroids, and oral contraceptives (49, 50). Rifampin and several anticonvulsant medications act as inducers of cytochrome P-450 IIIA and will reduce cyclosporine levels (50).

Cyclosporine exerts an effect on both humoral and cellular immunity (51). The major effect is on cell-mediated immunity and the early phase of CD4+, T-helper cell activation, in particular. Lymphokines such as interleukin (IL)-1, IL-2, IL-4, gamma interferon, macrophage inhibiting factors, and granulocyte–macrophage colony-stimulating factor are inhibited by cyclosporine. Inhibition of keratinocyte proliferation, macrophages, and antigen presenting cells has also been observed (52,53).

The adverse effect of major concern is the potential nephrotoxicity associated with cyclosporine use. This is usually reversible after discontinuation of cyclosporine. Bennett recommends that dosage be kept at or below 5 mg/kg/day and decreased by 25–50% if serum creatinine levels increase 30% over baseline (54). Concomitant use of drugs with known nephrotoxic potential, such as nonsteroidal anti-inflammatory drugs, aminogylcosides, trimethoprim–sulfamethoxazole, and amphotericin B, should be avoided.

Hypertension, a common problem in patients treated with cyclosporine, can be controlled by dose reduction or use of antihypertensive drugs such as calcium channel blockers and angiotensin I-converting enzyme inhibitors. Other common side effects include nausea, heartburn, fatigue, paresthesias, dysesthesias, hypertrichosis, gingival hyperplasia, and gingivitis. These are usually transient.

## VII. RETINOIDS

The antikeratotic properties of vitamin A were noted in 1932. Its use was discontinued due to the toxic effects of hypervitaminosis. In the early 1960s topical tretinoin was used in Europe to treat ichthyosis, actinic and seborrheic ketatoses, and basal cell carcinoma. Oral isotretinoin and etretinate were introduced in 1971 and 1975, respectively.

Three generations of retinoids have been developed based on modifications of the prototype vitamin A molecule (Fig. 6) (55). The first-generation retinoids, including isotretinoin and all-*trans*-retinoic acid, were synthesized by altering the polar end group and polyene side chain. The second generation, aromatic

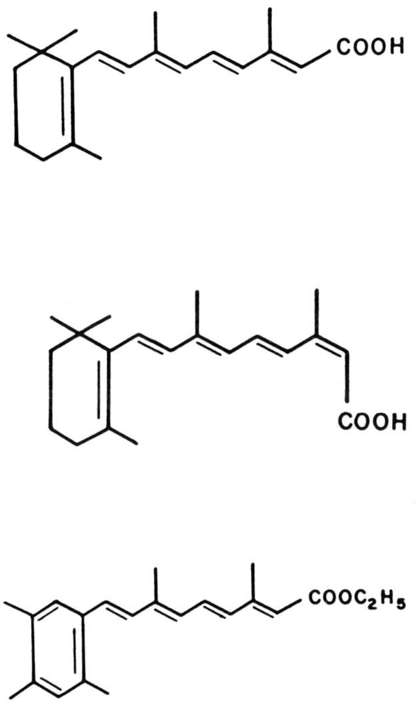


**Figure 6** Structure of the retinoids: tretinoin (top), 13-*cis*-retinoic acid (middle), and etretinate (bottom).

retinoids, etretinate and etretin, were formed by modifying the cyclic end group. Arotenoids, the third generation of retinoids were derived by cyclizing the polyene side chain.

Retinoids have been shown to have effects beyond those observed on keratinization. Keratinocyte differentiation and proliferation, sebum production, inflammation, immunomodulation, and tumor formation are influenced by retinoids (55).

Tretinoin (all-*trans* retinoic acid) is a natural metabolite of retinol (vitamin A). Tretinoin is not used orally due to side effects resembling hypervitaminosis A (56). Isotretinoin (13-*cis* retinoic acid), the other naturally occurring retinol metabolite, achieves maximum blood levels after 2–4 hr and reaches a steady-state concentration in 5–7 days. The elimination half-life ranges from 10.4 to 29.5 hr (57). The absorption of the drug is enhanced by food. The drug is primarily bound to albumin. It can be found in metabolized and unmetabolized forms in the bile and urine. When this teratogenic drug is used in women of childbearing potential, an adequate means of contraception is recommended during therapy, and for at least one menstrual cycle after. This provides a window of safety, since the elimination half-life of isotretinoin is 24 hr, and all the drug should be eliminated in 7 days.

Etretinate, the first aromatic retinoid, has a higher therapeutic index and greater immunomodulatory effect than isotretinoin (57). The absorption of etretinate is increased with milk and fatty foods. Peak plasma levels are reached in 4 hr and the half-life of the drug is approximately 10 hr. It is bound to serum lipoprotein. After repeated dosing, the drug accumulates in the fat and adrenal glands. Consequently, the elimination half-life is approximately 100 days. Detectable serum levels of etretinate were noted in some patients more than 2 years after the use of the drug was stopped (58).

Etretin, the carboxylic acid metabolite of etretinate, also reaches peak plasma levels in 4 hr, as does the parent compound. Etretin does not accumulate in any deep compartments and has an elimination half-life of 2 days (56).

The arotenoid family of retinoids consists of several drugs under investigation (56). RO 13-6298 and its free acid RO 13-7410 are more potent than etretinate, in that lower dosages are clinically efficacious. However, a similar incidence of toxic effects is noted. RO 15-1570 is 30 times more potent than 13-*cis*-retinoic acid and has few adverse effects in animals. These agents are potentially less toxic than second-generation retinoids.

The adverse effects of retinoids are similar to those seen in the hypervitaminosis A syndrome. These can be classified into acute effects, which are well tolerated, dose dependent, and reversible, and chronic effects, which may be irreversible (59). Acute effects involve various organs inluding the skin, gastrointestinal, central nervous, musculoskeletal, and reproductive systems. Laboratory abnormalities include elevated cholesterol and triglyceride levels, liver function tests, and levels of calcium, uric acid, and creatine phosphokinase. Hematologic and urinary abnormalities may be present as well.

Chronic effects may involve the skeletal system and the liver. Premature closure of the epiphyses, thinning of the long bones, fractures, and osteoporosis are among the reported side effects in children (60). Adults have developed a disorder resembling diffuse idiopathic skeletal hyperostosis; spinal hyperostoses, extraspinal ligament and tendon calcification, and spinal osteophytes. Liver toxicity may occur with chronic use of etretinate. A 3-year, 20-patient prospective study failed to demonstrate increased hepatotoxicity in high-risk patients treated with etretinate (61). However, in a retrospective study of 18 patients taking etretinate for more than 5 years, 4 patients had mild to severe portal fibrosis defined by formation of septa extending into the lobule or cirrhosis (62). Most patients in this study did not have pretreatment liver biopsies.

Variable results have been reported on photosensitivity induced by oral retinoids. Increased sensitivity to sunlight was reported in approximately 10% of patients treated with oral retinoids in one study (63). In another study, phototesting of 6 patients taking isotretinoin for at least 4 weeks and 9 patients taking etretinate for 3–12 months showed abnormal responses in 1 and 3 patients, respectively (64). In contrast, Diffey and colleagues (65) noted no significant change in minimal erythema dose in 5 patients after 3 weeks of isotretinoin use.

Retinoids are used in photomedicine predominantly for the treatment of photoaging, chemoprevention, and adjunctively with ultraviolet radiation. Investigators at the University of Michigan reported that tretinoin 0.1% cream ameliorated the effects of photoaging. After discontinuing topical tretinoin, clinical improvement was maintained for approximately 2 months but subsequently subsided (66).

Premalignant and malignant lesions have responded to topical and systemic retinoids (67). The most promising use for retinoids is in preventing or decreasing the rate of development of malignancies. Maintenance dosages of 0.25–1.5 mg/kg/day of isotretinoin have been successful. However, dosage of at least 1.5 mg/kg/day are needed to cause regression of pre-existing malignancies. Some patients have responded to etretinate at a dosage of 1 mg/kg/day (67). Different underlying causes responsible for multiple malignancies may influence the success of chemoprevention.

## REFERENCES

1. Kemeny L, Ruzicka T, Braun-Falco O. Dithranol: a review of the mechanism of action in the treatment of psoriasis vulgaris. Skin Pharmacol 1990; 3:1–20.
2. Ashton RE, Andre P, Lowe NJ, Whitefield M. Anthralin: historical and current perspectives. J Am Acad Dermatol 1983; 9:173–192.
3. Van dekerkhof PCM. Dithranol treatment for psoriasis after 75 years, still going strong! Eur J Dermatol 1991; 1:79–88.
4. Barr RM, Wong E, Cunningham FM, Mallet AE, Greaves MW. Effect of dithranol treatment on arachidonic acid and its lipoxygenase products in psoriasis. Arch Dermatol Res 1989; 280:474–476.

5. Fuchs J, Nitschmann WH, Packer L. The antipsoriatic compound anthralin influences bioenergetic parameters and redox properties of energy transducing membranes. J Invest Dermatol 1990; 94:71–76.
6. Kanerva L. Electron microscopy of the effects of dithranol on healthy and on psoriatic skin. Am J Dermatopathol 1990; 12:51–62.
7. Wilborn WH, Montes LF. Ultrastructural changes in psoriatic epidermis following anthralin treatment. J Cutan Pathol 1974; 1:132–150.
8. Ramsay B, Lawrence CM, et al. The effect of triethanolamine application on anthralin-induced inflammation and therapeutic effect in psoriasis. J Am Acad Dermatol 1990; 23:73–76.
9. Rogers MJ, Whitefield M, Marks VJ. Yellow hair discoloration due to anthralin. J Am Acad Dermatol 1988; 19:370–371.
10. Hjort N, Norgaard M. Tars. In: Roenigk HH Jr, Maibach HI, eds., Psoriasis, 2nd ed, New York: Marcel Dekker, 1991: 473–481.
11. Parrish JA, Morison WL, et al. Therapy of psoriasis by tar photosensitization. J Invest Dermatol 1978; 70:111–112.
12. Lowe NJ, Wortzman MS, Breeding J, Koudsi H, Taylor L. Coal tar phototherapy for psoriasis reevaluated: erythemogenic versus suberythemogenic ultraviolet with a tar extract in oil and crude coal tar. J Am Acad Dermatol 1983; 8:781–789.
13. Stern RS, Gange RW, Parrish JA, Tang SV, Arndt KA. Contribution of topical tar oil to ultaviolet B phototherapy for psoriasis. J Am Acad Dermatol 1986; 14:742–774.
14. Schoket B, Horkay I, Kosa A, Paldeak L, Hewer A, Grover PL, Phillips DH. Formation of DNA adducts in the skin of psoriasis patients, in human skin in organ culture, and in mouse skin and lung following topical application of coal-tar and juniper tar. J Invest Dermatol 1990; 94:241–246.
15. Rasmussen JE. The crudeness of coal tar. Prog Dermatol 1978; 12:23–29.
16. Maughan WZ, Muller SA, Perry HO, Pittelkow MR, O'Brien PC. Incidence of skin cancers in patients with atopic dermatitis treated with coal tar. J Am Acad Dermatol 1980; 3:612–615.
17. Pittelkow MR, Perry HO, Muller SA, Maughan WZ, O'Brien PC. Skin cancer in patients with psoriasis treated with coal tar. Arch Dermatol 1981; 117:465–468.
18. Stern RS, Zierler S, Parrish JA. Skin carcinoma in patients with psoriasis treated with topical tar and artificial ultraviolet radiation. Lancet 1980; 1:732–735.
19. Pathak MA. Molecular aspects of drug photosensitivity with special emphasis on psoralen photosensitization reaction. J Natl Cancer Inst 1982; 69:163–170.
20. deWolff FA, Thomas TV. Clinical pharmaocokinetics of methoxsalen and other psoralens. Clin Pharmacokinet 1986; 11:62–75.
21. Stolk LML, Siddiqui AH. Biopharmaceutics, pharmacokinetics and pharmacology of psoralens. Gen Pharmacol 1988; 19:649–653.
22. Brickl R, Schid J, Koss FW. Clinical pharmacology of oral psoralen drugs. Photodermatology 1984; 1:174–186.
23. Schmid J, Prox A, Reuter A, et al. The metabolism of 8-methoxypsoralen in man. Eur J Drug Metab Pharmocokinet 1980; 5:81–92.
24. Pathak MA, Dall'acqua F, et al. Metabolism of psoralens. J Invest Dermatol 1974; 62:347 (abstr).

25. Tanew A, Ortel B, Rappersberger K, Hönigsmann H. 5-Methoxypsoralen (Bergapten) for photochemotherapy. J Am Acad Dermatol 1988; 18:333–338.
26. Stern RS, et al. Benefits and risks of oral methoxsalen photochemotherapy a perspective based on the results from the sixteen-center PUVA follow-up study. In: Farber EM, Nall L, Morhenn V, Jacobs PH, ed. Psoriasis: Proceedings of the Fourth International Symposium. New York: Elsevier, 1987:188–196.
27. Stern RS, et al. Genital tumors among men with psoriasis exposed to psoralens and ultraviolet A radiation (PUVA) and ultraviolet B radiation. N Engl J Med 1990; 322:1093–1097.
28. Lindelof B, Sigurgeirsson B, Tegner E, et al. PUVA and cancer: a large epidermiological study. Lancet 1991; 338:91–93.
29. Henseler T, Christophers E, et al. Skin tumors in the European PUVA study. J Am Acad Dermatol 1987; 16:108–116.
30. Cloud TM, Hakin R, Griffin AC. Photosensitization of the eye with methoxsalen II. Chronic effects. Arch Opthalmol 1961; 66:689–694.
31. Stern RS, Parrish JA, Fitzpatrick TB. Ocular findings in patients treated with PUVA. J Invest Dermatol 1985; 85:269–273.
32. Ortel B, Tanew A, Honigsmann H. Treatment of vitiligo with khellin and ultraviolet A. J Am Acad Dermatol 1988; 693–701.
33. Warhurst DC. Antimalarial drugs: an update. Drugs 1987; 33:50–65.
34. Isaacson D, Elgart M, Turner ML. Anti-malarials in dermatology. Int J. Dermatol 1982; 21:379–395.
35. McChesney EW, Fitch CD. 4-Aminoquinolones. In: Peters W, Richards WHG, ed. Handbook of Experimental Pharmacology 68/II. New York: Springer-Verlag, 1984:3–60.
36. Peyron JL, Meynadier J. The pharmacological basis for the treatment of photodermatoses. Biochimie 1986; 68:899–904.
37. Scholnick PL, Epstein J, Marver HS. The molecular basis of the action of chloroquine in porphyria cutanea tarda. J Invest Dermatol 1973; 61:226–232.
38. Taljaard JJF, Schanley BC, Stewart-Wynne EG, Deppe WM, Joubert SM. Studies on low dose chloroquine therapy and the action of chloroquine in symptomatic porphyria. Br J Dermataol 1972; 87:261–269.
39. Rynes RI. Ophthalmologic safety of long term hydroxychloroquine. Am J Med 1983; 75:35–39.
40. Easterbrook M. Ocular effects and safety of antimalarial agents. Am J Med 1988; 85:23–29.
41. Bernstein HN. Ophthalamologic considerations and testing in patients receiving long-term antimalarial therapy. Am J Med 1983; 75:25–39.
42. Wallace DJ. The use of quinacrine (atabrine) in rheumatic diseases: a reexamination. Semin Arthritis Rheum 1989; 18:282–297.
43. Carr RE, Henkind P, Rothfield W, Siegel IM. Ocular toxicity of antimalarial drugs. Am J Opthalmol 1968; 66:738–744.
44. Mackenzie AH. Dose refinements in long-term therapy of rheumatoid arthritis with antimalarials. Am J Med 1983; 75(1A):40–45.
45. Clements PJ, Davis J. Cytotoxic drugs. Their clinical application to the rheumatic diseases. Semin Arthritis Rheum 1986; 15:231–254.

46. Singh G, Fried JF, et al. Toxic effects of azathioprine in rheumatoid arthritis. Arthritis Rheum 1989; 32:837–843.
47. Dalton A, Curtis D, Harrington CI. Synergistic effects of azathioprine and ultraviolet light detected by sister chromatid exchange. Cancer Genet Cytogenet 1990; 45:93–99.
48. Norris PG, Camp RDR, Hawk JLM. Actinic reticuloid: response to cyclosporine. J Am Acad Dermatol 1989; 21:307–309.
49. McMillan MA. Clinical pharmacokinetics of cyclosporin. Pharmacol Ther 1989; 42:135–156.
50. Watkins PB. The role of cytochromes P-450 in cyclosporine metabolism. J Am Acad Dermatol 1990; 23:1301–1311.
51. Gupta AK, Brown MD, Ellis CN, et al. Cyclosporine in dermatology. J Am Acad Dermatol 1989; 21:1245–1256.
52. Kanitakis J, Thivolet J. Cyclosporine: an immunosuppressant affecting epithelial cell proliferation. Arch Dermatol 1990; 126:369–375.
53. Cooper KD, Voorhees JJ, Fisher GB et al. Effects of cyclosporine on immunologic mechanisms in psoriasis. J Am Acad Dermatol 1990; 23:1318–1328.
54 Bennett WM. Renal effects of cyclosporine. J Am Acad Dermatol 1990; 23:1280–1287.
55. Bollag W. The development of retinoids in experimental and clinical oncology and dermatology. J Am Acad Dermatol 1983; 9:797–805.
56. Bollag W. New retinoids with potential use in humans. In: Saurat JH, ed. Retinoids: New Trends in Research and Therapy. Basel: Karger, 1985:274–288.
57. Allen JG, Bloxham DP. The pharmacology and pharmacokinetics of the retinoids. Pharmacol Ther 1989; 40:1–27.
58. DiGiovanna JJ, Zech LA, Ruddel ME, Gantt G, Peck GL. Etretinate persistent serum levels after long-term therapy. Arch Dermatol 1989; 125:246–251.
59. DiGiovanna JJ, Peck GL. Retinoid toxicity. Progr Dermatol 1987; 21:1–8.
60. White SI, Mackie RM. Bone changes associated with oral retinoid therapy. Pharmacol Ther 1989; 137–144.
61. Roenigk HH Jr. Liver toxicity of retinoid therapy. Pharmacol Ther 1989; 40:145–551.
62. Camuto P, Shupack J, Orbuch P, Tobias H, Sidhu G, Feiner H. Long-term effects of etretinate on the liver. Am J Surg Pathol 1987; 11:30–37.
63. Strauss JS, Rapini RP, Shalita AR, Konecky E, Pochi PE, Comite H, Exner JH. Isotretinoin therapy for acne: results of a multicenter dose-response study. J Am Acad Dermatol 1984; 10:490–496.
64. Ferguson J, Johnson BE. Retinoid associated phototoxicity and photosensitivity. Pharmacol Ther 1989; 40:123–135.
65. Diffey BL, Spiro JG, Hindson TC. Photosensitivity studies and isotretinoin therapy. J Am Acad Dermatol 1985; 12:119–120.
66. Ellis CN, Weiss JS, Hamilton TA, Headington JT, Zelickson AS, Voorhees JJ. Sustained improvement with prolonged topical tretinoin (retinoic acid) for photoaged skin. J Am Acad Dermatol 1990; 23:629–637.
67. Lippman SM, Meyskens FL Jr. Results of the use of vitamin A and retinoids in cutaneous malignancies. Pharmacol Ther 1989; 40:107–122.

**Table 1** Comparison of Effects of UVB and UVA on Human Skin

| | UVB | UVA |
|---|---|---|
| *Clinical features* | | |
| Erythema | No immediate phase 12–24 hr | Biphasic 6–24 hr |
| *Histopathologic features* | | |
| Keratinocytes | Dyskeratosis | Edema |
| Langerhans cells | 24 hr | 48 hr |
| Endothelial cells | 24 hr | 24–48 hr |
| Mast cells | 4–6 hr | 3–48 hr |
| Neutrophils | 14 hr | 3 hr |
| Mononuclear cells | 11–48hr | 3–38 hr |
| Fibrin | Absent | Present |
| *Biochemical alterations (blister aspirates and serum)* | | |
| Histamine | 6 hr | 15 hr |
| $PGD_2$ | 6 hr | 5–9 hr |
| $PGI_2$ | 6 hr | 5 hr |
| $PGE_2$ | 24 hr | 5–9 hr |
| IL-6 | 24 hr | No change |
| TGF$\alpha$ | 24 hr | No change |

60 and 86 years of age, although there is a decrease in the slopes of the UVB erythema dose–response curves in those 60–86 years of age (8). When compared with individuals under 30 years of age, the erythema in individuals over 60 years of age is less intense during the first 24 hours (9); the erythema also is more persistent in older individuals (10) (Table 2). In all age groups, erythema is followed by increased skin pigmentation resulting from augmented melanin synthesis in melanocytes and its transfer to keratinocytes.

## B. Histopathologic Alterations

The epidermis is the portion of the skin that is most affected by UVB (7, 11) (Table 1). In studies using 1 $\mu$m sections, in young persons under 30 years of age, damaged keratinocytes (so-called sunburn cells) were recognized as early as 30 min after irradiation and were most numerous at 24 hr (Fig. 1A). Although altered keratinocytes were found throughout the epidermis, they were most common in the lower half above the basal cell layer. By 72 hr, damaged keratinocytes were confined to the superficial layers, constituting a parakeratotic stratum corneum (Fig. 2A). Intercellular edema was an infrequent finding. A few

**Table 2** Effects of UVB in Individuals Over Age 60

| | |
|---|---|
| *Clinical features* | |
| Erythema | Less intense over 24 hr<br>More persistent |
| *Histopathologic features* | |
| Sunburn cells | Fewer at 24 hr<br>Present at 72 hr |
| Mast cells | Less extensive alterations |
| Endothelial cells | Less at 24 hr and more prominent at 72 hr |
| Langerhans cells | Less rapid decrease for 4 hr but reached similar low levels at 24 hr |
| *Biochemical alterations* | |
| Histamine | Decreased amounts in blister aspirates |
| $PGE_2$ | Slower generation with same peak levels by 24 hr |

lymphocytes were noted in the epidermis 24 and 72 hr after irradiation. The most prominent dermal changes involved the superficial vascular plexus. Endothelial cell enlargement was noted as early as 30 min after irradiation and became progressively more extensive during the next 24 hr. At 72 hr, venular changes were apparent only focally (Fig. 3A). Perivenular edema was present one hour after irradiation but was prominent at four hours. Endothelial cell activation and perivenular edema also were apparent in vessels of the deep reticular dermis and subcutaneous fat. Endothelial cell necrosis and fibrin deposition were not observed.

At the time erythema first developed, there was a reduction in the number of dermal mast cells. Although hypogranulation and degranulation were more prominent in the papillar dermis, these changes also were recognized in the deep reticular dermis. By 24 hr, mast cell number and morphology were normal.

In contrast, skin in individuals over the age of 60 contained fewer sunburn cells, especially at 24 hr (Fig. 1B); dyskeratotic keratinocytes were still present at 72 hr (Fig. 2B) (Table 2). The old subjects showed the same qualitative changes in mast cells, but the alterations were not so extensive. At 24 hr, hypogranulated mast cells were noted in old subjects. Minor vascular changes were present up to 24 hr, and endothelial cell alterations were prominent only at 72 hr (Fig. 3B).

As assessed by light and electron microscopic studies, in young individuals epidermal Langerhans cells began to decrease in number by 1 hr, were absent by 24 hr, and persisted at low levels up to 72 hr. The Langerhans cell number decreased less rapidly in older subjects during the first 4 hr but reached similar low levels by 24 hr (12). In subjects of all ages, the number of mononuclear cells exceeded the number of neutrophils at all times in the dermal infiltrate. Mononuclear cells gradually increased after irradiation to reach a plateau from 11 to 21 hr, with increased levels up to 48 hr. Dermal neutrophils appeared immediately after irradiation, reached peak levels at 14–24 hours, and decreased thereafter up to 48 hr (13, 14) without the accumulation of T lymphocytes (14). Macrophages increased with peak levels at 24 hr (14). Exposure of human skin to UVA or topical 0.15% 8-methoxypsoralen plus UVA inhibited the epidermal influx of neutrophils induced by the topical application of leukotriene $B_4$ $(LTB)_4$ (15).

## C. Biochemical Studies

Cutaneous suction blister aspirates, serum, skin biopsy specimens, and keratinocyte cultures have been used to study inflammatory mediators, cytokines, and adhesion molecules after exposure to UVB (Table 3). Elevated histamine levels were detected in blister fluid aspirates 4–8 hr after UVB irradiation and before the appearance of erythema. In tissue samples, increased histamine levels and decreased serotonin levels were noted between 3 and 48 hr (16). Prostaglandin (PG) $D_2$ levels were detected in suction blister aspirates at 6 hr, suggesting the mast cell as a potential source. $PGE_2$ and $PGF_{2\alpha}$ reached peak levels between 18 and 24 hr and returned to baseline levels by 48 hr (6, 17, 18). Delayed UVB erythema persisted beyond the return of the histamine and prostaglandin levels to normal. Alterations of 12-hydroxyeicosatetraenoic acid (12-HETE) levels showed a time course similar to that of $PGE_2$ (19). Leukotriene $B_4$ levels were unchanged (19). $PGI_2$ (prostacyclin) levels, as assessed by measuring its metabolite, 6-keto-$PGF_{1\alpha}$, increased within 6 hr after irradiation, were diminished by 24 hr, and returned to baseline levels by 48 hr (20). Histamine and certain arachidonic acid products appear to play a role in the early phases of UVB-induced erythema; the biological products involved in delayed erythema remain a mystery.

After UVB irradiation in humans, kinins were detected by a dermal perfusion technique within a few minutes and over a wide time course up to 48 hr after irradiation, suggesting that kinins probably have only a minor role (21). As yet, there are no data implicating the complement system in human UVB-induced inflammation; however, guinea pigs rendered complement deficient showed a weaker clinical response at 12–48 hr after UVB irradiation than did control animals (22).

After exposure of humans to one minimal erythema dose of UVB, serum interleukin (IL)-1 activity was elevated between 1 and 4 hr, with a return to baseline levels by 8 hr (23) (Table 3). After exposure of humans to two minimal erythema doses of UVB, serum IL-6 activity likewise reached peak levels at 12 hr and persisted over 72 hr (24).

Exposure of cultured human keratinocytes to UVB was associated with in-

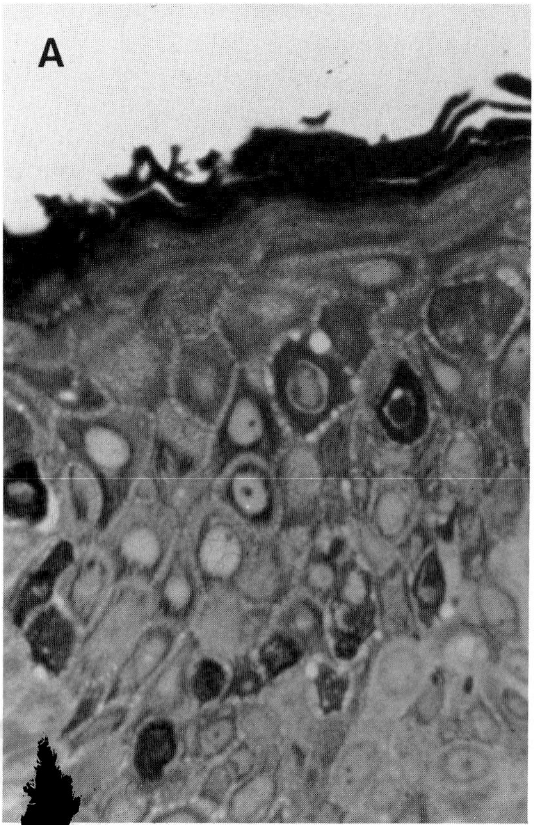

**Figure 1** Epidermal alterations at 24 hr after UVB exposure in individuals (A) under 30 years of age and (B) over 60 years of age. Note dyskeratotic cells on left and a rare dyskeratotic cell on right (1 μm sections; Giemsa, X196.5 in A; X78 in B).

creased levels of IL-1α- and β-mRNA at 6 hr (25). After exposure to ultraviolet light, intercellular adhesion molecule-1 (ICAM-1) appears on human keratinocytes at 48 hr (26). Moreover, supernatant fractions obtained from irradiated keratinocytes induced ICAM-1 on unirradiated keratinocytes as shown by Fluorescent activated cell sorter (FACS) analysis. This effect was blocked by antibodies to IL-1α but not tumor necrosis factor (TNF)α, suggesting that IL-1α is the principal mediator of the ultraviolet-induced expression of ICAM-1 on keratinocytes (26).

When exposed to both UVB plus UVA irradiation, ICAM-1 expression on the surface of cultured human keratinocytes as assessed by FACS analysis was biphasic. At 24 hr there was a dose-dependent suppression between 10 and 100

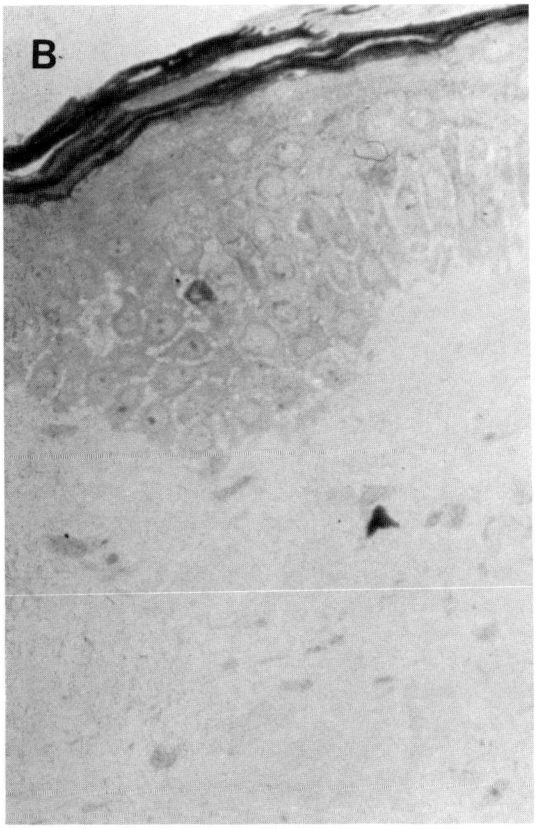


**Figure 1** *(continued)*

mJ/cm$^2$. Between 48 and 96 hr, there was a dose-dependent enhancement over the same dose range. At 100 mJ/cm$^2$, inhibition of interferon (INF)-$\gamma$-induction of ICAM-1 occurred at 24 hr, and this inhibition was dose-dependent over a range of 30–100 mJ/cm$^2$ between 48 and 96 hr (27). These results have been interpreted to suggest that early suppression is associated with therapeutic effects and late induction is associated with photosensitivity diseases (27).

After exposure to two minimal erythema doses of UVB, endothelial cell adhesion molecule-1 (ELAM-1) but not vascular cell adhesion molecule-1 (VCAM-1) was expressed on endothelial cells of the superficial venular plexus in skin biopsy specimens at 24 hr, with decreases by 72 hr (14). Exposure of cultured human dermal microvascular endothelial cells to UVB or reactive oxygen intermediates such as $H_2O_2$ resulted in an increase in ICAM-1 on the cell surface as assessed by FACS analysis and by mRNA expression (28). VCAM-1

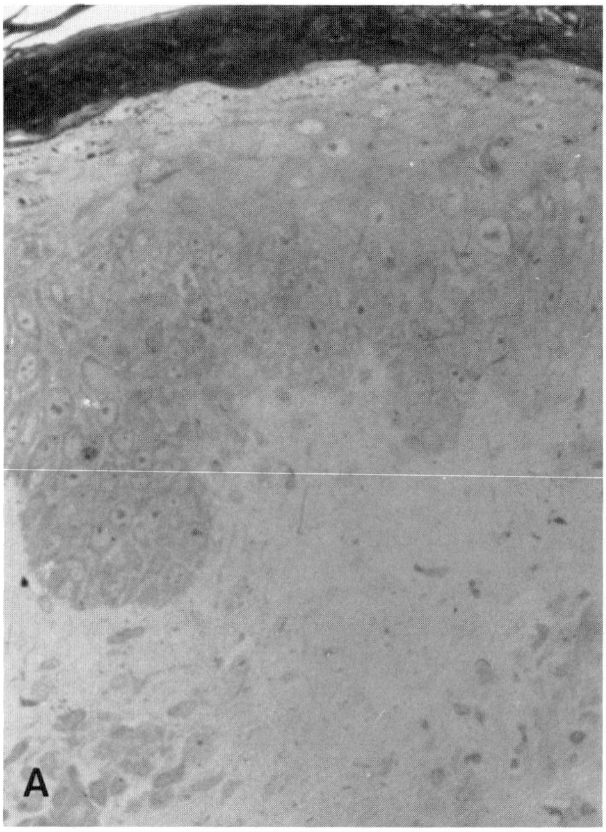

**Figure 2** Epidermal alterations at 72 hr after UVB exposure in individuals (A) under 30 years of age and (B) over 60 years of age. Note parakeratotic stratum corneum with an absence of dyskeratotic cells (A) and a loosely organized stratum corneum with dyskeratotic cells in the epidermis (B) (1 μm sections, Giemsa, X196.5).

and ELAM-1 were not induced. The induction of ICAM-1 was not blocked by anti-IL-1 antibodies.

Irradiation of human skin with three minimal erythema doses of UVB was associated with increased levels of transforming growth factor $\alpha$ (TGF$\alpha$) at 30 min, normal levels at four hours, and increased levels at 24 hr in exudates obtained using the skin chamber abrasion technique (29, 30), suggesting a role for this molecule in keratinocyte proliferation, epidermal hyperplasia, and angiogenesis. Increased levels of IL-1 and tumor necrosis factor were detected on epidermal cell membranes in skin biopsy specimens from subjects exposed to UVB radiation in vivo (31). The expression of ICAM-1 can be induced on

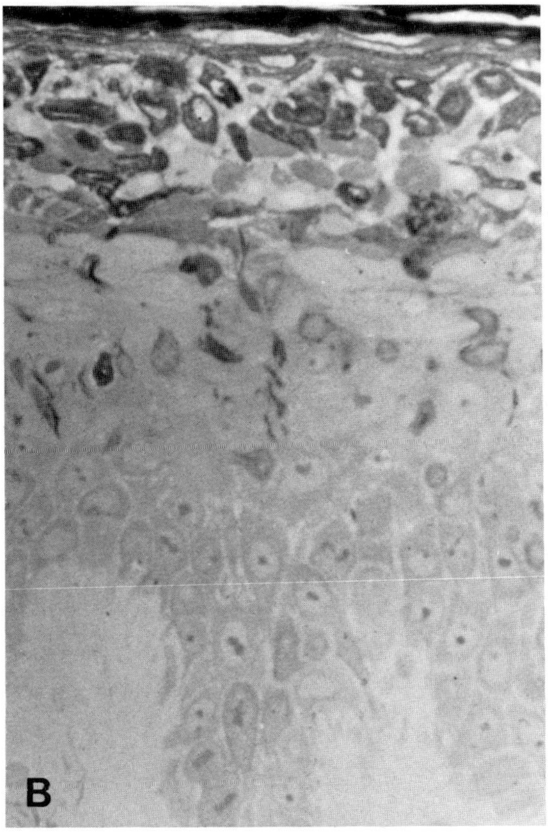


**Figure 2** *(continued)*

human keratinocytes after UVB irradiation for 16 hr (32). ELAM-1 was not induced on keratinocytes in skin biopsy specimens examined over a 72 hr period (14).

Normal human keratinocytes exposed to UVB but not UVA showed a dose-dependent release of IL-6 activity and enhanced IL-6 mRNA expression at 6 hr, which could be abrogated by hydrocortisone in physiological concentrations ranging from $10^{-7}$ to $10^{-9}$ M but not by prednisone (33). Exposure of human cultured keratinocytes and melanocytes to single or multiple doses of UVB, UVA, and psoralen plus UVA (PUVA) resulted in the modulation of the expression of the integrins A2, A3, A5, and B1 (34). Exposure of cultured melanocytes to UVB suppressed the ICAM-1 expression induced by TNF$\alpha$ and INF$\gamma$ (35).

Biochemical studies in old individuals are limited; a decreased amount of histamine and a slower generation of $PGE_2$ were noted when they were compared with young individuals (9) (Table 2).

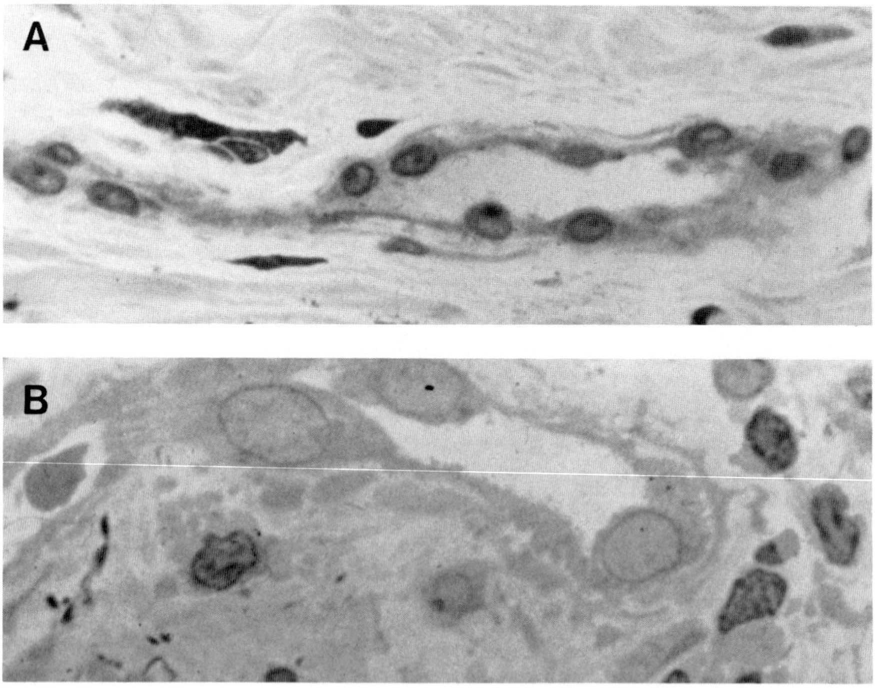


**Figure 3** Venular alterations at 72 hr after UVB exposure in individuals (A) under 30 years of age and (B) over 60 years of age. Note normal venules (A) and venule with enlarged endothelial cells (B) (1 μm sections, Giemsa, X1265).

## III. EFFECTS OF ULTRAVIOLET A

### A. Clinical Features

Although UVA (320–400 nm) is 1,000-fold less potent than UVB in producing erythema in human skin, it reaches the earth's surface in amounts 10–100-fold greater than UVB (5).

The time course of UVA erythema is biphasic. It is present after irradiation, decreases at 4 hr, and rises to a plateau between 6 and 24 hr (36, 37) (Table 1). The erythema induced by UVA in old individuals is more persistent than it is in young individuals (10). The studies considered have not compared the effects of UVA-II (320–340 nm) and UVA-I (340–400 nm).

### B. Histopathologic Alterations

After skin exposure to 2.5 minimal erythema doses of UVA, examination of 1 μm sections (38) demonstrated variation in the size of keratinocytes and their

**Table 3** Biochemical Effects of UVB on Human Skin

| | |
|---|---|
| *Mediators (blister aspirates)* | |
| Histamine | 4–8 hr |
| $PGD_2$ | 6 hr |
| $PGI_2$ | 6 hr |
| $PGE_2$ | 24 hr |
| $PGF_{2\alpha}$ | 24 hr |
| 12-HETE | 24 hr |
| $LTB_4$ | No change |
| $TGF_\alpha$ | 24 hr |
| *Cytokines (serum)* | |
| IL-1 | 1–4 hr |
| IL-6 | 12 hr |
| *Cytokines (skin samples and keratinocytes)* | |
| IL-1$\alpha$ and $\beta$ | |
| IL-6 | |
| TNF | |
| *Adhesion molecules (skin samples and keratinocytes)* | |
| ICAM-1 | |
| ELAM-1 | |
| *Adhesion molecules (endothelial cells)* | |
| ICAM-1 | |

staining properties, with many cells appearing swollen at 3 hr (Table 1). Perinuclear vacuoles and intercellular edema were prominent (Fig. 4). Dyskeratotic cells, characteristic of UVB-induced erythema reactions, were absent. Occasional lymphocytes were present, primarily in the lower one-third of the epidermis. Langerhans cells declined to approximately one-fifth of their previous number, with a statistically significant reduction achieved at 15 hr. There was no thickening of the stratum corneum (11).

Dermal neutrophils were present at all time points over 48 hr and were equally prominent in the papillary and reticular dermis. At 3 hr, most neutrophils were intravascular; subsequently, the majority were extravascular. Rare basophils and eosinophils also were observed. A lymphocytic infiltrate, far greater than the neutrophilic infiltrate, was present over 48 hr and involved the papillary and reticular dermis equally.

Endothelial cell enlargement was noted immediately after irradiation and became more extensive over the next 24 hr. At 24 and 48 hr, all of the venules throughout the dermis manifested changes. Hypogranulated mast cells accom-

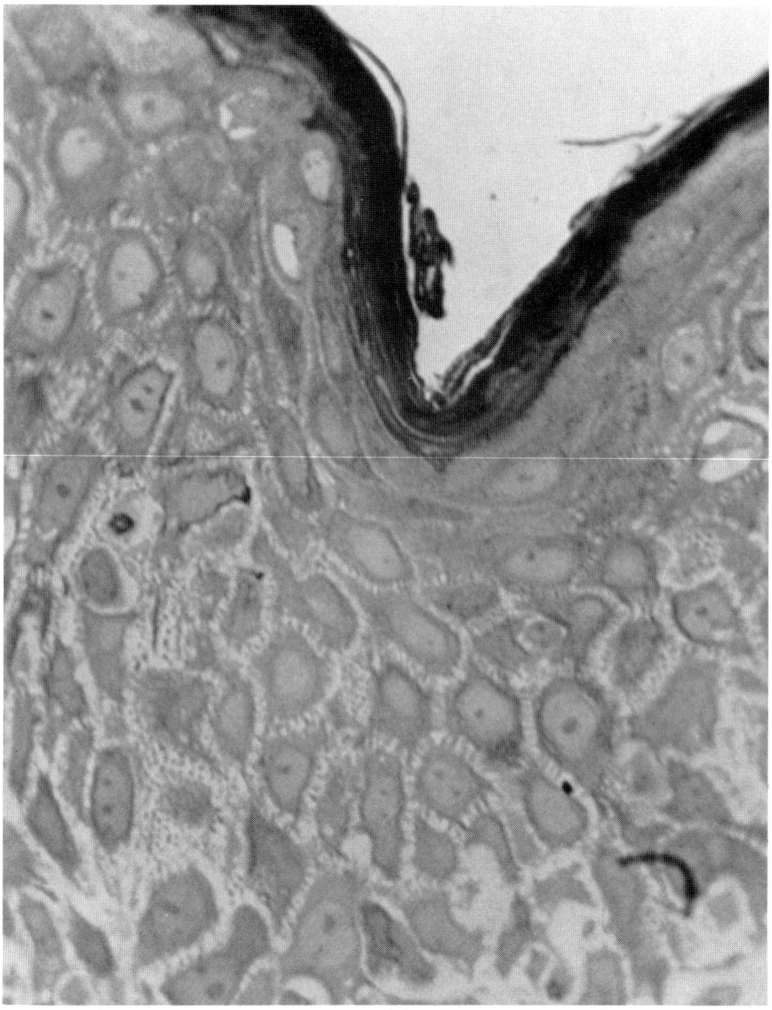

**Figure 4** Epidermal alterations at 24 hr after UVA exposure in individuals under 30 years of age. Note intercellular edema. Dyskeratotic cells are absent (1 μm section, Giemsa, X250).

panied by perivenular edema were observed over 48 hr. Foci of interstitial fibrin deposition in the reticular dermis were noted at 24 and 48 hr. After exposure to four minimal erythema doses of UVA, necrosis and loss of dermal endothelial cells of the superficial capillary venules and superficial venular plexus with a prominent infiltrate of neutrophils in the upper and mid dermis were noted (39).

There are no histologic studies of erythema induced by UVA in old individuals.

## C. Biochemical Studies

The biochemical studies of erythema induced by UVA are more limited than those of erythema induced by UVB. In suction blister aspirates examined over 48 hr (Table 4) (40), histamine concentrations rose between 9 and 15 hr, with a return to baseline levels by 24 hr. The concentrations of arachidonic acid, $PGD_2$, $PGE_2$, and 6-keto-$PGF_{1\alpha}$ were increased between 5 and 9 hr and returned to baseline levels by 24 hr. In exudates obtained from skin chambers after abrasion, three minimal erythema doses did not result in increased levels of $TGF\alpha$ at 4 or 24 hr (30).

Cultured keratinocytes exposed to UVA in vitro did not produce IL-6 (25). Exposure of human keratinocytes labeled with arachidonic acid to UVA in doses of 1–5 $J/cm^2$ resulted in the increased production of $PGE_2$ and 6-keto-$PGF_{1\alpha}$ but not $LTB_4$ between 1 and 6 hr (41). There are no studies implicating the complement system in UVA-induced inflammation. No biochemical studies have been performed in older individuals.

# IV. GENERAL CONSIDERATIONS

There are differences in the clinical features, histopathologic alterations, and biochemical studies when the erythema induced by UVB and UVA are compared (Table 1) and when young and old individuals are examined.

The skin contains many different types of cells with various capacities to produce inflammatory mediators. Human keratinocytes can produce cyclooxy-

**Table 4** Biochemical Effects of UVA on Human Skin

| | |
|---|---|
| *Blister aspirates* | |
| Histamine | 9–15 hr |
| $PGD_2$ | 5–9 hr |
| $PGI_2$ | 5 hr |
| $PGE_2$ | 5–9 hr |
| $TGF\alpha$ | No change |
| *Keratinocyte cultures* | |
| $PGE_2$ | 1–6 hr |
| $PGI_2$ | 1–6 hr |
| $LTB_4$ | No change |
| IL-6 | No change |

genase-derived and lipoxygenase-derived products from arachidonic acid as well as cytokines. Mast cells release a variety of vasopermeability molecules, such as histamine, prostaglandin $D_2$, and the sulfidopeptide leukotrienes. Cultured cutaneous microvascular endothelial cells produce $PGE_2$ and IL-1 but not $PGI_2$. Adhesion molecules can be induced on keratinocytes, melanocytes, and endothelial cells. Although the precise cell sources of the detected biological products in blister aspirates, serum, and skin samples in vivo is often unknown, erythema is likely a result of effects on each of these multiple types of cells with the production of various mediators, cytokines, and adhesion molecules.

## V. PIGMENTATION

Ultraviolet radiation leads to an increase in epidermal melanin pigmentation, which protects the skin against further radiation. The pigmentary response occurs in immediate and delayed phases that have different action spectra. Pigment responses to UVA, but not to UVB, can be abolished by making the skin hypoxic during exposure (42).

### A. Immediate Pigment Darkening

Immediate pigment darkening begins during ultraviolet radiation, is maximal immediately afterward, and may fade within minutes after exposure to small doses. After larger exposures it is more persistent and blends with delayed tanning. UVA and visible light both can cause immediate pigment darkening, which results from photo-oxidation of existing melanin and the distribution of melanosomes within melanocytes into the dendrites. It is most conspicuous in subjects who already have some degree of either constitutive or facultative pigmentation. Immediate pigment darkening does not seem to protect the skin against subsequent UVB radiation.

### B. Delayed Tanning

Delayed tanning becomes visible after about 72 hr. The major action spectrum for delayed tanning is UVB, but UVA and visible light also may cause delayed tanning. Dosage thresholds for erythema and delayed tanning differ depending on the wavelength. In individuals who tan well, suberythemal doses of UVA are melanogenic, whereas erythemogenic doses of UVB are required to induce delayed tanning. Delayed tanning from multiple exposures of either UVA or UVB is associated with increases in the number of melanocytes, the number of melanosomes synthesized, the degree of melanization, and the number of melanosomes transferred to keratinocytes (43).

The main function of delayed tanning is photoprotection, although the protection conferred by increased melanin alone without stratum corneum thickening is moderate. On histologic examination, there are fewer neutrophils and less severe endothelial cell venular alterations and less epidermal damage in tanning induced by UVA as compared to tanning induced by UVB (39). In the depigmented skin of patients with vitiligo, substantial photoprotection following ultraviolet exposure can occur as a result of stratum corneum thickening alone (44).

In light-skinned races, skin thickening is probably the more important photoprotective response, whereas in dark-skinned races constitutive pigmentation is the more important barrier. Although the skin melanin content is much greater in dark-skinned races, the numbers of melanocytes per unit area of skin does not differ significantly among the races.

## VI. SKIN THICKENING

UVB radiation causes the epidermis to thicken. This response, which is associated with upward transfer of melanin, increases tolerance to subsequent radiation and is the result of increased epidermal mitosis for up to 7 days. Preceding these responses, there is a transient depression of mitosis as well as of DNA, RNA, and protein synthesis (45). There is no epidermal thickening with UVA-induced delayed tanning, and the increased melanin is predominantly in the basal-cell layer. After exposure to suberythemogenic doses of either UVB or UVA for 3 weeks, there was an increase in stratum corneum ceramides, suggesting an improvement in barrier function (46).

The pathway through which stimulation of epidermal proliferation is mediated is not known. In hairless mice, UVB induces an increase in epidermal ornithine decarboxylase at 2 hr that reaches maximum levels by 28 hr (47). This is the rate-limiting enzyme in the synthesis of the polyamines putrescine, spermidine, and spermine, which are increased in amount in proliferative states. The induction of this enzyme by UVA has not been demonstrated except in the presence of the photosensitizers anthracene and 8-methoxypsoralen (48).

## VII. IMMUNOLOGIC CHANGES (49)

Ultraviolet B and UVA irradiation of human skin alters epidermal Langerhans cell morphology and impairs function (see Chap. 8). After exposure to UVB, epidermal non-Langerhans $CD1^{-}DR^{+}$ antigen-presenting cells are induced as well as circulating antigen-specific $CD_4{}^{+}2H_4{}^{+}$ T lymphocytes and $CD8^{+}$ T lymphocytes, thus enhancing immunologic suppression (50). After exposure to two minimal erythema doses of UVB in vivo, human epidermis showed decreased levels of superoxide dismutase immediately, and conjugated diene double bands were increased at 3 and 24 hr, suggesting impaired detoxification of oxidative enzymes (51).

## VIII. VITAMIN D SYNTHESIS

Ultraviolet B irradiation converts 7-dehydrocholesterol (provitamin $D_3$) to previtamin $D_3$ throughout the epidermis, with the highest concentration present in the basal and malpighian layers (52). Isomerizatin to vitamin $D_3$ occurs at physiological temperatures with thermal equilibrium reached at 4 days. Vitamin $D_3$ is transported into the circulation by plasma vitamin D binding protein. The selective removal of vitamin $D_3$ by vitamin D binding protein allows the efficient use of small quantities of provitamin $D_3$ that are generated in the skin during sun exposure.

## IX. THERAPEUTIC CONSIDERATIONS

A variety of therapeutic agents have been used in experimental studies in vivo in both experimental animals and humans to modify the erythema response to different wavebands of ultraviolet light. These agents include H1-type antihistamines, nonsteroidal anti-inflammatory agents, and corticosteroid preparations. The resulting experimental data have not allowed definitive recommendations for the treatment of human erythema responses to different wavebands of light.

## REFERENCES

1. Freeman SE, Hacham H, Gange RW, Maytum DJ, Sutherland JC, Sutherland BM. Wavelength dependence of pyrimidine dimer formation in DNA of human skin irridiated *in situ* with ultraviolet light. Proc Natl Acad Sci USA 1989; 86:5605–5609.
2. Arbabi L, Gange RW, Parrish JA. Recovery of skin from a single suberythemal dose of ultraviolet radiation. J Invest Dermatol 1983; 81:78–82.
3. Parrish JA, Zaynoun S, Anderson RR. Cumulative effects of repeated subthreshold doses of ultraviolet radiation. J Invest Dermatol 1981; 76:356–358.
4. Kaidbey KH, Kligman AM. Cumulative effects from repeated exposures to ultraviolet radiation. J Invest Dermatol 1981; 76:352–355.
5. Harber LC, Bickers DR. Photosensitivity Diseases. Principles of Diagnosis and Treatment, ed 2. Toronto, Canada: Decker, 1989:115–117.
6. Gilchrest BA, Soter NA, Stoff JS, Mihm MC Jr. The human sunburn reaction: histologic and biochemical studies. J Am Acad Dermatol 1981; 5:411–422.
7. Diffey BL, Oakley AM. The onset of ultraviolet erythema. Br J Dermatol 1987; 116:183–187.
8. Cox NH, Diffey BL, Farr PM. The relationship between chronological age and the erythemal response to ultraviolet B radiation. Br J Dermatol 1992; 126:315–319.
9. Gilchrest BA, Stoff JS, Soter NA. Chronologic aging alters the response to ultraviolet-induced inflammation in human skin. J Invest Dermatol 1982; 79:11–15.
10. Guarrera M. Aging and skin response to ultraviolet radiation. J Cutan Aging Cosmet Dermatol 1988; 1:135–144.

11. Rosario R, Mark GJ, Parrish JA, Mihm MC Jr. Histological changes produced in skin by equally erythemogenic doses of UV-A, UV-B, UV-C and UV-A with psoralens. Br J Dermatol 1979; 101:299–308.
12. Gilchrest BA, Murphy GF, Soter NA. Effect of chronologic aging and ultraviolet irradiation on Langerhans cells in human epidermis. J Invest Dermatol 1982; 79:85–88.
13. Hawk JLM, Murphy GM, Holden CA. The presence af neutrophils in human cutaneous ultraviolet-B inflammation. Br J Dermatol 1988; 118:27–30.
14. Norris P, Poston RN, Thomas DS, Thornhill M, Hawk J, Haskard DO. The expression of endothelial leukocyte adhesion molecule-1(ELAM-1), intercellular adhesion molecule-1(ICAM-1), and vascular cell adhesion molecule-1(VCAM-1) in experimental cutaneous inflammation: a comparison of ultraviolet B erythema and delayed hypersensitivity. J Invest Dermatol 1991; 96:763–770.
15. Chang A, Alkemade JAC, Van de Kerkhof PCM. PUVA and UVB inhibit the intra-epidermal accumulation of polymorphonuclear leukocytes. Br J Dermatol 1988; 119:281–287.
16. Jeanmougin M, Haimart M, Launay JM, Gareau-Demaray H, Civatte J. Étude dynamique de l'histamine et de la sérotonine tissulaires au cours de l'érythème actinique. Sem Hôp Paris 1985; 61:1208–1212.
17. Kobza Black A, Greaves MW, Hensby CN, Plummer NA. Increased prostaglandins $E_2$ and $F_{2\alpha}$ in human skin at 6 and 24 H after ultraviolet B irradiation (290–320 nm). Br J Clin Pharmacol 1978; 5:431–436.
18. Black AK, Fincham N, Greaves MW, Hensby CN. Time course changes in levels of arachidonic acid and prostaglandins $D_2$, $E_2$, $F_{2\alpha}$ in human skin following ultraviolet B irradiation. Br J Clin Pharmacol 1980; 10:453–457.
19. Greaves MW. Ultraviolet erythema: causes and consequences. Curr Prob Dermatol 1986; 15:18–24.
20. Black AK, Hensby CN, Greaves MW. Increased levels of 6-oxo-$PGF_{1\alpha}$ in human skin following ultraviolet B irradiation. Br J Clin Pharmacol 1982; 13:351–354.
21. Greaves MW, Sondergaard J. Pharmacologic agents released in ultraviolet inflammation studied by continuous skin perfusion. J Invest Dermatol 1970; 54:364–367.
22. Torinuki W, Tagami H. The role of complement in UVB-induced inflammation. Acta Derm Venereol (Stockh) 1986, 66.386–390.
23. Granstein RD, Sauder DN. Whole-body exposure to ultraviolet radiation results in increased serum interleukin-1 activity in humans. Lymphokine Res 1987; 6:187–193.
24. Urbanski A, Schwarz T, Neuner P, Krutmann J, Kirnbauer R, Köck A, Luger TA. Ultraviolet light induces circulating interleukin-6 in humans. J Invest Dermatol 1990; 94:808–811.
25. Kupper TS, Chua AO, Flood P, McGuire J, Gubler U. Interleukin 1 gene expression in cultured human keratinocytes is augmented by ultraviolet irradiation. J Clin Invest 1987; 80:430–436.
26. Krutmann J, Trefzer U, Kapp A, Schöpf E, Luger TA. Interleukin(IL)1$\alpha$, but not tumor necrosis factor (TNF)$\alpha$, mediates ultraviolet (UV) radiation-induced ICAM-1 expression in human keratinocytes. J Invest Dermatol 1992; 98:578.

27. Norris DA, Lyons B, Middleton MH, Yohn JJ, Kashihara-Sawami M. Ultraviolet radiation can either suppress or induce expression of intercellular adhesion molecule 1(ICAM-1) on the surface of cultured human keratinocytes. J Invest Dermatol 1990; 95:132–138.
28. Cornelius LA, Li LJ, Sepp N, Lawley TJ, Caughman SW. The effect of ultraviolet B (UVB) and reactive oxygen intermediates (ROI) upon dermal microvascular endothelial cell (EC) adhesion molecule expression. J Invest Dermatol 1992; 98:596.
29. James LC, Moore AM, Wheeler LA, Murphy GM, Dowd PM, Greaves MW. Transforming growth factor $\alpha$: in vivo release by normal human skin following UV irradiation and abrasion. Skin Pharmacol 1991; 4:61–64.
30. Murphy GM, Quinn DG, Camp RDR, Hawk JLM, Greaves MW. In-vivo studies of the action spectrum and time course for release of transforming growth factor-$\alpha$ by ultraviolet irradiation in man. Br J Dermatol 1991; 125:566–568.
31. Oxholm A, Oxholm P, Staberg B, Bendtzen K. Immunohistological detection of interleukin I-like molecules and tumour necrosis factor in human epidermis before and after UVB-irradiation in vivo. Br J Dermatol 1988; 118:369–376.
32. Krutmann J, Köck A, Schauer E, Parlow F, Möller A, Kapp A, Förster E, Schöpf E, Luger TA. Modulation of ICAM-1 expression on human keratinocytes by cytokines and UVB radiation. J Invest Dermatol 1989; 92:465.
33. Kirnbauer R, Köck A, Neuner P, Förster E, Krutmann J, Urbanski A, Schauer E, Ansel JC, Schwarz T, Luger TA. Regulation of epidermal cell interleukin-6 production by UV light and corticosteroids. J Invest Dermatol 1991; 96:484–489.
34. Hogan PA, Morelli JG, Norris DA. Effect of ultraviolet radiation on human melanocyte and keratinocyte integrin expression. J Invest Dermatol 1992; 98:653.
35. Kirnbauer R, Schauer E, Charvat B, Urbanski A, Förster E, Neuner P, Micksche M, Schwarz T, Luger TA. Regulation of ICAM-1 expression on human melanocytes and melanoma cells by cytokines and UV-light. J Invest Dermatol 1991; 96:609.
36. Gilchrest BA, Soter NA, Hawk JLM, Barr RM, Black AK, Hensby CN, Mallet AI, Greaves MW, Parrish JA. Histologic changes associated with ultraviolet A-induced erythema in normal human skin. J Am Acad Dermatol 1983; 9:213–219.
37. Diffey BL, Farr PM, Oakely AM. Quantitative studies on UVA-induced erythema in human skin. Br J Dermatol 1987; 117:57–66.
38. Gilchrest BA, Soter NA, Hawk JLM, Barr RM, Black AK, Hensby CN, Mallet AI, Greaves MW, Parrish JA. Histologic changes associated with ultraviolet A-induced erythema in normal human skin. J Am Acad Dermatol 1983; 9:213–219.
39. Margolis RJ, Sherwood M, Maytum DJ, Granstein RD, Weinstock MA, Parrish JA, Gange RW: Longwave ultraviolet radiation (UVA, 320–400 nm)-induced tan protects human skin against further UVA injury. J Invest Dermatol 1989; 93:713–718.
40. Hawk JLM, Black AK, Jaenicke KF, Barr RM, Soter NA, Mallet AI, Gilchrest BA, Hensby CN, Parrish JA, Greaves MW. Increased concentrations of arachidonic acid, prostaglandins $E_2$, $D_2$, and 6-oxo-$F_{1\alpha}$, and histamine in skin following UVA irradiation. J Invest Dermatol 1983; 80:496–499.
41. Hanson D, DeLeo V. Long-wave ultraviolet light induces phospholipase activation in cultured human epidermal keratinocytes. J Invest Dermatol 1990; 95:158–163.

42. Kaidbey KH, Kligman AM. The acute effects of long-wave ultraviolet radiation on human skin. J Invest Dermatol 1979; 72:253–256.
43. Pathak MA, Jimbow K, Parrish JA, Kaidbey KH, Kligman AL, Fitzpatrick TB. Effect of UV-A, UV-B, and psoralen on in vivo human melanin pigmentation: cellular and subcellular characterization on delayed tanning reactions induced by single or multiple exposures to UV-A, UV-B or UV-A plus 8-methoxypsoralen. In: Riley V, ed. Pigment Cell, Vol 3, Unique Properties of Melanocytes, Proceedings of the 9th Internatinal Pigment Cell Conference, Houston, TX, 1975 (Part II). Basel: S. Karger. 1976: 291–298.
44. Everett MA: Protection from sunlight in vitiligo. Arch Dermatol 1961; 84:997–998.
45. Epstein JH, Fukuyama K, Fye K. Effects of ultraviolet radiation on the mitotic cycle and DNA, RNA and protein synthesis in mammalian epidermis *in vivo*. Photochem Photobiol 1970; 12:57–65.
46. Wefers H, Melnik BC, Flür M, Bluhm C, Lehmann P, Plewig G: Influence of UVC irradiation on the composition of human stratum corneum lipids. J Invest Dermatol 1991; 96:959–962.
47. Verma AK, Lowe NJ, Boutwell RK. Induction of mouse epidermal ornithine decarboxylase activity and DNA synthesis by ultraviolet light. Cancer Res 1979; 39:1035–1040.
48. Gange RW. Epidermal ornithine decarboxylase activity and thymidine incorporation following treatment with ultraviolet A combined with topical 8-methoxypsoralen or anthracene in the hairless mouse. Br J Dermatol 1981; 105:247–255.
49. Granstein RD. Photoimmunology. Semin Dermatol 1990; 9:16–24.
50. Baadsgaard O, Fox DA, Cooper KD. Human epidermal cells from ultraviolet light-exposed skin preferentially activate autoreactive $CD4^{+}2H4^{+}$ suppressor-inducer lymphocytes and $CD8^{+}$ suppressor/cytotoxic lymphocytes. J Immunol 1988; 140:1738–1744.
51. Punnonen K, Autio P, Kiistala U, Ahotupa M. *In-vivo* effects of solar-simulated ultraviolet irradiation on antioxidant enzymes and lipid peroxidation in human epidermis. Br J Dermatol 1991; 125:18–20.
52. Holick MF, MacLaughlin JA, Parrish JA, Anderson RR. The photochemistry and photobiology of vitamin $D_3$. In Regan JD, Parrish JA, eds. The Science of Photomedicine. New York: Plenum, 1982: 195–218.

# 6

# Photoaging

**Barbara A. Gilchrest and Gary S. Rogers**
*Boston University School of Medicine, Boston, Massachusetts*

Photoaging is the term used to describe the clinical, histologic, and functional changes in chronically sun-exposed skin of middle-aged and elderly adults. Terms previously used in the medical literature and lay press as synonyms include premature aging, accelerated aging, dermatoheliosis, heliodermatitis, chronic sun damage, and even aging itself. The confusion between intrinsic or chronologic aging and photoaging is understandable, as the two processes have much in common, although detailed comparison reveals both quantitative and qualitative differences (1).

The importance of photoaging lies in its cosmetic and hence psychosocial impact on a large percentage of older adults, its intimate but incompletely understood relationship to skin cancer, as well as various functional losses in skin, and the model it provides for studies of adverse environmental effects on human health. In this latter context, chronic sun exposure may be considered a "gerontogen," or extrinsic agent that accelerates the apparent intrinsic aging process in one or more organ systems (2). The rapidly lengthening average lifespan, the public's persistent habit of extensive life-long unprotected sun-exposure, and the continued loss of atmospheric ozone with consequent increase in terrestrial ultraviolet (UV) irradiation combine to guarantee that photoaging will be an ever-greater dermatologic concern in the decades to come.

## I. FUNCTIONAL CHANGES

Photoaging is widely presumed to have an adverse effect on essentialy all skin functions, although very few controlled data exist to support this belief. A decrease in the cutaneous manifestation of delayed hypersensitivity to standard recall antigens in adjacent sun-exposed vs. sun-protected skin sites has been reported (3). Other physiological consequences of chronic sun exposure are presumptive and mentioned below in the context of specific clinical or histologic changes.

## II. CLINICAL FEATURES

At a clinical level, differences between habitually sun-exposed and sun-protected skin of elderly individuals are readily apparent, often with abrupt transitions in the skin appearance corresponding to the neckline or sleeveline. Characteristic changes in the exposed skin vary with complexion and degree of damage.

Young and middle-aged fair-skinned adults (skin types I and II) initially display freckling, followed by fine wrinkling and telangiectases, classically on the cheeks and nose. With increasing age and severity of damage, such individuals often develop markedly irregular pigmentation including freckling, lentigines (age spots), guttate hypomelanosis, and amelanotic pseudoscars. A tight, shiny quality of the skin surface may be observed as a consequence of diffuse epidermal atrophy (Fig. 1). Multiple actinic keratoses are also common. Lesions vary from minute minimally erythematous rough patches to well-demarcated plaques usually 4–10 mm in diameter, most numerous in the maximally sun-exposed areas (Fig. 2).* Basal cell carcinomas, squamous cell carcinomas, lentigo maligna, and some melanomas classically occur in this context. Focal and diffuse atrophy, particularly of the dorsal forearms and hands, may be severe and is often associated with skin ecchymoses, fragility, and slowly healing erosions at sites of minor trauma (Fig. 3). In other individuals, laxity and fine wrinkling predominate (Fig. 4, see color plate).

Follicular changes are also common in sun-damaged facial skin, but difficult to differentiate with certainty from intrinsic aging, because there is no comparable sun-protected control skin. However, sebaceous hyperplasia, clinically manifest as yellow perifollicular papules, appears to be far more common in moderately to severely photoaged fair skin than in darker skin or even in less photodamaged fair skin. Prominent follicular orifices are likewise common in photoaged skin, possibly as a consequence of UV damage to perifollicular connective tissue.

Darker-skinned persons (skin types III, IV, and V) with extensive sun exposure develop clinical changes somewhat later in life, and the more prominent abnormalities are of a somewhat different character. Rather than being telangiectatic and diffusely erythematous, the skin tends to become sallow, with a dull yellow cast attributed to decreased blood flow and possible changes in the

*Fig. 2, see color plate.

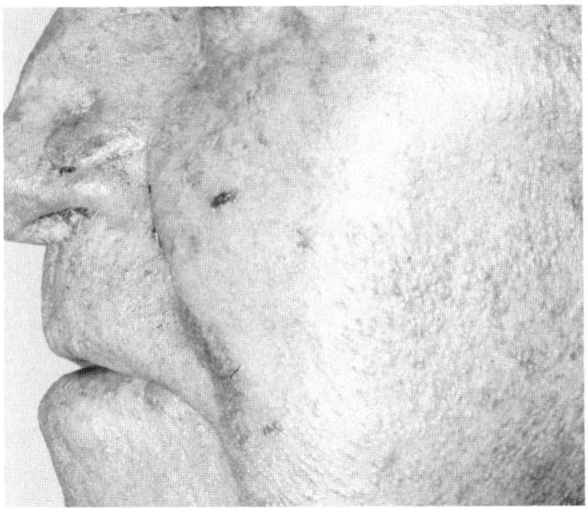

**Figure 1** Severe photoaging in a fair-skinned man, manifested primarily as increased surface reflectance (atrophy), irregular epidermal pigmentation, and telangiectases.

dermal matrix. Elastosis, manifested clinically by a pebbly, thickened character of the skin, is also characteristic of severely sun-damaged dark skin (Fig. 5, see color plate). The only prominent epidermal changes are diffuse hyperpigmentation and other pigmentary disturbances, particularly lentigines.

The relative severity of the above findings varies considerably among individuals, even those of the same age and skin type, undoubtedly reflecting inherent differences in vulnerability and cutaneous repair capacity for the solar insult, as well as differences in sun exposure over many decades. The patient's sex also influences the character of photoaging, with different distributions of lesions attributable at least in part to different hair styles, patterns of dress, and nature of sun exposure (occupational vs. recreational) between men and women over the past generations. Other sex differences such as epidermal thickness and sebaceous gland activity may also influence the character of photoaging. At least some lesions appear to be site-specific. For example, sebaceous hyperplasia and telangiectases are common only on the face; guttate hypomelanosis occurs on the extremities but not on the face.

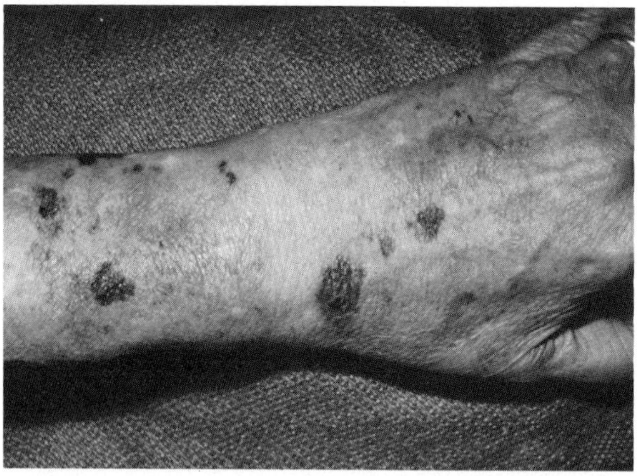

**Figure 3** Actinically induced atrophy of the dorsal forearm and hand, manifested by easy bruising and prominent dermal vasculature. Irregular epidermal pigmentation including depigmented pseudoscars is also evident.

## III. HISTOLOGIC FEATURES

Photodamaged skin has a variable but characteristic and indeed diagnostic histologic appearance, which differs quantitatively and qualitatively from sun-protected skin of the same individual (Fig. 6).

### A. Epidermal Changes

The stratum corneum often appears normal, but hyperkeratosis may be present. The epidermis may be hypertrophic, atrophic, or of normal thickness. Nuclear atypia and often some degree of disordered keratinocyte maturation are present in moderate to severe photoaging, but are not early markers. The basement membrane zone is thickened, presumably reflecting repeated injury to basilar keratinocytes. Variable keratinocyte size and loss of rete ridges are usually observed, but these changes are also features of old sun-protected skin (1).

Melanocytes are usually irregularly distributed along the basement membrane and are highly variable in size, dendricity, and degree of pigmentation (4,5). Melanocyte density (number per unit skin surface area) is approximately twice as great in habitually sun-exposed old skin as in sun-protected skin of the same

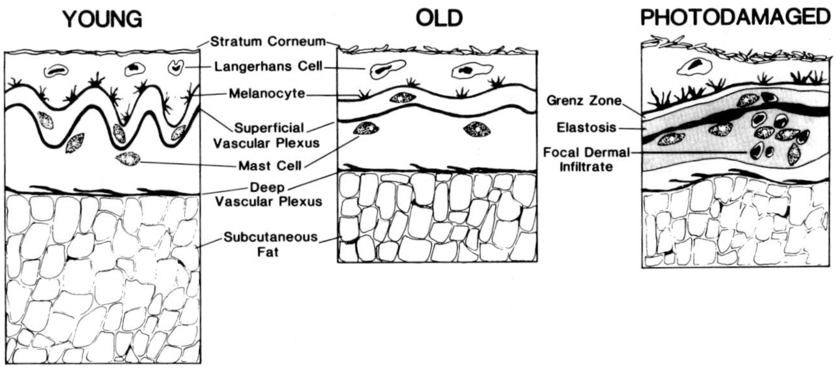


**Figure 6** Histologic changes associated with aging and photoaging. With age, there is a reduction in number of epidermal Langerhans cells and melanocytes, as well as effacement of epidermal rete ridges with a reduction in the proportion of basilar keratinocytes. There is also mild dermal atrophy with reduction in the area of the superficial vascular plexus. In old sun-exposed skin, there is variable epidermal atrophy or hyperplasia, irregular hyperkeratoses, and atypia of keratinocytes and melanocytes. Langerhans cell number is further reduced, while melanocyte number is increased relative to paired sun-protected skin. The most striking change is dermal elastosis, often with a mild perivascular infiltrate and focal vascular dilatation. (From Ref 2.)

individuals (4) and, unlike the increase observed after single tanning doses of UV irradiation in young adult subjects, does not return to baseline values after prolonged sun avoidance. This response is assumed to represent a protective adaptation to UV damage, although dendritic connections to keratinocytes and resultant pigment transfer often appear disturbed. With continued sun exposure, melanocytes are focally lost, undoubtedly compromising the UV melanin barrier.

Among the best-studied histologic aspects of photoaging is loss of epidermal Langerhans cells (LC). Studies comparing different sun-exposed vs. sun-protected sites have shown a decrease of approximately 50% in LC density in habitually sun-exposed skin of older adults (6,7). Because a single UV exposure is known to decrease epidermal LC density markedly for several days (8,9), the long-term decrease in the LC population of photoaged skin appears also to be a failure of acute UV changes to return to normal, as described for melanocytes above. The functional consequences of decreased LC density are unknown, but it is postulated that this reduction in immune recognition capacity may contribute to photocarcinogenesis by permitting emergence of otherwise antigenic malignantly transformed clones.

## B. Dermal Changes

In general, there is a vertical gradient of dermal damage, consistent with the progressive attenuation of UV radiation in skin. The depth and severity of the changes are dependent on skin type and the degree of photoaging, as noted above. The most striking and pathognomonic dermal change in photoaged skin is elastosis (Fig. 7). The term refers histologically to altered elastic filters occupying a variable portion of the dermis, beginning approximately at the junction of the papillary and reticular dermis (10). Old sun-protected skin does not manifest this change. Recent studies suggest that the precise character of elastosis may vary somewhat with anatomical site, being more diffuse in facial skin and more focal, with nodular aggregates, in arm skin (11). At the electron microscopic level, elastin fibers in mildly photodamaged skin are noted to be hypertrophied, while in severely damaged skin, they are "moth-eaten" and virtually amorphous

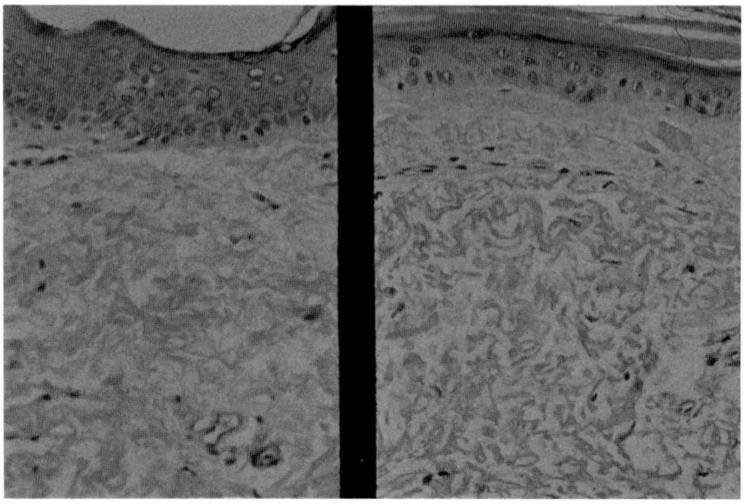

**Figure 7** Habitually sun-exposed young adult compared with old adult facial skin. Left panel: Biopsy specimen of a fair-skinned 24-year-old shows increased melanocytes and some effacement of rete ridges. At this magnification, the dermis appears normal. Right panel: In contrast, biopsy specimen of a 73-year-old shows more advanced epidermal atrophy and atypia, accompanied by hyperkeratosis. Below a narrow Grenz zone of pink-staining collagen, elastosis extends to the base of the specimen, giving a bluish discoloration to the hematoxylin and eosin-stained tissue section (original magnification, 40; photomicrographs courtesy of Jag Bhawan, M.D.).

(12). In terms of function, the material is assumed to have lost the mechanical property that characterizes normal elastin, specifically the ability to return stretched tissue rapidly to its original conformation.

Above the elastosis, extending to the basement membrane, is a narrow strip of relatively normal-appearing papillary dermis termed the Grenz, or border, zone. Ultrastructural studies have demonstrated, however, that the region is probably not truly spared, but represents a subtle scar (13). Fibroblasts in this superficial portion of the dermis are characteristically large and activated, containing abundant endoplasmic reticulum, indicative of protein synthesis, presumably reflecting attempts to repair repeated UV irradiation. The lack of elastosis in the Grenz zone may thus be the result of different fibroblast synthetic repertoires in the superficial vs. mid-dermis, rather than insufficient injury.

As noted in the discussion of clinical photoaging changes, the degree of vascular damage varies markedly among individuals. In general, in mildly photodamaged skin, there is venular wall thickening and basement membrane reduplication, while in severely damaged skin the vessel walls are thinned and supporting perivascular veil cells are reduced in number (14). The number of vascular cross-sections is reduced, sometimes markedly so (15), presumably in proportion with the total area of the superficial vascular plexus. There are local dilations, corresponding to clinical telangiectases that may compromise nutrient and oxygen transfer in the tissue.

Modest increases in perivenular mast cells and a minimal to mild superficial, predominantly lymphocytic, inflammatory infiltrate are common (16), particularly in facial skin (11). The presence of these cells may be a response to the UV-altered dermal environment as well as to individual acute sun exposures, and has been postulated to contribute to the photoaging process through chronic release of collagenases, elastases, and other injurious enzymes (16).

## IV. PATHOGENESIS

By definition, photoaging is due to chronic sun exposure. Ultraviolet B (UVB) (sunburn spectrum UV) is presumed to be primarily responsible, as for photocarcinogenesis, although the more deeply penetrating and more abundant ultraviolet A (UVA) photons are believed by many authorities to play a substantial role also. In the rhino mouse model of photoaging, both UVB and UVA can produce elastosis (17).

One hypothesis to explain the different character of photoaging in light vs. dark skin is depicted in Figure 8. In lightly pigmented skin, the small and sparse epidermal melanin granules absorb very few UV photons. The average photon thus penetrates deeply into the skin, with approximately 10% of UVB energy and 50% of UVA energy entering the dermis. In the absence of melanin, an excellent absorber of UV energy, intracellular DNA is a major target for these UV

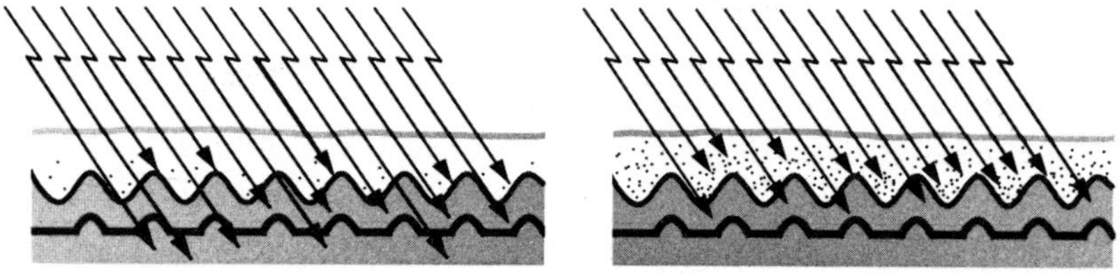


**Figure 8** Localization of UV photon absorption and consequent injury in fair-skinned (type I–II) vs. dark-skinned (type III–IV) persons. The lesser amount of epidermal melanin in fair-skinned persons permits the average photon to penetrate more deeply, thus reaching cellular targets in the epidermal basal layer and in the dermis far more often than in darker skin.

photons. Photons absorbed by DNA in cells above the epidermal basal layer have no permanent impact, because the cells ultimately desquamate. However, photons cumulatively absorbed by basal keratinocytes or melanocytes, or by fibroblasts, endothelial cells, or perivascular veil cells in the dermis, may be expected to contribute to photaging. In lightly pigmented skin, relatively few photons are absorbed in the epidermis and many enter the dermis. The greater the number of photons reaching the dermis, the deeper a "biologically significant" number will penetrate. It is assumed that many reach the superficial vasculature and may thus directly contribute to UV-associated vascular ectasia. In darkly pigmented skin, in contrast, large melanin granules throughout the epidermis absorb UV photons, permitting relatively few to penetrate to nuclei in the basal layer or beyond. Only a very small fraction of incident photons cross the basement membrane and thus hypothetically the "biologically significant" dose is smaller and generally confined to the superficial papillary dermis, above the vascular plexus.

The above considerations alone, however, account only partially for the different character of photoaging in fair vs. dark skin or indeed for the existence of this distinct clinical syndrome. Sunburn and tanning are transient UV responses, lasting a few days to a few weeks respectively. Why is photoaging cumulative and, for practical purposes, irreversible? Permanent UV-induced changes in cell behavior, leading to altered cutaneous appearance and function, must logically reflect either mutation (unrepaired or incorrectly repaired DNA damage) or an altered state of differentiation in response to the UV injury. The former mechanism is well supported by experimental data both in vivo and in vitro and appears to be a least partially responsible for photocarcinogenesis. The latter mechanism, a permanent UV-induced change in cellular behavior, equivalent to an altered state of differentiation, is suggested by the quite stable, characteristic, and predictable features of photoaged skin in the population. Random errors in the DNA would seem less likely to result in so consistent a clinical syndrome that affects essentially all skin compartments and evolves simultaneously, in a dose-dependent manner, over the entire skin surface. Consistent with altered state of differentiation as a mechanism for photoaging is the fact that keratinocytes and fibroblasts cultured from habitually sun-exposed skin of adult volunteers consistently differ from paired cell cultures derived from sun-protected skin of the same donors in having a decreased response to growth modulators, a decreased confluent density, and a decreased culture lifespan (18–20). Thus, the individual cells in photoaged skin have stereotypically altered behavior, compared to cells from nonphotoaged skin of the same donor, and this altered behavior persists for the lifespan of the cells in culture.

How the interaction of UV photons with the genome might trigger a differentiation-like process is presently unknown. The gene response might be specific to UV injury. As an alternative, UV-induced cell death might simply trigger compensatory division in the surrounding undamaged cell population, leading ultimately to "senescence" through consumption of the programmed potential total lifetime divisions for the progenitor cells. This possibility is suggested by the epidermal and dermal atrophy that often accompanies severe photoaging and is consistent with data comparing early and late passage fibroblasts in the Hayflick model of in vitro aging (1).

The different characteristics of mild, moderate, and severe photoaging in lightly pigmented vs. darkly pigmented skin may also reflect the characteristic response pattern of many cells to increasing injury. Initially cells respond by hypertrophy, hyperplasia, and/or increased function; and later by reduced function and/or proliferative capacity (Fig. 9). Thus, hyperpigmentation, either diffuse or focal as in lentigines, may be considered a manifestation of earlier or milder damage than islands of permanent depigmentation. The prominent coarse pebbly elastosis observed in dark-skinned individuals with extensive sun exposure may likewise reflect increased fibroblastic activity, while the dermal atrophy and pseudoscars commonly observed in those of Celtic ancestry may reflect decreased fibroblast number and protein synthetic capacity secondary to severe photodamage.

## V. TREATMENT

### A. Sun Avoidance

The cornerstone of management for photoaging is avoidance of further sun exposure. Patients should avoid not only intermittent "excessive" sun exposure, such as experienced during intentional sunbathing, but should also minimize cumulative casual daily sun exposure. Sun-seeking behaviors should be dis-

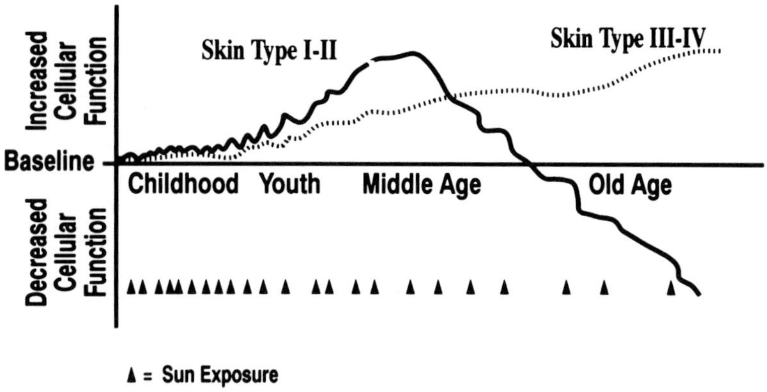


**Figure 9** Time course of lifetime sun exposures and consequent changes in irradiated cell function in fair-skinned (type I–II) vs. dark-skinned (type III–IV) persons. In general, excessive sun exposures occur disproportionately in childhood and photoaging, other than freckling, is first appreciated decades later. For many years, repeated UV injury induces increased proliferation and/or protein synthesis. Eventually, at least in fair skin, this adaptive cellular response is exhausted and further sun exposures result in cell loss and/or decreased function of remaining cells. In most dark-skinned individuals, this limit is never reached. Thus, in older adults, sun-damaged fair skin is often atrophic, while sun-damaged dark skin is hypertrophic.

couraged and outdoor activities should not be scheduled for midday (10:00 am–2:00 pm) whenever practical. Use of protective clothing, such as hats and long sleeves, should be encouraged within the limits of comfort and style. Exposed areas, including the ears, neck, and dorsa of the hands, as well as the face, should be protected with a sunscreen if more than minimal exposure is anticipated.

The action spectrum for photoaging is presently unknown. Available data suggest that UVB wavelengths are primarily responsible, but the prominent dermal changes associated with photoaging, such as elastosis and telangiectases, raise the possibility that the more deeply penetrating UVA portion of the spectrum may also contribute. Sunscreens with a high sun protection factor (SPF) provide excellent UVB protection but in general only modest UVA protection, leading some authorities to recommend additional use of a second product with preferential absorption in the UVA range. Complete protection can be obtained only with opaque sunscreens, among which only heavy make-up is practical for everyday use.

Prudent lifelong sun avoidance is a daunting prospect for many patients. Compliance may be enhanced by the knowledge that sun avoidance not only prevents further cumulative damage but also permits repair of photodamaged skin. This has been amply demonstrated in animal models of photoaging (21,22). Recent multicenter trials of tretinoin therapy (see below) have also documented a statistically significant improvement in both clinical and histologic appearance of skin within 6 months, when compared to a placebo regimen that consisted of sun protection and moisturizing cream only (23,24).

## B. Topical Tretinoin

Treatment of photoaging was revolutionized by the observation that daily use of topical tretinoin (all *trans* retinoic acid) for 4–6 months improves fine and coarse wrinkling, mottled pigmentation, skin texture, and sallowness associated with chronic sun damage (23,25–27). Multicenter studies involving over 500 patients established a dose–response of photoaged skin to tretinoin therapy at both the clinical (23) and histologic (24) levels. At the highest tretinoin concentration studied, maximum improvement is achieved within the first year of treatment and can be maintained by less frequent maintenance applications (28). At lower concentrations, improvement is more gradual and continues for a longer time. All beneficial results of treatment are assumed to be reversible with cessation of tretinoin therapy, although this has not been rigorously established. Indefinite maintenance therapy, two or three applications per week, is recommended once treatment goals have been met.

Of the various components of photodamage, skin texture (surface smoothness), irregular pigmentation, and sallowness are probably most responsive to tretinoin therapy. Fine and coarse wrinkling also responds, dramatically in a few patients, but more slowly. Telangiectasia is not reliably improved, and in some patients unwanted ruddiness and dilated vessels appear to be exacerbated by tretinoin therapy, possibly due to retinoid-induced neovascularization. Changes in skin appearance attributable to tretinoin therapy are accompanied by striking and consistent dose-dependent histologic changes, particularly in the epidermis within the first 3–6 months (24), but most then regress despite continued clinical improvement (29). This suggests that the such early histologic changes as altered stratum corneum morphology, and increased thickness of the granular layer and viable epidermis may be markers of retinoid effect only, with ultimate accommodation, rather than being causally related to the desired effects. An exception is the decreased and more evenly distributed epidermal melanin, which does appear to correlate well with tretinoin's correction of irregular pigmentation (24, 29). It is now believed, however, that subtle retinoid effects on the dermis, such as increased numbers of anchoring fibrils below the basement membrane (30,31) and new collagen deposition (22), appreciated only by immunofluorescence staining and/or electron microscopy, are responsible for the decreased wrinkling.

Retinoid treatment also appears effective in preventing and/or in reducing the number and size of established actinic keratoses in photodamaged skin (32–35). This is consistent with the recently documented ability of oral isotretinoin to prevent photocarcinogenesis in patients with xeroderma pigmentosum (36), a genetic disorder of impaired DNA repair capacity for UV damage with greatly enhanced skin cancer risk, and with extensive studies both in vivo and in vitro demonstrating a reversal of the malignant phenotype by retinoids (37). Thus, topical tretinoin therapy may reduce the risk of photocarcinogenesis in otherwise normal chronically sun-damaged skin, as well as improve its appearance.

Considerable data now suggest that tretinoin, bound to one of a family of nuclear receptors, acts directly on specific response elements in the DNA to alter gene transcription and thus cell behavior (37). In experimental model systems other than skin, tretinoin has been shown definitively to alter normal cellular differentiation programs and to cause reversion toward normal of gene expression in malignantly transformed cells (37). Possible changes in gene regulation responsible for tretinoin's beneficial effects on photoaged skin are an area of active investigation.

Theoretic concerns regarding long-term safety of topical tretinoin therapy have been raised. Nearly 30 years ago, some investigators found that topical tretinoin enhanced UV-induced skin cancer formation in hairless mice during the course of daily carcinogenic UV exposures, while others found that tretinoin exerted a protective effect (34). These opposite results have been attributed to differences in strains of mice, UV-irradiation protocols, and tretinoin preparations, but never resolved. In addition, tretinoin has been widely used by cancer-prone older adults for far less than a decade. Thus, the more than 20 years of topical tretinoin use as an acne medication without any suggestion of accelerated photocarcinogenesis can offer only partial reassurance regarding its safety. A second theoretic concern is the teratogenicity of retinoids. However, available data suggest that the amount of tretinoin absorbed into the circulation and thus available to the fetus after either short-term or long-term application to the face (38) is far less than the normal endogenous level derived from the diet and essential for health.

At a practical level, the major impediment to topical tretinoin therapy of photoaging (other than cost) is skin irritation, also termed retinoid dermatitis (23,25–27). Especially during the initial weeks of treatment, the skin is dry, easily irritated, and often pruritic. Superficial desquamation and erythema are common. These side effects can be minimized by regular use of emollients and, if necessary, by a short course of topical corticosteroid therapy in areas of symptomatic dermatitis. The skin usually accommodates to topical tretinoin within several weeks and the retinoid dermatitis resolves, although continued use of a moisturizer is recommended. Because the severity of retinoid dermatitis is dose dependent, many authorities suggest that the patient begin treatment at a low dose (e.g., 0.025%), and gradually increase to 0.05% and 0.1% cream over several months, as the skin accommodates. Other authorities recommend that the patient be encouraged to begin treatment at the high concentration, despite the virtual inevitability of dermatitis, so as to achieve a beneficial response more rapidly. Available evidence does strongly indicate that the rate of patient improvement is dose dependent.

## C. Surgical Procedures

A wide range of out-patient procedures is employed in the treatment of photoaged skin, although virtually no published data document their success rates, relative efficacy, or frequency of complications. On an anecdotal level, there appears to be considerable variation among practitioners in technique and cosmetic results. No single procedure addresses all aspects of photoaging.

### *1. Cryotherapy*

Liquid nitrogen, applied briefly, is routinely used to treat actinic keratoses and solar lentigines. Successful cryotherapy presumably results from destruction of mutated basilar keratinocytes (actinic keratoses) or melanocytes (lentigines), with sparing of surrounding normal cellular elements that then repopulate the skin.

### *2. Laser Therapy*

Both the continuous-wave argon laser (39) and, more recently, the flashlamp-pulsed tunable dye laser (40) have been extensively used to treat facial actinic telangiectases and so-called poikiloderma of Civatte, the reticulated hyperpigmentation and telangiectases that classically involve the sun-exposed lateral neck. Laser light (488 nm and 514 nm for argon, usually 585 nm for tunable dye), directed at the involved vessels through a fiberoptic cable, is specifically absorbed in the skin by intravascular hemoglobin, which is complementary in color to the light, and is immediately transformed to heat energy that then destroys the targeted ectatic vessels. New vessels formed in the ensuing healing period are smaller in diameter and thus invisible from the skin surface, as in nonphotodamaged skin. Results are usually excellent. However, permanent if subtle depigmentation is common in the treated areas following argon laser treatment, due to inadvertent melanocyte ablation; scarring (sclerosis) may occur, particularly on the neck. The tunable dye laser has a much smaller risk of scarring, attributed to the confinement of thermal damage to the immediate perivascular area by virtue of its short pulse duration. Persistence or recurrence of individual telangiectases may be observed after treatment with either instrument. Other lasers, such as carbon dioxide (10,600 nm) and the copper vapor (511 nm and 578 nm) lasers, as well as conventional electrocautery with a fine needle inserted into the offending vessel, are also employed with good to excellent results in most patients.

Lasers, particularly the flashlamp-pulsed tunable dye laser set at 504 nm (41), have also been used to treat epidermal hyperpigmentation, including solar lentigenes, and other stigmata of photoaging. In this instance, melanin granules within melanocytes are the presumed target or chromophore, and absorbed laser energy is presumed to alter the abnormal program of melanogenesis permanently in the melanocyte without eliminating pigment production totally. Long-lasting normalization of skin color can be achieved in at least some patients. More clinical experience will be required to judge the overall clinical efficacy of this modality.

### 3. *Lifting Procedures*

The face lift, blepharoplasty, and brow lift redistribute and excise sagging skin and subcutaneous tissue. The goal is to remove redundant folds of skin and fat to define better the esthetic landmarks of the face. However, this type of surgery has no effect on the quality, texture, or appearance of photodamaged skin and has little effect on expression lines and wrinkles of the central face.

### 4. *Filler Materials*

These can be injected in a simple, office-based, low-morbidity procedure to soften expression lines and wrinkles. They have no effect on skin folds resulting from laxity or on the changes in color and texture associated with photoaging. Filler materials include bovine collagen, porcine collagen mixed with the patient's fibrin, autologous fat, and polydimethylsiloxane (silicone).

### 5. *Dermabrasion and Chemical Peels*

Facial skin involved with actinic keratoses, irregular pigmentation, and/or fine wrinkling (lesions of the epidermis and superficial dermis) may be superficially destroyed by either mechanical or chemical means in an attempt to reconstitute more attractive skin from deeper and hence less photodamaged progenitor cells in follicular remnants and deeper dermis. Results are often good, but clinically apparent scarring is a definite risk. The primary indication for chemical exfoliation is treatment of sun-damaged skin (42,43).

## VI. SUMMARY

Photoaging, the combined effect of intrinsic aging and chronic sun damage on human skin, is a well-defined clinical and histologic syndrome. Ultraviolet irradiation, particularly UVB, is believed to act directly on cells in both the epidermis and dermis to alter their function and ultimately to deplete their populations. Unrepaired UV-induced DNA damage, perhaps compounded by decreased immunosurveillance and other functional impairments due to photoaging, results in photocarcinogenesis. Photoaging is completely preventable. Institution of effective sun protection even after severe damage may permit limited repair, but most changes do not show spontaneous regression. Topical tretinoin has been shown to improve many aspects of photaging when applied daily for at least 4–6 months, and various surgical procedures address individual features with good to excellent results. However, heightened public and professional awareness of the cumulative adverse effects of habitual sun exposure is needed to decrease the burden of photoaging on future generations.

## REFERENCES

1. Gilchrest BA. Skin and Aging Processes. Boca Raton, FL: CRC Press, 1984: 130 pp.
2. Gilchrest BA. Physiology and pathophysiology of aging skin. In: Goldsmith LA, ed. Biochemistry and Physiology of the Skin. England: Oxford University Press, 1991: 47–56.
3. O'Dell BL, Jessen T, Becker LE, Jackson RT, Smith EB. Diminished immune response in sun-damaged skin. Arch Dermatol 1980; 116:559–561.
4. Gilchrest BA, Blog FB, Szabo G. The effects of aging and chronic ultra-violet irradiation on melanocytes in human skin. J Invest Dermatol 1979; 73:141–143.
5. Breathnach AS, Wylie LM. Electron microscopy of melanocytes and melanosomes in freckled human epidermis. J Invest Dermatol 1964; 42:389–394.
6. Deleo VA, Dawes L, Jackson R. Density of Langerhans cells (LC) in normal vs. chronic actinically damaged skin (CADS) of humans. J Invest Dermatol 1981; 76:330–334.
7. Gilchrest BA, Murphy GF, Soter NA. Effect of chronologic aging and ultraviolet irradiation on Langerhans cells in human skin. J Invest Dermatol 1982; 79:85–88.
8. Aberer W, Schuler G, Stingl G, Honigsmann H, Wolff K. Ultraviolet light depletes surface markers of Langerhans cells. J Invest Dermatol 1981; 76.202–210.
9. Gilchrest BA, Stoff JS, Soter NA. Chronologic aging alters the response to UV-induced inflammation of human skin. J Invest Dermatol 1982; 79:11–15.
10. Kligman AM. Early destructive effect of sunlight on human skin. JAMA 1969; 210:2377–2380.
11. Bhawan J, Chil-Hwan Oh, Lew R, Nehal K, Labadie R, Gilchrest BA. Histopathologic differences in the photoaging process in facial versus arm skin. Amer J Dermatopath 1992; 14:224–230.
12. Braverman IM, Fonferko E. Studies in cutaneous aging. I. The elastic fiber network. J Invest Dermatol 1982; 78:434–443.
13. Lavker RM. Structural alterations in exposed and unexposed aged skin. J Invest Dermatol 1979; 73:59–66.
14. Braverman IM, Fonferko E. Studies in cutaneous aging. II. The microvasculature. J Invest Dermatol 1982; 78:444–448.
15. Kligman AM. Perspectives and problems in cutaneous gerontology. J Invest Dermatol 1979; 73:39–46.
16. Lavker RM, Kligman AM. Chronic heliodermatitis: a morphologic evaluation of chronic actinic dermal damage with emphasis on the role of mast cells. J Invest Dermatol 1988; 90:325–330.

17. Kligman LH, Akin FJ, Kligman AM. The contributions of UVA and UVB to connective tissue damage in hairless mice. J Invest Dermatol 1985; 84:272–276.
18. Gilchrest BA. Relationship between actinic damage and chronologic aging in keratinocyte cultures of human skin. J Invest Dermatol 1979; 72:219–223.
19. Gilchrest BA. Prior chronic sun exposure decreases the lifespan of human skin fibroblasts in vitro. J Gerontol 1980; 35:537–541.
20. Stanulis-Praeger BM, Gilchrest BA. Effect of donor age and prior sun exposure on growth inhibition of cultured human dermal fibroblasts by all trans-retinoic acid. J Cell Physiol 1989; 139:116–124.
21. Kligman LH, Akin FJ, Kligman AM. Sunscreens promote repair of ultraviolet radiation-induced dermal damage. J Invest Dermatol 1983; 81:98–102.
22. Kligman LH, Kligman AM. The nature of photoaging: its prevention and repair. Photodermatol 1986; 3:215–227.
23. Weinstein GD, Nigra TP, Pochi PE, Savin RC, Allan A, Benik K, Jeffes E, Lufrano L, Thorne EG. Topical tretinoin for treatment of photodamaged skin: a multicenter study. Arch Dermatol 1991; 127:659–665.
24. Bhawan J, Gonzalez-Serva K, Nehal K, Labadie L, Lufrano L, Thorne EG, Gilchrest BA. Effects of tretinoin on photodamaged skin: a histologic study. Arch Dermatol 1991; 127:666–672.
25. Kligman AM, Grove GL, Hirose R, Leyden JJ. Topical tretinoin for photoaged skin. J Am Acad Dermatol 1986; 15:836–859.
26. Weiss JS, Ellis CN, Headington JT, Tincoff T, Hamilton TA, Voorhees JJ. Topical tretinoin improves photoaged skin: a double-blind vehicle-controlled study. JAMA 1988; 259:527–532.
27. Leyden JJ, Grove GL, Grove MJ, Thorne EG, Lufrano L. Treatment of photodamaged facial skin with topical tretinoin. J Am Acad Dermatol 1989; 21:638–644.
28. Ellis CN, Weiss JS, Hamilton TA, Headington JT, Zelickson AS, Voorhees JJ. Sustained improvement with prolonged topical tretinoin (retinoic acid) for photoaged skin. J Am Acad Dermatol 1990; 23:629–637.
29. Bhawan J, Lufrano L, Thorne EG, Gilchrest BA. Some histologic changes in tretinoin-treated skin regress from 24 to 48 weeks despite continued clinical improvement. Clin Res 1991; 39:502A.
30. Zelickson AS, Mottaz JH, Weiss JS, Ellis CN, Voorhees JJ. Topical tretinoin in photoaging: an ultrastructural study. J Cutan Aging Cosmet Dermatol 1988; 1:41–47.
31. Woodley DT, Zelickson AS, Briggaman RA, Hamilton TA, Weiss JS, Ellis CN, Voorhees JJ. Treatment of photoaged skin with topical tretinoin increases epidermal-dermal dermoepidermal anchoring fibrils: preliminary report. JAMA 1990; 263:3057–3059.
32. Moriarty M, Dunn J, Darragh A, Lambe R, Brick I. Etretinate in the treatment of actinic keratosis. Lancet 1982; 1:364–365.
33. Peck GL. Topical tretinoin in actinic keratosis and basal cell carcinoma. J Am Acad Dermatol 1986; 15:829–835.
34. Epstein JH. All-trans-retinoic acid and cutaneous cancers. J Am Acad Dermatol 1986; 15:772–778.
35. Thorne EG, Roach JB Jr, McElroy JA. Topical tretinoin in the treatment of actinic

(solar) keratoses. Atlanta, Georgia: Exhibit at 49th Annual Meeting of the American Academy of Dermatology, 1990.
36. Kraemer KH, DiGiovanna JJ, Moshell AN, Tarone RE, Peck GL. Prevention of skin cancer inxeroderma pigmentosum with the use of oral isotretinoin. N Engl J Med 1988; 318:1633–1637.
37. Gilchrest BA. Retinoid pharmacology and skin. In: Mukhtar H, ed. Pharmacology of the Skin. Boca Raton, FL: CRC Press, 1992:167–181.
38. Worobec SM, Wong FA, Tolman EL, Abrams LS, Parker GR, Thorne EG. Percutaneous absorption of $^3$H-tretinoin in normal volunteers. Clin Res 1990; 38: 786A.
39. Apfelberg DB, Maser MR, Lash H. The argon laser for cutaneous lesions. JAMA 1981; 245:2073–2075.
40. Polla LL, Tan OT, Garden J, Parrish JA. Tunable pulsed dye laser for the treatment of cutaneous vascular ectasia. Dermatologia 1987; 174:11–17.
41. Kurban AK, Cheong WF. Laser treatment of pigmented lesions. In: Achauer BM, Vander Kam V, eds. Lasers in Plastic Surgery and Dermatology. New York: Thieme Medical Publishers, 1991.
42. Baker TJ, Gordon HL. Chemical face peeling and dermabrasion. Surg Clin North Am 1971; 51:387–401.
43. Stegman ST, Tromovitch TA, Glogau TG. Chemical peels. In: Stegman ST, Tromovitch TA, Glogau RG, eds. Cosmetic Dermatologic Surgery. Chicago: Year Book Medical Publishers, 1990:35–38.

# 7

# Skin Cancer

**Robin Marks**
*Anti-Cancer Council of Victoria, Melbourne, Victoria, Australia*

**Arthur J. Sober**
*Harvard Medical School, Boston, Massachusetts*

Most skin cancers (nonmelanoma skin cancer and melanoma) appear to be directly or indirectly related to previous sun exposure. The relationship between cause and solar exposure appears different for the two forms. The evidence linking solar exposure with these cancers and the dose–response relationships will be explored in this chapter.

Nonmelanoma skin cancer includes basal cell carcinoma and squamous cell carcinoma. Solar keratoses, while not being true invasive carcinomas, have an intimate relationship to the development of nonmelanoma skin cancer (either being a precursor or an indicator of high risk). Thus, any discussion of nonmelanoma skin cancer needs to take into account the nature and extent of solar keratoses as well.

## I. NONMELANOMA SKIN CANCER

In 1894 Paul Gerson Unna reported that there appeared to be a relationship between exposure to sunlight and the development of nonmelanoma skin cancer (1). Almost 100 years later, we are still making the same observation, but the exact nature of the relationship is not yet entirely clear.

Two main factors contribute to the incidence of nonmelanoma skin cancer in any given population: the physical characteristics of individuals in the population, and the sunlight to which the individuals in the population are exposed.

These two are not entirely separate: the physical characteristics of individuals appear to predispose them to be sensitive to sunlight. Thus they both merit attention in a discussion of nonmelanoma skin cancer.

## A. Physical Characteristics

### *1. Skin Color*

The major physical characteristic predisposing individuals to the development of nonmelanoma skin cancer is skin color. The darker the natural untanned color of the skin, the lower the incidence of nonmelanoma skin cancer. For example, the annual incidence rate in black South Africans has been reported as being less that 1:100,000 persons whereas in white Australians, whose sunlight exposure is similar to persons in South Africa, the annual incidence rate is greater than 800:100,000 persons (2,3). The incidence rate of basal cell carcinoma for white men in New Mexico in 1977–1978 was likewise 495:100,000/year, while the rate for Hispanic men living in the same area was 64.3:100,000/year (4).

The untanned color of skin is used as a measure of risk here, not suntan. However, ability to tan when exposed to strong sunlight is a predictor of risk of nonmelanoma skin cancer in fair-skinned persons. There is a gradient of risk, from the highest in those persons who "burn only" when exposed to strong sunlight through to moderate risk for those who burn first and then tan. The lowest risk is among those who state that they "tan only" and "never burn" when exposed (3) (Fig. 1). The relative risk of those who "burn only" compared to those who "tan only" when exposed to strong sunlight is around 3. This has led to the perception that a tan may be protective against nonmelanoma skin cancer. What is even worse is that some people have actually advocated the seeking of a "safe" tan as a way of reducing personal risk. Extrapolating from case–control studies and relative risks into the development of public health messages is fraught with danger.

In a population-based incidence study from Australia, the rate of nonmelanoma skin cancer among people, who stated that they "burn only" when exposed to strong sunlight was 1,764:100,000 persons/year. In contrast, the rate for those who stated that they "tan only" when exposed was 616:100,000 persons/year (3). This latter rate is higher than any other previously reported nonmelanoma skin cancer rate anywhere else in the world. In general, the proportion of fair-skinned populations reporting a response to exposure to strong sunlight of "tan only" and "never burn" is 20–25%. Thus a recommendation to seek a "safe" tan is likely to lead to substantially higher rates for the 75–80% of the population who are more sensitive to sunlight.

Other characteristics associated with increased risk of nonmelanoma skin cancer include blue eyes compared with brown eyes and light hair compared with dark hair (5). Similar physical characteristics have also been associated with de-

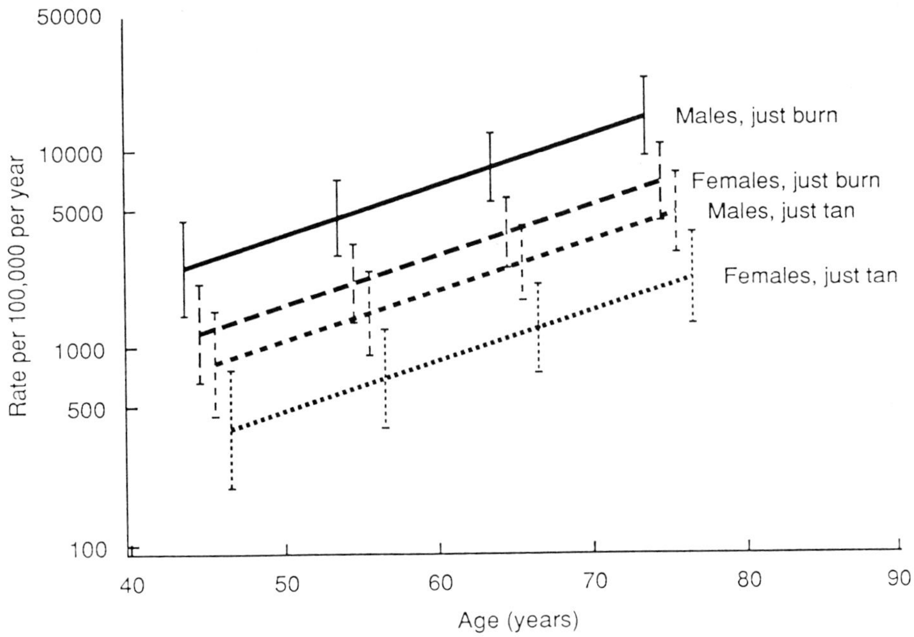


**Figure 1** Incidence of nonmelanoma skin cancer in Maryborough, Australia, shows a higher rate in men than women and in those people who "burn only" when exposed to sunlight compared to those who "tan only." (From Ref. 20.)

velopment of solar keratoses (6). Since skin, eye, and hair color are all inherited characteristics, it follows that there is an inherited predisposition to be sensitive to sunlight and thus to be at increased risk of nonmelanoma skin cancer. This is believed to be highest among those with Celtic ancestry who have red hair, blue eyes, who cannot tan when exposed to sunlight, and who have a tendency to develop freckles. Whether or not these people have other characteristics apart from their skin type that predispose them to skin cancer has been questioned but, as yet, no clear evidence has been put forward to support this.

### 2. Gender

Gender is another important physical characteristic that is a risk factor for nonmelanoma skin cancer. The incidence rate is substantially higher in men than women, particularly for squamous cell carcinoma (4,5). This can be almost totally explained by behavioral and occupational characteristics of men compared with women. More men tend to work outdoors than women and the rate of nonmelanoma skin cancer has been shown to be higher in outdoor than in indoor workers (7) (Fig. 2). A recent study from the Centre for Behavioural Research in

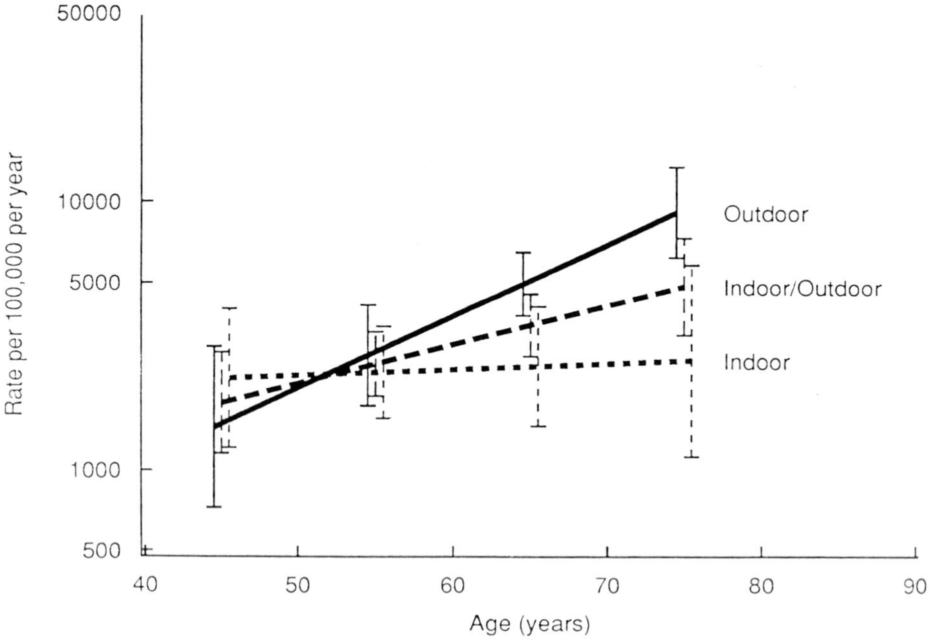


**Figure 2** Incidence rate of nonmelanoma skin cancer in men in Maryborough, Australia, shows a higher rate in those who work outdoors than in those who work indoors. (From Ref. 20.)

Cancer at the Anti-Cancer Council of Victoria demonstrated in Australia that men tend to take part in more outdoor leisure activities than women, particularly during the middle of the day (8). This study also noted that men tend to be outdoors for longer periods during their leisure hours than women.

### *3. Immune Competence*

Immunocompetence is an important factor that can influence individual risk of squamous cell carcinoma. The ratio of basal cell carcinoma to squamous cell carcinoma, normally around 4:1, is substantially reduced in immunosuppressed patients who have had renal transplants and may even reverse as squamous cell carcinomas become increasingly common among these patients with time (9). The role of sunlight in affecting both local skin and systemic immune responses is being increasingly elucidated, as reflected in this book. Its exact role in the development of nonmelanoma skin cancer in people who have an apparently normal immune response is not yet clear (see Chap. 8).

The role of immunity in the control of human papilloma virus (HPV) infection is also not entirely clear. Recurrent HPV infection of skin (i.e. warts) in renal transplant patients can become a major management problem. Because of the association of squamous cell carcinoma of the skin in immunocompromised people, the increased risk of squamous cell carcinoma in patients with epidermodysplasia verruciformis, and the recent association of HPV infection of the cervix with dysplasia and cervical carcinoma (squamous cell carcinoma), a possible role for HPV in the development of solar keratoses and squamous cell carcinoma of apparently normal people is being explored (10). Its role as a cocarcinogen with sunlight is being specifically questioned. Until now HPV has been mainly associated in individuals with an intact immune system with dysplasias and squamous cell carcinomas of the anogenital region and mucosae. If the presence of HPV is found in association with actinic-related tumors, a further step to demonstrate causality rather than mere association will need to be demonstrated.

Other examples of physical factors contributing to the risk of nonmelanoma skin cancer include xeroderma pigmentosum and patients who have been treated with ionizing radiation. In each of these populations, the rate of nonmelanoma skin cancer is higher than would be expected from exposure to sunlight alone. In xeroderma pigmentosum, which involves defective DNA repair following exposure to ultraviolet light, the tumors occur at sites that are normally sun-exposed (11). With ionizing radiation, the tumors occur following ultraviolet light exposure at the sites of radiation exposure (12).

### *4. Age*

With increasing age, there tends to be a log-linear increase in the incidence of both basal cell and squamous cell carcinomas (4). This could be due to the accumulation of total sunlight with time, but age itself is a risk factor for many cancers. Whether for nonmelanoma skin cancer this is cumulative exposure, increased sensitivity to the carcinogen with advanced age, decreased responsiveness of the immune system, or some other factor is almost impossible to determine from epidemiologic studies. Separating out increasing age alone from increasing cumulative dose of sunlight is extremely difficult. Some other method to look specifically at the effect of age on skin carcinogenesis will probably be necessary before this question can be answered.

## B. Exposure to Sunlight

There is little doubt that sunlight plays an important part in the development of nonmelanoma skin cancer. But, as yet, we do not have precise answers to questions such as how much sunlight is necessary to develop lesions; how many different wavelengths of sunlight are involved; in what way exposure to sunlight is important (e.g., are high doses over short periods likely to have the same effect as the identical cumulative dose delivered over a long period); does age at which sunlight is received make any difference; and is there a threshold dose of sunlight necessary before tumors will appear. However, there are partial answers to some of these questions.

### 1. *Site Distribution*

Between 50 and 75% of squamous cell carcinomas occur on the head and neck. The rest occur mainly on the dorsa of the hands, forearms, and light-exposed areas of the legs (Fig. 3). Nonmelanoma skin cancers of both types are more common on the ears in men than women, possibly related to hair fashions. In the past, it has been suggested that up to 90% of basal cell carcinomas occur on the head and neck. More recent studies have shown an increasing proportion of tumors occurring on the trunk. A 1990 study from Western Australia reported 53% of basal cell carcinomas in men and 43% in women were present on the trunk. Most of these occurred on the upper back and shoulders (13). If this relative change in distribution is due to increased sunlight exposure with change in clothing and bathing habits, it still remains to be explained why a similar change in anatomical distribution has not been seen with squamous cell carcinoma. Whether it means that basal cell carcinomas need less exposure to induce a lesion than squamous cell carcinomas is not clear. Certainly the mean age of people with basal cell carcinoma tends to be around 10 years younger than the mean age for squamous cell carcinoma (5).

It is interesting, however, that around one-third of basal cell carcinomas that occur on the head and neck appear in sites not those of maximal exposure (e.g, behind the ear, or in inner canthi). In fact, these areas tend to receive around only 20% of the maximum exposure possible (14). This is in contrast to squamous cell carcinomas, which tend to occur on sites of maximal exposure on the head and neck.

The difference in exposure characteristics between squamous cell carcinoma and basal cell carcinoma highlights the fact that these tumors have quite different biological charactistics. Most studies tend to suggest a more clear relationship between squamous cell carcinoma and sunlight than there is for basal cell carcinoma.

Solar keratoses are also being noted more frequently on the trunk. In a recent study from Queensland, 61% of people who had a full body examination as part of a population study had lesions on sites apart from the head and neck, hands, and forearms (15). Solar ketaroses appear to be a sensitive indicator of sunlight exposure in people who are predisposed to nonmelanoma skin cancer, both basal cell carcinoma and squamous cell carcinoma (16). Although actinic keratoses have been thought to be precursors of squamous cell carcinoma, a prospective study suggested that fewer than 1:1000 of these lesions is likely to change to a squamous cell carcinoma within a year (17). Up to 25% of these lesions may remit spontaneously within 12 months, particularly in people who have the ability to reduce their sunlight exposure (18). Actinic keratoses are also a clinical

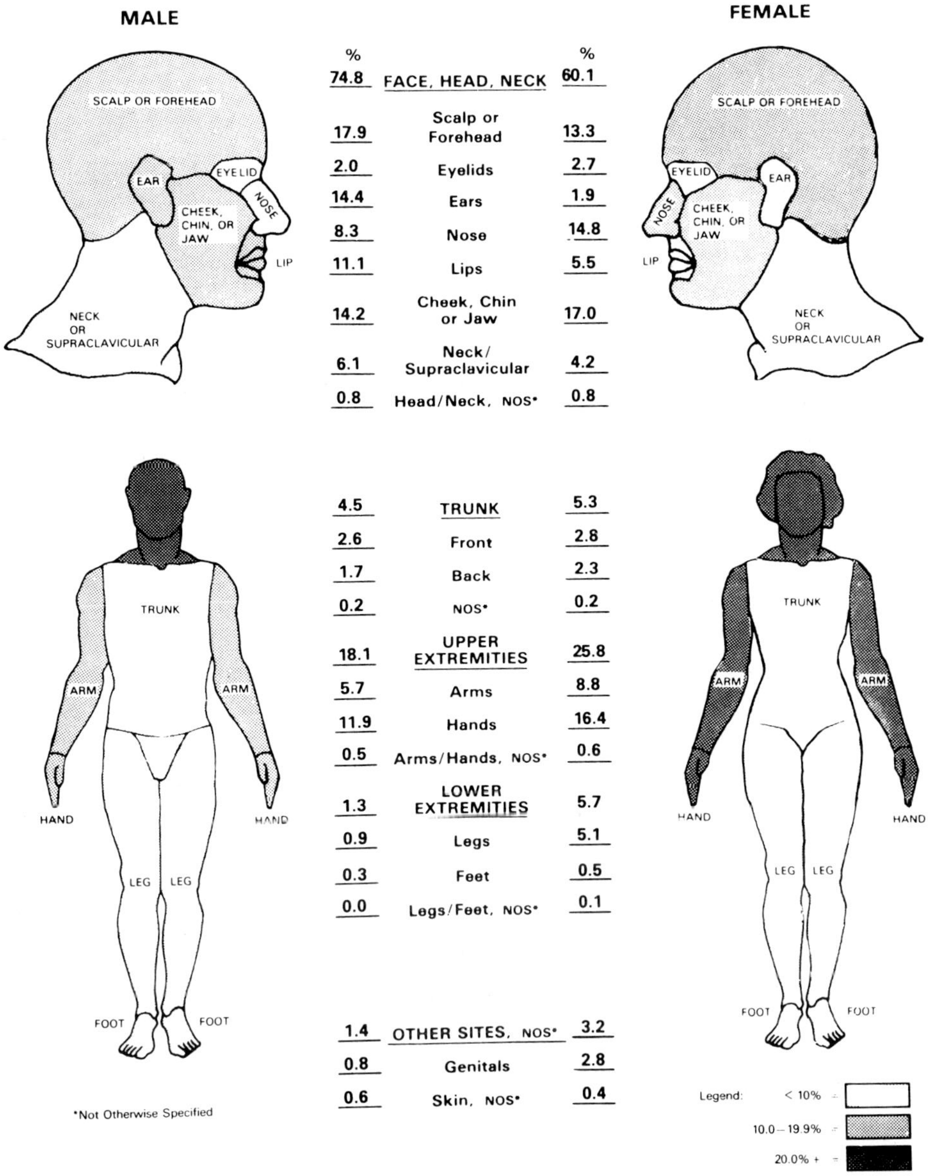


| Male % | Site | Female % |
|---|---|---|
| 74.8 | FACE, HEAD, NECK | 60.1 |
| 17.9 | Scalp or Forehead | 13.3 |
| 2.0 | Eyelids | 2.7 |
| 14.4 | Ears | 1.9 |
| 8.3 | Nose | 14.8 |
| 11.1 | Lips | 5.5 |
| 14.2 | Cheek, Chin or Jaw | 17.0 |
| 6.1 | Neck/ Supraclavicular | 4.2 |
| 0.8 | Head/Neck, NOS* | 0.8 |
| 4.5 | TRUNK | 5.3 |
| 2.6 | Front | 2.8 |
| 1.7 | Back | 2.3 |
| 0.2 | NOS* | 0.2 |
| 18.1 | UPPER EXTREMITIES | 25.8 |
| 5.7 | Arms | 8.8 |
| 11.9 | Hands | 16.4 |
| 0.5 | Arms/Hands, NOS* | 0.6 |
| 1.3 | LOWER EXTREMITIES | 5.7 |
| 0.9 | Legs | 5.1 |
| 0.3 | Feet | 0.5 |
| 0.0 | Legs/Feet, NOS* | 0.1 |
| 1.4 | OTHER SITES, NOS* | 3.2 |
| 0.8 | Genitals | 2.8 |
| 0.6 | Skin, NOS* | 0.4 |

*Not Otherwise Specified

**Figure 3** Site distribution of squamous cell carcinoma in the United States shows the tumors occurring predominantly in areas of maximum sunlight exposure. (From Ref. 4.)

indicator of high risk of nonmelanoma skin cancer and may be useful as a surrogate for study of the behavior of these tumors, particularly squamous cell carcinoma (16).

### 2. *Latitude*

The difference in incidence of nonmelanoma skin cancer with difference in latitude of residence is another strong point in favor of sunlight causing these tumors. A recent population-based incidence rate from South Wales, United Kingdom, was calculated to be 83:100,000 persons/year for basal cell carcinoma and 19:100,000 persons/year for squamous cell carcinoma, giving a combined rate of 102 (19). A rate for an equivalent population in Victoria, Australia which has a similar genetic makeup to the Welsh population was calculated as 672:100,000 persons/year for basal cell carcinoma, 201:100,000 persons/year for squamous cell carcinoma, and a combined rate of 873 (20).

Within countries there are similar differences in incidence with latitude in the same populations. This has been well demonstrated in both the United States and Australia. It is likely that differences in rates between different parts of the same country and different countries are not only due to differences in the latitude. Areas at lower latitudes also tend to have warmer climates as well as increased radiation, thus encouraging people to be outdoors with less clothing and to undertake more outdoor leisure activity. Therefore behavioral differences leading to different exposure patterns need to be accounted for when comparing different populations. A study revealing a substantially increased prevalence of solar ketatoses at all ages between two different populations in Victoria, Australia, revealed that some of the differences in radiation between the two areas were a result of differences not only in latitude but also in altitude (a 15% increase in ultraviolet radiation for each 1000 m above sea level) and in cloud cover (21) (Fig. 4). Cloud cover, although absorbing some ultraviolet radiation, also scatters it, so that the total reduction in ultraviolet radiation at the earth's surface may not be as great as expected. Cloud (water vapor) absorbs proportionally more radiation in the infrared spectrum than the ultraviolet spectrum, thus reducing the heat of the sun more than the ultraviolet radiation. This frequently leads to a false sense of security on cloudy summer days, when the resultant ultraviolet radiation-induced erythema (sunburn) is attributed to "windburn" rather than its real case.

### 3. *Ozone Depletion*

Differences in incidence rate with different latitudes in the United States and Australia have been used as the basis for estimating the amount of ultraviolet radiation necessary to develop nonmelanoma skin cancer. This has occurred recently in attempts to forecast the effects of ozone depletion. In general, the estimates suggest an increase of 1.3–4% in nonmelanoma skin cancer for each 1% increase in total lifetime exposure to ultraviolet B (UVB) (290–320 nm) (22,23). The estimates are remarkably similar between the United States and Australia, with both showing that for the same increase in UVB, a proportionally

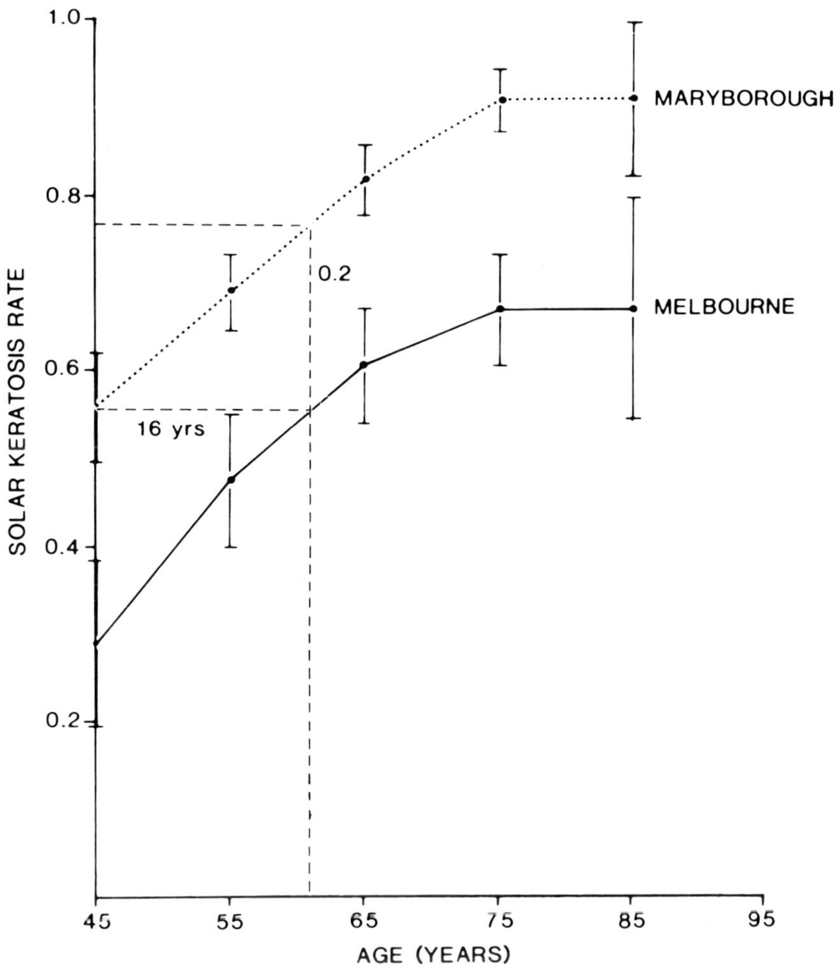


**Figure 4** A comparison of solar keratosis prevalence rate (1.0= 100%) in Australian-born male outdoor workers in Melbourne and Maryborough shows a significant difference between the two areas (Z=4.065, $p<0.001$). On the linear part of the curve, there is a difference of approximately 16 years in the age of persons who have a particular prevalence rate in Maryborough compared with the same rate in Melbourne. There is a 14% higher level of UVB in Maryborough than in Melbourne. (From Ref. 21.)

greater increase in incidence rate is likely to occur in those populations with the highest incidence rates already due to current high background UVB levels. This means that areas like the southern United States or northern Australia will be affected more by the effects of ozone depletion than the northern United States or southern Australia. These estimates are based on total lifetime exposure. It is difficult to calculate the lag period effect of short-term exposures. However, calculations based on studies such as those of Scotto et al. across the United States have suggested that the amount of sunlight necessary to cause nonmelanoma skin cancer is predominantly related to the amount accumulated over time (4). The fact that the incidence rate of nonmelanoma skin cancer (particularly squamous cell carcinoma) increases in a logarithmic fashion with increasing age suggests the cumulative nature of the exposure necessary to develop these tumors. These age-specific incidence curves tend to be similar in slope for different countries, differing mainly in the slope at the young age groups when more tumors occur in younger age groups in countries with high background isolation, (e.g., Australia) (14). Thus attempts to estimate the effects of increased exposure to sunlight on the incidence rate of nonmelanoma skin cancer suggest that not only will the incidence of the tumors rise but the age of onset will also fall. The slopes with latitude appear to be steeper for squamous cell carcinoma than for basal cell carcinoma.

Calculations of the increased risk to British people associated with recreational ultraviolet light exposure over a 20 year period (Table 1) (24) show that for relatively small additions to normal exposure occurring in the United Kingdom, there is a consistent increase in risk, particularly when the increased exposure occurs at a young age.

### 4. *Age of Exposure*

A number of studies of the incidence and mortality due to nonmelanoma skin cancer in British immigrants to Australia show significantly lower rates in the migrant population compared to the Australian-born population (25). A recent study comparing the prevalence of solar keratoses in Australian-born people with

**Table 1** Risk of nonmelanoma skin cancer at age 70 years from recreational ultraviolet light exposure for 20 years

| | Annual Exposure from Sunbathing | | | |
|---|---|---|---|---|
| Age at Sunbathing (years) | None | 1 week UK Sunbathing (~ 20 MED) | 2 weeks UK Sunbathing (~ 40 MED) | 3 weeks Mediterranean Sunbathing (~ 80 MED) |
| 5–25 | 1.0 | 1.4 | 2.1 | 4.2 |
| 20–40 | 1.0 | 1.2 | 1.5 | 2.5 |
| 35–55 | 1.0 | 1.1 | 1.3 | 1.7 |
| 50–70 | 1.0 | 1.1 | 1.2 | 1.4 |

MED, Minimal erythema dose.
Source: Ref. 24.

British migrants showed that those people who migrated to Australia at less than 20 years had, by the age of 50 years, rates equalling or exceeding those in the Australian-born population (26). Those people who migrated from Britain after the age of 20 years never reached the rates in Australian-born people, even though they may have lived in Australia for many years after they migrated (Fig. 5). These data do not explain whether it is merely that a large proportion of total lifetime sunlight is received within the first 20 years of life (i.e., nonmelanoma skin cancer is due to cumulative exposure), or whether there is some particular sensitivity of childhood skin that puts it at risk when exposed during that age period. The former reason has been postulated by Stern et al., who used this assumption in a computer model that predicted that over 75% of nonmelanona skin cancer could be prevented by the correct use of sunscreens in childhood and adolescence (27).

### 5. *Wavelengths of Ultraviolet Radiation*

The use of sunscreens to prevent nonmelanoma skin cancer introduces the last point for discussion: what particular wavelengths of sunlight are responsible for the development of these tumors. It has been postulated that UVB is the major

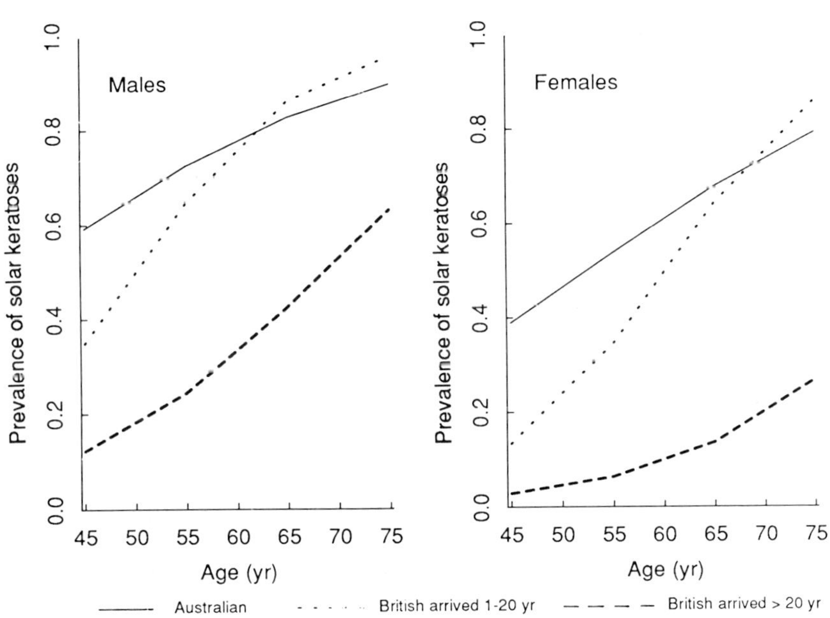


**Figure 5** The prevalence of solar keratoses (1.0=100%) in Australian-born persons compared to persons who migrated to Australia from Great Britain before and after the age of 20 years. (From Ref. 26.)

offender, on the basis of animal data and the fact that this range is most clearly identified with the ability to induce erythema in humans (28,29). A large proportion of sunscreens in some countries, the United States and United Kingdom included, contain agents that block out only that range of ultraviolet radiation. More recent data have suggested that ultraviolet A (UVA) (320–400 nm) may have a contributory, although not as powerful, role in nonmelanoma carcinogenesis. This is based on animal tumor induction studies as well as human data (30,31). Use of UVA in association with psoralens for treatment of psoriasis has been shown to change the normal basal cell carcinoma to squamous cell carcinoma ratio, with an increase in the number of squamous cell carcinomas in patients undergoing this treatment (32). On the basis of these data, it does seem prudent to recommend that UVA-blocking agents should be included in sunscreens, in addition to those that block UVB to reduce the risk of skin cancer. Finally, although there have been suggestions that other wavelengths of sunlight, including infrared radiation and even the visible spectrum, may make some contribution to carcinogenesis, the evidence for these claims is very scanty at this time.

## II. CUTANEOUS MELANOMA

Although much more extensively studied than nonmelanoma skin cancer, the relationship of sunlight exposure to melanoma appears more indirect and is less well quantitated. Several recent review articles summarize this area (33–35). Exploring the relationship between sunlight and melanoma and, more specifically, between the specific wavelengths of sunlight and melanoma to determine an action spectrum for tumor production, has been hampered by the lack of an animal model. Until recently, there has been a report of only one mouse in which a melanoma had been induced by UVB (36). A potentially significant advance is a study on the South American opossum, *Monodelphis domestica*, which develops a melanomalike tumor upon exposure to UVB (34). This tumor develops without the addition of any other chemical carcinogen.

### A. Epidemiologic Evidence in Humans

Table 2 lists the observations for and against a role for solar exposure in cutaneous melanoma. Perhaps the most direct evidence linking solar exposure to melanoma occurs with lentigo maligna melanoma, the least common type of melanoma (5% frequency) and the first to be differentiated from the other forms of melanoma. With lentigo maligna melanoma, there is clear evidence that chronic cumulative sun exposure, similar to the nonmelanoma skin cancers, is the most likely causative factor. These tumors occur primarily on sun-exposed surfaces, especially the nose and cheek of older individuals (mean age 70), in chronically sun-damaged skin. The other types of melanoma, by and large, do not seem to be linked to chronic sun exposure, and, in fact, individuals with the

**Table 2** Sunlight and cutaneous melanoma: evidence for and against causation

| |
|---|
| *Positive Evidence* |
| Inverse latitude gradient worldwide: melanoma incidence increases in areas closer to the equator. |
| Melanoma occurs primarily in lightly pigmented Caucasians and is rare in more heavily pigmented people (e.g., Asians, Blacks, and Hispanics). |
| Migration studies suggest that melanoma risk increases in those moving to an area of more intense sun exposure. |
| Melanoma risk increases with the number of melanocytic nevi, which are concentrated on sun-exposed areas, and rare on "double protected" body sites. |
| Melanoma occurs most commonly on body areas exposed to the sun during recreational exposure (e.g., back and legs). |
| Patients with melanoma have increased rates of severe sunburns and sun sensitivity. |
| Patients with melanoma have lower minimal erythema doses and higher rates of skin types I (burn, never tan) and II (burn, then tan) than controls. |
| Direct dose–response relationship between ultraviolet light radiation and lentigo maligna melanoma. |
| Melanoma rates are increased in xeroderma pigmentosum, a disease entity characterized by the inability to repair ultraviolet light damaged DNA. |
| *Negative Evidence* |
| Lower rates of melanoma in persons with outdoor occupations and higher rates in white-collar, educated, indoor workers. |
| Melanoma is not common on the face and hands, except for the lentigo maligna melanoma subtype. |
| Little solar elastosis in skin in the vicinity of melanoma, in contrast to nonmelanoma skin cancer. |
| The latitude gradient for melanoma is not clear in studies in western Australia and central Europe. |
| Melanoma cannot consistently be induced in animals using ultraviolet radiation alone. |

Source: Ref. 35.

highest cumulative sun exposure often have lower relative risks than individuals who have had less sun exposure (38). Studies linking increased melanoma risk to chronic cumulative sun exposure (39) have noted a direct relationship between total sun exposure throughout life and increasing risk of developing melanoma.

What is the indirect evidence that sunlight is a factor in the production of melanoma?

### 1. *Race and Skin Type*

Melanoma is primarily a disease of the lighter-complexioned Caucasian races. The frequency in Blacks, Asians, and very dark-skinned Caucasians (natives of

India, Puerto Rico, etc.) is about one-tenth the rate noted in Caucasians of light complexion. When melanomas occur in these darker-complexioned individuals, the sites of distribution are different from those occurring in lighter-skinned Caucasians. Palms, soles, nailbeds, and mucous membranes are the preferred sites in darker-skinned individuals. One recent study has suggested that the site-specific rates for melanoma on the sole is the same in Blacks and Caucasians (40).

In the four skin types occurring in whites, there is a gradient of up to 3 times for increased melanoma risk noted in individuals with either type I and type II skin as compared to individuals of higher skin types. Those individuals who "always burn" and "never tan," who freckle, who have blond or red hair, blue or grey eyes, are those at greater risk for melanoma in contrast to individuals lacking these characteristics.

### 2. *Latitude Gradient*

Early reports of increased melanoma rates as the equator is approached were noted in Australia (41), Norway (42), and from the Third National Cancer Survey in the United States. In the United States the latitude gradient is the least steep for melanoma in comparison to the two forms of nonmelanoma skin cancer. In 1980, the incidence rate in Tucson, Arizona, was about two-and-one-half times higher than the incidence rate for Connecticut (43). Not only is Tucson, Arizona, closer to the equator, it is also at substantially higher elevation than Connecticut, which is mostly at sea level. Latitude gradients appear to exist within countries; however, in Western Europe, there is a paradoxical relationship noted for melanoma incidence with latitude. Incidence rate in the Mediterranean areas are lower than those in the Scandinavian countries. An explanation for this paradox is the obvious difference in phenotype and skin type as well as potential differences in recreational behavior patterns.

### 3. *Site Distribution*

Figure 6 shows the anatomical distribution of primary melanoma according to gender for patients in the Melanoma Clinical Cooperative Group. This distribution is similar to that observed in Queensland, Australia, and western Canada. Of note is an increased number of melanomas on the upper back in both genders, relative sparing of the breast and abdomen in women, and of the area covered by underpants in both genders. There is relative sparing of the legs in men and an increased number of tumors noted in women from knee to ankle. Differences in clothing patterns and the intermittent exposure to sunlight that occurs with the use of bathing suits has been offered to explain the distribution noted. In western Australia, the relationship between relative risk for superficial spreading melanoma of the torso was studied in women by the type of bathing suit worn from late adolescence into early adult life. If a relative risk of 1 was assigned to those who wore a one-piece high-back suit, then a fourfold increase was observed with one-piece low-back suits and a 13-fold increase was noted for those who wore bikinis or who bathed nude (44). Melanomas are three to fivefold more common on the ears of men than women (45).

#### 4. Relationship Between Sunburn and Melanoma

Almost all epidemiologic studies have shown an association between history of sunburn and the increased risk for melanoma later in life (34). Traumatic sunburn as measured by the development of blistering (second-degree burn) during adolescence was noted to be associated with a doubled risk of melanoma later in life (46). Since melanoma is noted from the midteens on, sun exposure in childhood and adolescence has been examined (46,47). Sun exposure prior to 20 years of age is more closely associated with melanoma risk than sun exposure after 30 years of age (47).

#### 5. Migrant Studies

Strong evidence linking solar exposure to melanoma comes from the studies of melanoma rates in populations of similar genetic makeup in their country of origin and destination country. A striking increase in melanoma is noted in individuals in Australia, in contrast to individuals in the United Kingdom of similar ethnic background. Also of note is the several-fold higher melanoma rates for Japanese in Hawaii compared with those in Japan (48). Perhaps the earliest migrant study lending support for the role of solar exposure early in life or continuously through life is in native-born Israelis, who had higher age-specific melanoma rates than immigrants prior to 1948, who, in turn, had higher rates than immigrants after 1948 (49). This relationship, by and large, has held up with subsequent studies of the Israeli population (50). In more detailed studies from western Australia, melanoma rates of individuals born in western Australia or who migrated there by age 10 are four times higher than those of individuals who migrated to Australia after the age of 15 (51). A similar relationship of higher rates in native-borns or early childhood migrants is noted in New Zealand (52). These data taken in aggregate suggest that either the childhood period has increased vulnerability or that spending long periods of life in sunny climates is a relevant factor in the production of melanoma.

## B. Xeroderma Pigmentosum

In this rare genetically transmitted disorder of DNA repair, there is also a clear relationship between solar exposure and the production of cutaneous melanoma. In xeroderma pigmentosum, there is failure to repair the errors introduced into DNA from ultraviolet damage. Many of these patients die in adolescence or early adult life from metastatic melanoma. A recent summary (53) of the anatomical locations for cutaneous melanoma in xeroderma pigmentosum patients within the United States showed an increased frequency of melanomas on the exposed areas

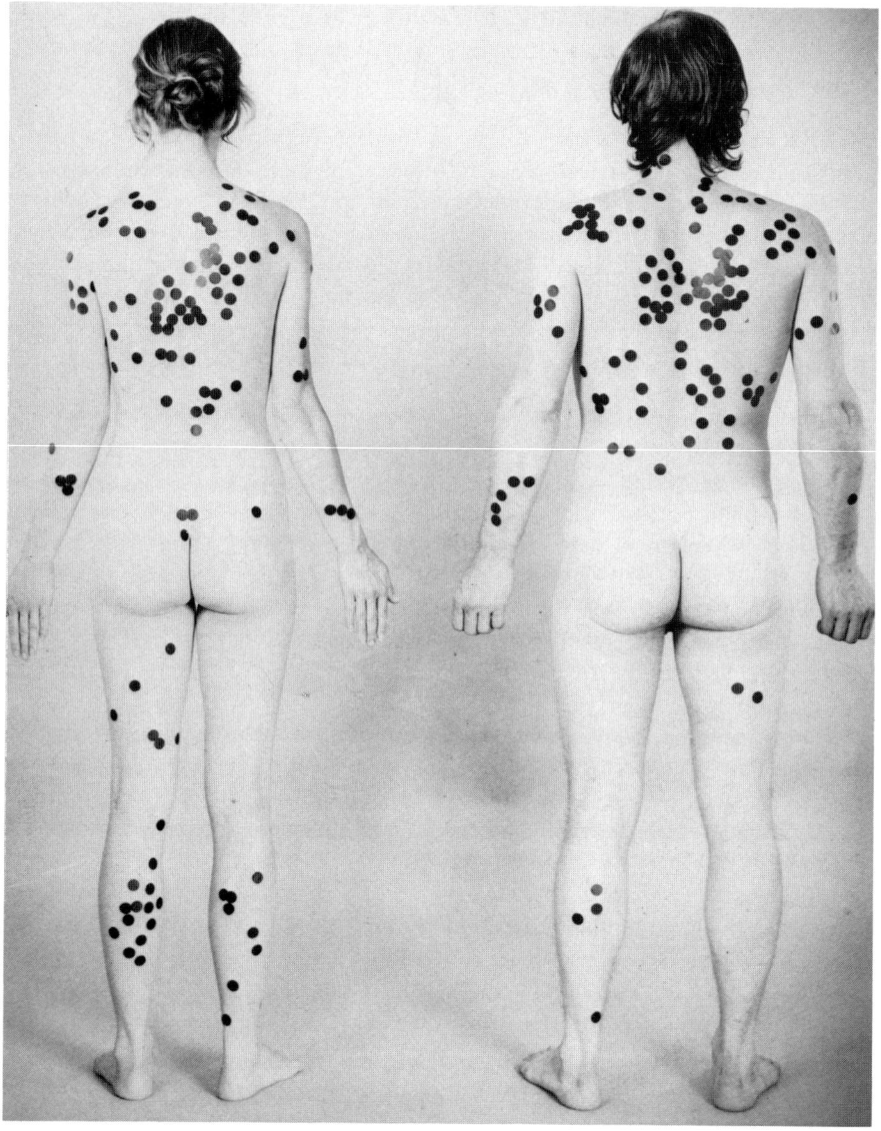

**Figure 6** Anatomical distribution of primary melanomas from the Melanoma Clinical Cooperative Group (National Academy of Sciences, with permission).

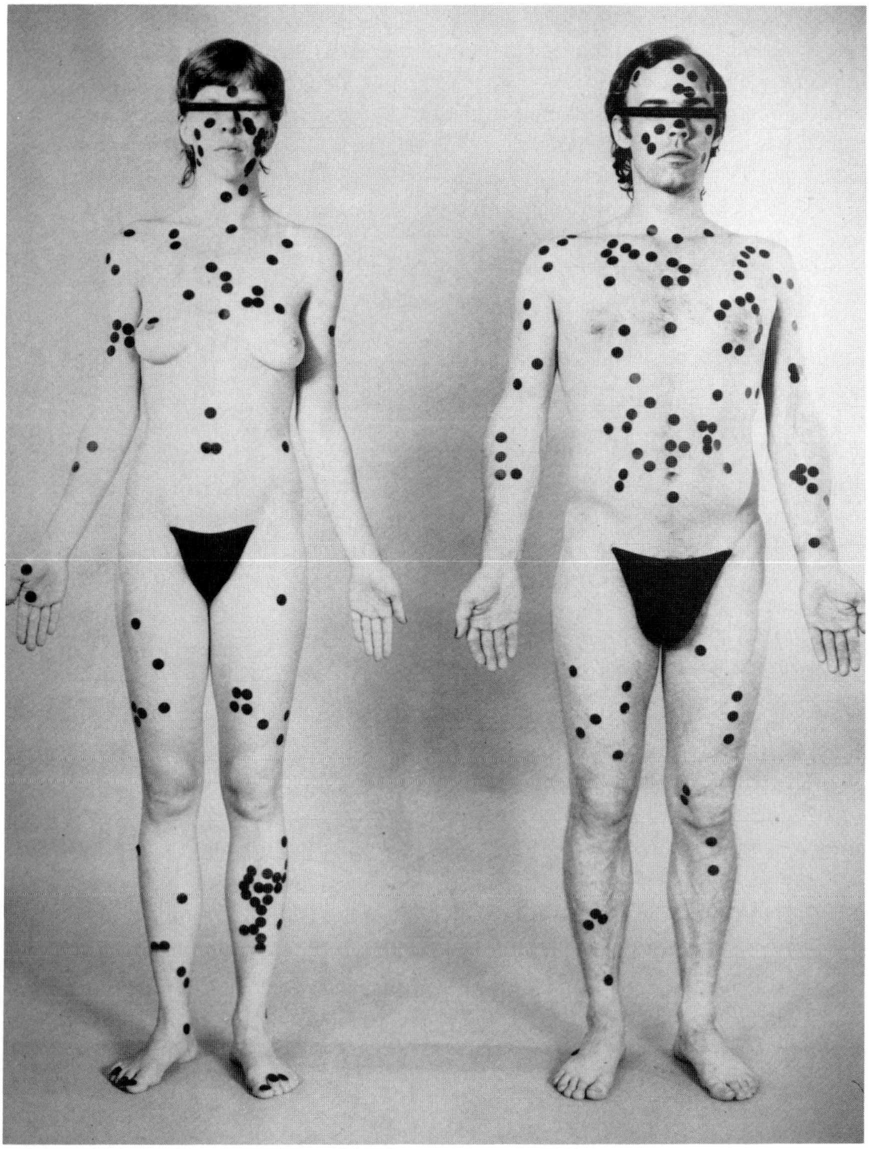

**Figure 6** *(continued)*

of the face, head, and neck, in contrast to the distribution in the general population (Fig. 6). As more data on melanoma accumulate in patients with xeroderma pigmentosum, the distribution shows an increased frequency of torso lesions, more similar to that noted in the general population (54). Patients with xeroderma pigmentosum have more than a thousand-fold increased risk of developing cutaneous melanoma (55). A reduction in DNA repair was noted in melanocytes and also in nevus cells of a patient with xeroderma pigmentosum (55).

## C. Relationship of Benign Acquired Nevi and Dysplastic Nevi to Solar Exposure

Benign acquired nevi and dysplastic nevi appear to be increased in number at anatomical sites of greater solar exposure (56,57). These studies provide an additional hypothesis of the way in which sunlight may be a factor in the production of melanoma. Sunlight may induce targets in nevi that may spontaneously degenerate into melanoma, or, as an alternative, sunlight might act on these targets in nevi to further promote their degeneration. Sunlight may also act as an immunosuppressive agent to allow neoplastic cells to escape the normal immune surveillance destruction mechanisms. Sunlight could also be acting directly on cutaneous melanocytes to transform them directly into cells with malignant capacity.

## D. Artificial Ultraviolet Exposure and Melanoma

A 2.9-fold relative risk for melanoma with the use of ultraviolet lamps and sunbeds (58) was reported from the United Kingdom. The 95% confidence interval was 1.3–6.4. The risk was significantly related to duration of use. Those individuals who had used ultraviolet beds or lamps at least 5 years before the presentation of melanoma had a relative risk of 9.1 (95% confidence intervals, 2.0–40.6) (58). In a large case–control study from Canada, odd ratios of 1.88 for males and 1.45 for females were noted for those who had sunbed or sunlamp exposure (59). The risk increased with duration and amount of use (59). These authors concluded that artificial tanning device use appeared to be a risk factor for cutaneous melanoma (59).

## E. Action Spectrum

Even though some paradoxes still exist relating solar exposure to melanoma, the compelling evidence is that for a substantial numbers of melanomas, exposure to the sun appears to be an important factor. Evidence for UVB as important wavelengths of exposure is presented in Table 3. The increased melanoma risk associated with UVA sunbeds raises the issue of whether UVA wavelengths may also be of causative importance.

**Table 3** Evidence favoring an association of UVB exposure with melanoma

| |
|---|
| The latitude gradient for UVB as measured on the earth's surface fits the changes with latitude of melanoma incidence within countries. |
| UVB wavelengths are responsible for sunburn; blistering sunburns in adolescence are associated with doubled risk of melanoma. |
| UVB flux, as measured by Robertson-Berger meters, correlates directly with melanoma risk in the United States. |
| UVB exposures are immunosuppressive in some strains of mice. |
| Changes in squamous cell carcinoma rates fit a chronic UVB exposure model. |
| After UVB exposure, the number of functioning melanocytes increases. |
| Cell transformation and chromosomal changes (sister chromated exchanges) are evidence for the mutagenic affects of UVB. |
| Of solar energy reaching the earth's surface, UVB wavelengths are the most damaging to cellular DNA. |

Source: Ref. 33.

## F. Paradoxes

It is difficult to propose that all melanomas are related to solar exposure. Certainly the development of melanoma in giant congenital nevi appears to have little to do with sun exposure. The development of melanoma in areas that are not sun exposed, such as the mucosal melanomas of anal or genital skin, is likewise difficult to relate to solar exposure. Melanomas of the palms and soles may also have little relationship with solar exposure. The paradoxic relationship of the latitude gradient as seen in Europe can be explained in part by differences in phenotype and potential differences in sun exposure practices. Table 2 lists the negative evidence for an association between sunlight and melanoma. Lower rates of melanoma are noted in individuals with outdoor occupations. Higher rates are noted for educated white collar indoor workers. The most common anatomical site for melanoma is not the maximally exposed surfaces of head and neck and dorsa of hands. In most patients there is limited evidence for chronic solar damage in the vicinity of melanoma, although "sunburn freckles" may be noted across the shoulders and nose as evidence of at least one severe traumatic solar exposure. Additional studies to evaluate the relationships between sunlight and melanoma are needed.

## G. Ozone Depletion

The estimated effect of ozone depletion on increasing the rates of nonmelanoma skin cancer is 1.3–4% increase for each 1% decrease in ozone (22,23). If one assumes a significant relationship between sunlight and melanoma, there are anticipated consequences as well for melanoma from the potential ozone depletion currently being monitored. There have been recent revisions of the risk estimates to determine what the affects of ozone depletion would have on increasing UVB and what the biological multiplier effect would subsequently be on melanoma rates (60). A 5% reduction in the ozone layer has been estimated to result in a 10% increase in biologically effective radiation. This is estimated to result in an increase of 5–8% in the incidence of malignant melanoma over the next 20 years (60).

## III. SUMMARY

The evidence is overwhelming that chronic solar exposure is the most important single factor in the production of basal and squamous cell carcinomas. Ultraviolet exposure also appears to participate in the genesis of some melanomas. Depletion of the ozone layer cannot be blamed for the increase in number of cases of skin cancer to date, but surely will make the situation worse in the future. Until some method is achieved to reverse the changes induced by ultraviolet exposure, emphasis must be placed on prevention through reduction of ultraviolet exposure, especially early in life (protective clothing, sunblocks of high sun protection factor and broad spectrum, avoiding midday sun, shade tree planting programs), and through early detection by the implementation of screening programs for individuals at elevated risk and education programs both for the public and health care professionals.

## ACKNOWLEDGMENT

Supported in part by a grant from the Marion Gardner Jackson Trust, Bank of Boston (Trustee).

## REFERENCES

1. Unna P. Histopathologie der Hautkrankheiten. Berlin: August Hirschwald, 1894.
2. Isaacson C. Cancer of the skin in urban blacks in South Africa. Br J Dermatol 1979; 100:347–350.
3. Giles GG, Marks R, Foley P. Incidence of non-melanocytic skin cancer treated in Australia. Br Med J 1988; 296:13–17.
4. Scotto J, Fears TR, Fraumeni JF. Incidence of non-melanoma skin cancer in the United States. NIH Publication No. 83-2433. Washington DC: National Cancer Institute, 1983:15–104.
5. Skin cancer in man. In: Laerum OD, Iversen OH, eds. Biology of Skin Cancer (Excluding Melanomas). Geneva: UICC Technical Report Series. 1981; 63:58–86.
6. Marks R, Ponsford MW, Selwood TS, Goodman G, Mason G. Non-melanotic skin cancer and solar keratoses in Victoria. Med J Aust 1983; 2:619–622.

7. Beral V, Robinson N. The relationship of malignant melanoma, basal and squamous cell cancers to indoor and outdoor work. Br J Cancer 1982; 44:886–891.
8. Hill D, Borland R. ACCV Sun Survey. Centre for Behavioural Research in Cancer, Anti-Cancer Council of Victoria, 1988.
9. Hartvelt MM, Bouwes Bavinck JN, Kootte AMM, Vermeer BJ, Vandenbroucke JP. Incidence of skin cancer after renal transplantation in the Netherlands. Transplantation 1990; 49:506–509.
10. Dyall-Smith D, Varigos G. The malignant potential of papilloma virus. Aust J Dermatol 1985; 26:102–107.
11. Robbins JH, Kraemer KH, Lutzner MA. Xeroderma pigmentosum: an inherited disease with sun sensitivity, multiple cutaneous neoplasms and abnormal DNA repair. Ann Intern Med 1974; 80:221–248.
12. Davis MM, Hanke CW, Zollinger TW, Montebello JF, Hornback NB, Norins AL. Skin cancer in patients with chronic radiation dermatitis. J Am Acad Dermatol 1989; 20:608–616.
13. Kricker A, English DR, Randell PL. Skin cancer in Geraldton, Western Australia: a survey of incidence and prevalence. Med J Aust 1990; 152:399–407.
14. Gordon D, Silverstone H. Worldwide epidemiology of premalignant and malignant cutaneous lesions. In: Andrade R, Gumport SL, Popkin GL, Rees TD, eds. Cancer of the Skin. Philadelphia: WB Saunders, 1976:405–434.
15. Green A, Beardmore G, Hart V, Leslie D, Marks R, Staines D. Skin cancer in a Queensland population. J Am Acad Dermatol 1988; 19:1045–1052.
16. Marks R, Rennie G, Selwood TS. The relationship of basal cell carcinomas and squamous cell carcinomas to solar keratoses. Arch Dermatol 1988; 124:1039–1042.
17. Marks R, Rennie G, Selwood TS. Malignant transformation of solar keratoses to squamous cell carcinoma in the skin: a prospective study. Lancet 1988; 1:795–797.
18. Marks R, Foley P, Goodman G, Hage BH, Selwood TS. Spontaneous remission of solar keratoses: the case for conservative management. Br J Dermatol 1986; 115:649–655.
19. Roberts DL. Incidence of non-melanoma skin cancer in West Glamorgan, South Wales. Br J Dermatol 1990; 122:399–403.
20. Marks R, Jolley D, Dorevitch AP, Selwood TS. The incidence of non-melanocytic skin cancer in an Australian population: results of a five year prospective study. Med J Aust 1989; 150:475–478.
21. Marks R, Selwood TS. Solar keratoses—the association with erythemal ultraviolet radiation in Australia. Cancer 1985; 56:2332–2336.
22. Fears TR, Scotto J. Estimating increases in skin cancer morbidity due to increases in ultraviolet radiation exposure. Cancer Invest 1983; 1:119–126.
23. Health effects of ozone layer depletion. A report of the NHMRC Working Party, Melbourne, 1989. Canberra: Australian Government Publishing Service, 1989:70–95.
24. Diffey BL. Analysis of the risk of skin cancer from sunlight and solaria in subjects living in Northern Europe. Photodermatology 1987; 4:118–126.
25. Armstrong BK, Woodings TL, Stenhouse NS. Mortality from cancer in migrants to Australia—1962 to 1971. Perth: The University of Western Australia, 1983:81–83.

26. Marks R, Jolley D, Lecatsass S, Foley P. The role of childhood sunlight exposure in the development of solar keratoses and non-melanocytic skin cancer. Med J Aust 1990; 152:62–65.
27. Stern RS, Weinstein MC, Baker SG. Risk reduction for non-melanoma skin cancer with childhood sunscreen use. Arch Dermatol 1986; 122:537–545.
28. Forbes PD, Davies RE, Urbach F. Experimental ultraviolet photocarcinogenesis: wavelength interactions and time–dose relationships. Natl Cancer Inst Monogr 1978; 50:31–38.
29. Strickland PT, Vitasa BC, West SK, Rosenthal FS, Emmett EA, Taylor HR. Quantitative carcinogenesis in man: solar ultraviolet B dose dependence of skin cancer in Maryland watermen. J Natl Cancer Inst 1989; 81:1910–1913.
30. Young AR, Gibbs NK, Thomson RJ. Modification of 5-methoxypsoralen phototumorigenesis in hairless albino mice by sunscreen filters. Br J Dermatol 1983; 190:705.
31. Epstein JH. Phototherapy and photochemotherapy. N Engl J Med 1990; 322:1149–1151.
32. Stern RS, Lange R. Non-melanoma skin cancer occurring in patients treated with PUVA five to ten years after first treatment. J Invest Dermatol 1988; 91:120–124.
33. Sober AJ. Solar exposure in the etiology of cutaneous melanoma. Photodermatology 1987; 4:23–31.
34. Armstrong BK. Epidemiology of malignant melanoma: intermittent or total accumulated exposure to the sun? J Dermatol Surg Oncol 1988; 14:835–849.
35. Koh HK, Kligler BE, Lew RA. Sunlight and cutaneous malignant melanoma: evidence for and against causation. Photochem Photobiol 1990; 51:765–779.
36. Kripke ML. Speculations on the role of ultraviolet radiation in the development of malignant melanoma. J Natl Cancer Inst 1979; 63:541–548.
37. Ley RD, Applegate LA, Padilla RS, Stuart TD. Ultraviolet radiation-induced malignant melanoma in *Monodelphis domestica*. Photochem Photobiol 1989; 50:1–5.
38. Graham S, Marshall J, Haughey B, et al. An inquiry into the epidemiology of melanoma. Am J Epidemiol 1985; 122:606–619.
39. Green A. Sun exposure and the risk of melanoma. Aust J Dermatol 1984; 25:99–102.
40. Stevens NG, Liff JM, Weiss NS. Plantar melanoma; is the incidence of melanoma of the sole of the foot really higher in blacks than white? Int J Cancer 1990; 45:691–693.
41. Lancaster HO. Some geographical aspects of the mortality from melanoma in Europeans. Med J Aust 1956; 2:1082–1087.
42. Magnus K. Incidence of malignant melanoma of the skin in the five Nordic countries: significance of solar radiation. Int J Cancer 1977; 20:477–485.
43. Schreiber MM, Moon TE, Bozzo PD. Malignant melanoma in Southern Arizona: increasing incidence and sunlight as an etiologic factor. Arch Dermatol 1981; 117:6–11.
44. Holman CDJ, Armstrong BK, Heenan PJ. Relationship of cutaneous malignant melanoma to individual sunlight exposure habits. J Natl Cancer Inst 1986; 76:403–414.

45. Hudson DA, Krige JEJ, Strover RM, King HS. Malignant melanoma of the external ear. Br J Plast Surg 1990; 43:608–611.
46. Lew RA, Sober AJ, Cook N, Marvell R, Fitzpatrick TB. Sun exposure habits in patients with cutaneous melanoma: a case control study. J Dermatol Surg Oncol 1983; 9:981–988.
47. Weinstock MA, Colditz GA, Willett WC, Stampfer MJ, Bronstein BR, Mihm MC Jr, Speizer FE. Nonfamilial cutaneous melanoma incidence in women associated with sun exposure before 20 years of age. Pediatrics 1989; 84:199–204.
48. Muir C, Waterhouse J, Mack T, et al, eds. Cancer Incidence in Five Continents, Vol 5. Lyon: International Agency for Research on Cancer, 1987.
49. Movshovitz M, Modan B. Role of sun exposure in the etiology of malignant melanoma; epidemiologic influence. J Natl Cancer Inst 1973; 51:777–779.
50. Anaise D, Steinitz R, Ben Hur N. Solar radiation: a possible etiologic factor in malignant melanoma in Israel; a retrospective study (1960–1972). Cancer 1978; 42:299–304.
51. Holman CDJ, Armstrong BK. Cutaneous malignant melanoma and indicators of total accumulated exposure to the sun: an analysis separating histogenetic types. J Natl Cancer Inst 1984; 73:75–82.
52. Cooke KR, Fraser J. Migration and death from malignant melanoma. Int J Cancer 1985; 36:175–178.
53. Kraemer KH, Lee MM, Scotto J. Xeroderma pigmentosum, cutaneous, ocular, and neurologic abnormalities in 830 published cases. Arch Dermatol 1987; 123:241–250.
54. Kraemer KH. Personal communication, 1990.
55. Kraemer KH, Herlyn M, Yuspa SH, Clark WH Jr, Townsend GK, Neises GR, Hearing VJ. Reduced DNA repair in cultured melanocytes and nevus cells from a patient with xeroderma pigmentosum. Arch Dermatol 1989; 125:263–268.
56. Kopf AW, Lazer M, Bart RS. Prevalence of nevocytic nevi in lateral and medial aspects of arms. J Dermatol Surg Oncol 1978; 4:153–158.
57. Kopf AW, Lindsay AC, Rogers GS, Friedman RJ, Rigel DS, Levenstein M. Relationship of nevocytic nevi to sun exposure in dysplastic nevus syndrome. J Am Acad Dermatol 1985; 12:656–662.
58. Swerdlow AJ, English JSC, MacKie RM, O'Doherty CJ, Hunter JAA, Clark J, Hole DJ. Fluorescent lights, ultraviolet lamps and risk of cutaneous melanoma. Br Med J 1988; 297:647–650.
59. Walter SD, Marrett LD, From L, Hertzman C, Shannon HS, Roy P. The association of cutaneous malignant melanoma with the use of sunbeds and sunlamps. Am J Epidemiol 1990; 131:232–243.
60. Longstreth JD, Pitcher HM. Skin cancer: revising the risk estimates due to ozone depletion. Skin Cancer Found J 1990; 8:27–69.

# III

# Effects of Light on the Immune System

# 8

# Photoimmunology: Effects of Ultraviolet B Radiation on Cutaneous Photocarcinogenesis and Allergic Contact Sensitivity

**Ponciano D. Cruz, Jr., and Paul R. Bergstresser**
*University of Texas Southwestern Medical Center, Dallas, Texas*

## I. INTRODUCTION

The study of effects of nonionizing radiation, principally of the ultraviolet (UV) and visible spectra, on the immune system has been termed photoimmunology. Although this definition is an acceptable description, it leaves out an important intermediate concept, which is that for nonionizing radiation to exert any influence on effector lymphocytes of the immune system, it must first transmit its effects through skin (Fig. 1). This concept takes on greater importance when one considers that skin contains cells of several different lineages that can initiate and implement immune responses. Keratinocytes, for example, can secrete a wide variety of immunoregulatory cytokines, and epidermal Langerhans cells (and dermal dendrocytes and macrophages) can present foreign antigens to T lymphocytes. The outcome of many immune responses initiated through skin can therefore be attributed directly to photoinduced perturbation of the function of these immunocompetent cutaneous cells.

The focus of this chapter is on the influence of ultraviolet B (UVB) radiation on T-cell-mediated immune responses initiated in skin. The first observations in photoimmunology were made in humans when it was realized that a number of photodermatoses involved interactions between sunlight and the immune system. Two such disorders are solar urticaria and photoallergy, in which sunlight is

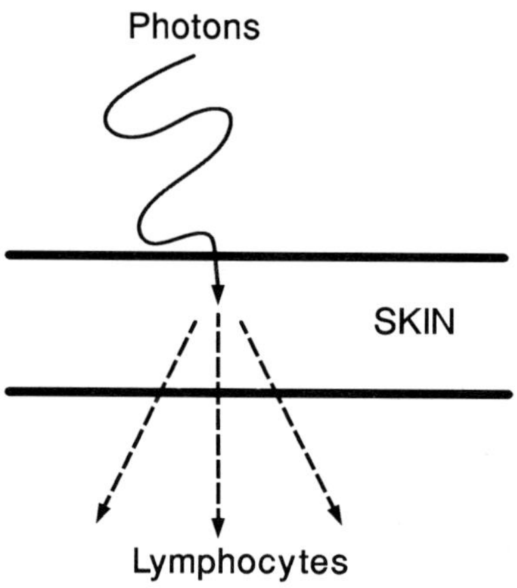


**Figure 1** For photons to exert their influence on effector lymphocytes of the immune system, they must first transmit their effects through skin.

responsible for triggering immune responses. In terms of impact, however, the importance of photoimmunology became firmly established following the recognition that the immune system plays an important role in the development of UVB-induced skin cancers.

## II. UVB RADIATION PROMOTES THE DEVELOPMENT OF SKIN CANCER BY SUPPRESSING T-CELL RESPONSES GENERATED TO PROTECT AGAINST SUCH TUMORS

UVB radiation is a carcinogen. UVB-induced carcinogenesis in humans is supported by substantial epidemiologic and clinical evidence (1, 2). This evidence is buttressed by experimental studies in animals, which have shown that chronic exposure to high dosages of UVB radiation ($>10^8$ $J/m^2$) not only initiates the growth but also promotes the development of epidermal cell-derived cancers (3, 4). This promotional influence of UVB radiation on skin carcinogenesis, in turn, is due to adverse UVB effects on the immune system.

The idea that immunologic mechanisms play an important role in curtailing the development of UVB-induced skin cancers was derived originally from observations concerning the antigenicity of UVB-induced cancers (5). UVB-induced skin cancers transplanted into genetically identical normal mice were readily rejected (6) (Fig. 2). By contrast, skin cancers engrafted onto immunocompromised animals, including mice treated with sublethal doses of $\gamma$ radiation or subcarcinogenic doses of UVB radiation, became established (7). A

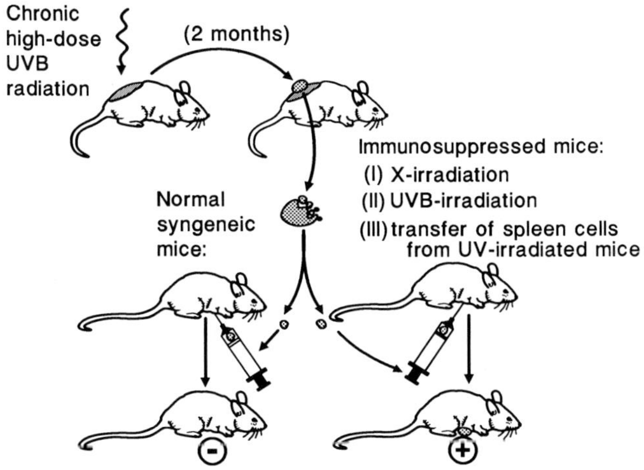


**Figure 2** Exposure to chronic high-dosage UVB radiation leads to growth of skin cancer. When transplanted onto normal, syngeneic (immunocompetent) mice, these skin cancers are readily rejected. By contrast, these skin cancers become established when engrafted onto immunocompromised mice.

similar protective role of immunity in humans is supported by the observation that patients who receive prolonged immunosuppressive treatment during the course of organ transplantation have a high incidence of cutaneous squamous cell carcinomas (8, 9). Furthermore, such tumors develop primarily in sites of sun exposure (8, 9).

The capacity of UVB radiation to deter immunosurveillance for skin cancers has been attributed to the generation of "suppressor" T cells that are specific for UVB-induced tumor antigens. This was first suggested by the finding that growth of transplanted UVB-induced skin cancers (but not of other types of skin cancers) in unirradiated mice was facilitated when recipients were first infused with spleen cells from UVB-irradiated mice (Fig. 2). In further studies, it was revealed that a tumor-specific population of T-cells was responsible for suppressing the immunologic destruction of the skin cancers (10). Generation of these "suppressor" T cells was demonstrated to predate visible growth of the skin cancers, which in mice were squamous cell carcinomas in virtually all cases (7). Finally, the UVB dose required for generation of these "suppressor" T cells was small compared to that necessary for the induction of skin cancer (7). It is important to note that UVB protocols used to induce skin cancer in rodents do not cause a complete shutdown of the immune system (11, 12). It has already been mentioned that UVB-irradiated mice are capable of rejecting non-UV-induced skin tumors. In addition, UVB-irradiated mice also exhibit normal humoral responses. By contrast, T-cell-mediated immune responses are suppressed by UVB radiation (11, 12). Among experimental models of T-cell-mediated immunity, UVB-induced suppression of allergic contact sensitivity has been the most extensively investigated.

## III. UVB RADIATION SUPPRESSES ALLERGIC CONTACT SENSITIVITY THROUGH MECHANISMS INITIATED WITHIN EPIDERMIS

Allergic contact sensitivity represents an animal model of a delayed-in-time, T-cell-mediated reactivity that is produced by cutaneous application of certain reactive chemicals (haptens) capable of binding directly to soluble and to cell-associated proteins (13). Figure 3 portrays a representative set of assays used to assess allergic contact sensitivity responses in mice. In experimental settings, two assays are usually performed: a test for responsiveness and a test for regulation. In testing for responsiveness, immunization is performed on day 0 by painting a sensitizing dose of hapten on shaved abdominal skin. Elicitation is

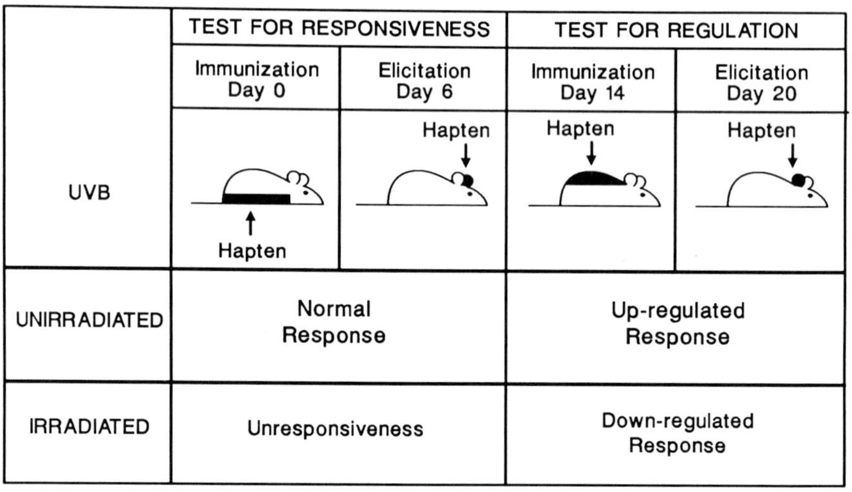


**Figure 3** Assays for allergic contact sensitivity include an initial test for responsiveness and a subsequent test for regulation. Immunization with hapten painted on unirradiated abdominal skin leads to normal ear swelling responses; reimmunization with hapten painted on unirradiated back skin results in augmented (up-regulated) responses. By contrast, immunization with hapten painted on UVB-irradiated abdominal skin produces markedly reduced ear swelling responses (unresponsiveness); more important is the observation that after reimmunization with hapten painted on unirradiated back skin, these mice exhibit less than normal (down-regulated) ear swelling responses.

performed on day 6 by challenging one ear with a more dilute concentration of the hapten; ear swelling is then measured. In testing for regulation, the same mice are reimmunized at a later date (e.g., day 14) by painting hapten at a different site (e.g., the dorsal back). Subsequently (e.g., day 20), the other ear is challenged, and ear swelling is again measured.

In general, unirradiated mice exhibit good ear swelling responses (Fig. 3); furthermore, when these mice are resensitized with hapten at a different site, they display augmented (upregulated) responses. By contrast, UVB-irradiated mice exhibit reduced ear swelling responses (unresponsiveness) (Fig. 3); in addition, even upon subsequent resensitization with hapten at an unirradiated site, these mice produce considerably less than normal (downregulated) responses.

UVB-induced suppression of allergic contact sensitivity in mice has been linked to the development of hapten-specific "suppressor" T cells (14, 15). It should be noted, however, that although all strains of mice tested to date with UVB irradiation prior to skin painting with the hapten, dinitrofluorobenzene (DNFB), were found to generate DNFB-specific "suppressor" T cells (16), it has been reported that not all strains are susceptible to the actual suppression induced by UVB radiation as judged by reduced ear swelling responses (17).

A similar form of UVB-induced suppression of allergic contact sensitivity was demonstrated in humans. Localized exposure to low-dosage UVB (14 $J/m^2$/day for 4 consecutive days) radiation impaired induction of allergic contact sensitivity to dinitrochlorobenzene (DNCB) in 40% of normal individuals and in 92% of patients with UVB-induced skin cancer (18). In addition, 45% of the patients with skin cancer developed specific tolerance to DNCB, whereas none of the normal individuals demonstrated tolerance (18). The existence in humans of populations and in mice of strains that are either sensitive or resistant to the immunosuppressive effects of UVB radiation suggests a genetic basis for susceptibility to UVB-induced immunosuppression (17).

Concurrent with the capacity to suppress allergic contact sensitivity, UVB radiation can produce immunologic alterations in epidermal Langerhans cells and keratinocytes. These parallel findings have led investigators to propose several scenarios in which such changes may explain the suppression of T-cell-mediated immune responses by UVB radiation.

## IV. ALTERATIONS IN EPIDERMAL LANGERHANS CELLS

One scenario highlights the capacity of UVB radiation to alter the normal antigen presenting function of Langerhans cells. The first clue to this possibility arose from the finding that erythemogenic (or even suberythemogenic) dosages of UVB radiation ($<10^3$ $J/m^2$) cause Langerhans cells in irradiated mouse or human skin to lose their dendricity and some surface markers, including major histocompatibility complex (MHC) class II molecules and ATPase activity (19)

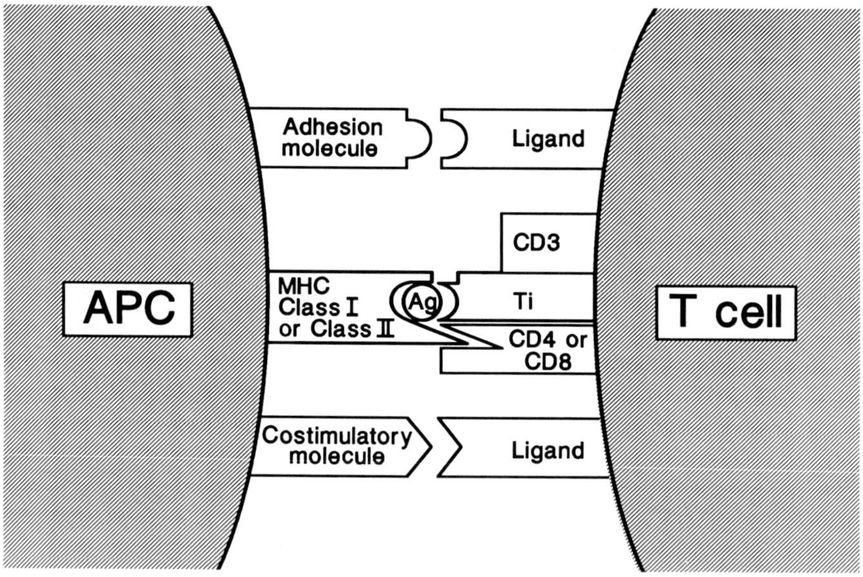


**Figure 4** Antigen-presenting cells (APC) present antigens to the T-cell receptor complex (TCR: Ti in association with CD3) of T cells in the context of MHC molecules. To achieve optimal engagement and communication between APC and T cells, adhesion molecules and costimulatory molecules on APC bind to their appropriate ligands on T cells.

(Fig. 4). In vitro studies have subsequently shown that UVB irradiation causes Langerhans cells to lose the capacity to present antigens to T cells (20).

## A. Depletion of Langerhans Cells Allows "Suppressor" Antigen-Presenting Cells in Epidermis to Evoke Immunosuppression

Since Langerhans cells are the principal (if not the sole) antigen presenting cells in epidermis, it has been reasoned that depletion of these cells by UVB radiation would preclude antigens (such as contact sensitizers) from being presented to T cells, thus preventing the initiation of an immune response against such antigens. It was hypothesized further that in the absence of Langerhans cells, the inherent suppressive effect of other immunocompetent cells within epidermis becomes dominant (Fig. 5). Among the possible "suppressive" antigen-presenting cells in epidermis are: two resident epidermal cells in mice, dendritic epidermal T cells (DETC) that bear $\gamma\delta$ T-cell receptors (21, 22) and I-$J^+$ epidermal cells (23, 24); and in humans $OKM5^+DR^+CD1^-$ cells that appear to migrate into epidermis 72 hr after UVB irradiation (25–28).

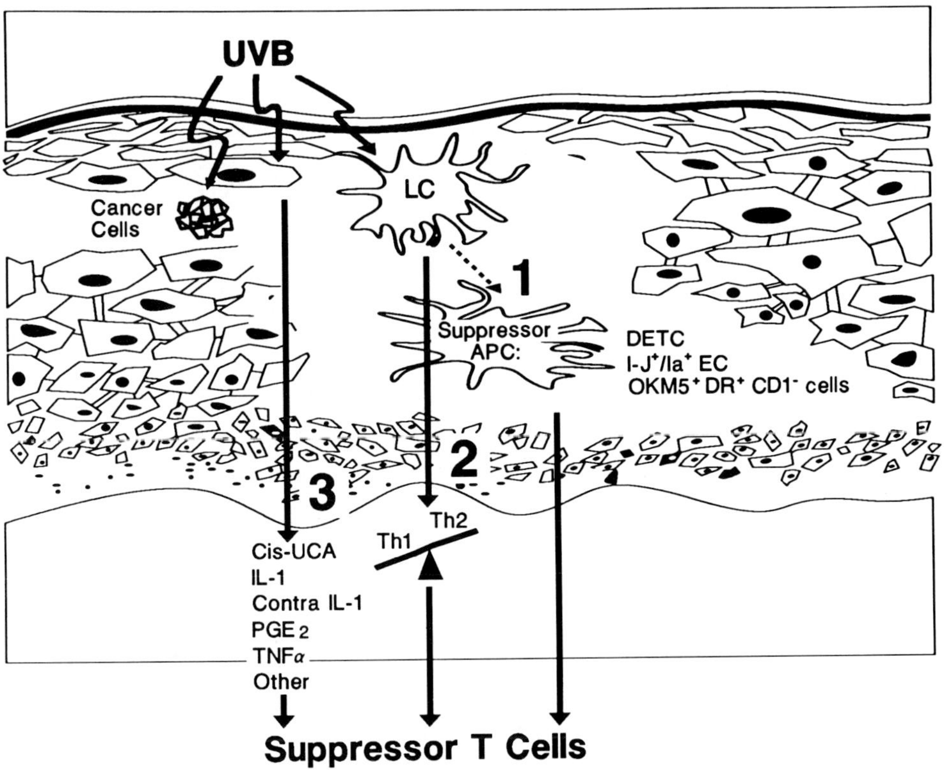


**Figure 5** In association with its direct effects on epidermal cells as a carcinogen, UVB radiation also promotes the development of skin cancer by activating "suppressor" T cells that prevent the immunologic destruction of these cancers. These suppressor T cells may be activated as a result of: abrogation of the antigen-presenting function of Langerhans cells, thus allowing "suppressor" antigen-presenting cells in epidermis to evoke unopposed effects; conversion of Langerhans cell function from an immunogenic to a tolerogenic antigen-presenting cell via the induction of clonal anergy in Th1 and the preferential activation of Th2; and induction of keratinocytes to secrete soluble cytokines, which lead to activation of "suppressor" T cells. (UVB, ultraviolet B radiation; LC, Langerhans cells; DETC, dendritic epidermal $\gamma\delta$ T-cell receptor-bearing T cells; I-J$^+$/Ia$^+$EC, I-J-bearing and Ia$^+$ epidermal cells distinct from Langerhans cells; OKM5$^+$DR$^+$CD1$^-$ cells, dermal melanophages that migrate into epidermis 72 hr after UVB exposure; Th1, CD4$^+$ T cells that mediate delayed hypersensitivity responses through production of interferon $\gamma$; Th2, CD4$^+$ cells that help B cells produce antibodies through production of IL-5; Cis-UCA, *cis* isomer of urocanic acid; PGE$_2$, prostaglandin E$_2$, TNF$\alpha$, tumor necrosis factor $\alpha$.)

## B. Altered Function Enables Langerhans Cells to Initiate Immunosuppression

Recent studies in mice have shown that erythmogenic and suberythmogenic dosages of UVB are not, in fact, lethal for Langerhans cells. Rather, such relatively low dosages of UVB convert the antigen-presenting function of Langerhans cells from one responsible for triggering T-cell-mediated immunity to one capable of inducing T-cell-mediated suppression (29, 30).

Unlike B cells, which recognize antigens free in solution by their native conformation, T cells require antigen to be presented to its T-cell receptor in the form of oligopeptides (10–20 amino acids long) bound to MHC molecules (class I in the case of $CD8^+$ T cells and class II in the case of $CD4^+$ T cells) on the surfaces of antigen-presenting cells (31) (Fig. 4). Although engagement of the oligopeptide/MHC complex with the T-cell receptor is essential for antigen-specific T-cell activation and proliferation, binding of other molecules on antigen-presenting cells to appropriate ligands on T cells also contributes to antigen-specific activation (Fig. 4). These accessory molecule-ligand interactions may strengthen the engagement of antigen presenting cells and T cells (adhesion molecules), modify the signal initiated through the T-cell receptor, or initiate signaling events distinct from those of the T-cell receptor (costimulatory molecules) (31). Two subsets of $CD4^+$ T cells in mice have been identified: Th1 cells mediate delayed-in-time hypersensitivity responses and tumor cytolysis, whereas Th2 cells help B cells produce primary antibody responses. In addition, Th1 secrete interleukin (IL)-2, interferon $\gamma$ (IFN$\gamma$), and lymphotoxin, whereas Th2 produce IL-4, IL-5, IL-7, and IL-10 (31).

Murine Langerhans cells irradiated in vitro with a single dose of 200 $J/m^2$ UVB have been shown to induce in $CD4^+$ Th1 cells a form of unresponsiveness called clonal anergy (29, 30). The mechanism by which UVB radiation achieves this effect appears to be the inactivation of an accessory function expressed by Langerhans cells (30). In this context, inactivation of the accessory function precludes the initiation of the hypersensitivity response. Since UVB-irradiated Langerhans cells lose the capacity to stimulate proliferation of Th1, but maintain the ability to induce proliferation of Th2 (29) (Fig. 5), it has been speculated that the UVB-sensitive accessory function is required for antigen-driven proliferation of Th1, but not for analogous proliferation of Th2 (29). In addition, Th2 have been postulated to serve as the "suppressor" cells responsible for the tolerance induced by UVB radiation in vivo (32), based primarily on observations that Th2 can secrete the cytokines, IL-4 and IL-10, which can inhibit Th1 proliferation and/or Th1 cytokine production.

Studies with the South American opossum (whose cells contain photoreactivating enzyme, which can repair spontaneously UV-induced DNA damage) have demonstrated that both UVB-induced morphologic alterations in Langerhans cells and UVB-induced suppression of allergic contact sensitivity can be reversed by exposure to photoreactivating light (33). These findings suggest that damage to DNA is involved in these UVB effects, and that such DNA damage can be repaired by photoreactivation in the opossum. On the other hand, photoreactivation studies in humans have failed to show reversal of both of these UVB effects (34); it should also be noted that the existence of photoreactivating enzyme in humans remains controversial.

## V. SOLUBLE FACTORS DERIVED FROM KERATINOCYTES IMPLICATED AS IMMUNOSUPPRESSIVE AGENTS

The second principal scenario invoked to explain suppression of allergic contact hypersensitivity following UVB irradiation is the production in skin, principally by keratinocytes, of soluble factors that modify the normal immunologic pathway, resulting in activation of "suppressor" T cells (Fig. 5). Serum from UVB-irradiated mice as well as culture supernatants collected from UVB-irradiated epidermal cells have each been shown to suppress allergic contact sensitivity when transferred into normal recipients (35, 36). Based on independent work by different laboratories, several keratinocyte-derived soluble factors have been proposed as candidates for the putative immunosuppressive soluble factors.

### A. *Cis*-Urocanic Acid

Urocanic acid is synthesized normally in skin in the *trans*-isomerized form via deamination of the amino acid histidine within keratinocytes. Following absorption of UVB radiation, *trans*-urocanic acid is converted to the *cis*-isomer. In vivo studies in mice and in vitro systems have indicated that *cis*-urocanic acid can induce suppression of T-cell-mediated responses directed at contact sensitizers and viral antigens (37–39). In a report in hairless mice, it was also suggested that *cis*-urocanic acid can act as a promoter for UVB-induced squamous cell cancer (40).

### B. Interleukin-1 (IL-1) and Contra-IL-1

IL-1 refers to a family of molecules that exhibit a wide array of biological activities including, among others, the mediation of fever, secretion of acute phase reactants, and induction in T lymphocytes of IL-2 responsiveness and cytokine synthesis (41). Normal epidermis contains large amounts of functionally active IL-1 (42), and UV light has been shown to induce even greater secretion of IL-1 by keratinocytes (43, 44). In addition, the fever response following sunburn reactions has been shown to be due to UV-induced increase in epidermal IL-1 production (45). IL-1 itself and a specific inhibitor of IL-1 (contra-IL-1), which is also produced by keratinocytes, have each been implicated as an immunosuppressive factor. In particular, contra-IL-1 was shown to inhibit the induction of allergic contact sensitivity in mice (36, 46).

## C. Prostaglandins

Treatment of mice with the cyclooxygenase inhibitor, indomethacin, has been shown to abrogate the capacity of UVB radiation to induce suppression of allergic contact sensitivity in mice (47), thus indicating that products of the cyclooxygenase pathway, possibly prostaglandins, are somehow involved in generating immunosuppression. UVB exposure can enhance prostaglandin secretion by epidermal cells. This is particularly true of prostaglandin $E_2$, which is synthesized either directly as an effect of UVB stimulation or as a consequence of IL-1 effects on keratinocytes (48, 49). In view of the latter finding, it is not surprising that IL-1-treated mice are markedly compromised in their capacity to elicit normal allergic contact sensitivity responses (47).

## D. Tumor Necrosis Factor $\alpha$

Tumor necrosis factor $\alpha$ (TNF$\alpha$) is a cytokine that regulates several important physiological and immunologic responses; UVB radiation has been shown to induce secretion of TNF$\alpha$ by keratinocytes (50). More important is that recent studies have demonstrated that the intracutaneous injection of TNF$\alpha$ mimics two effects of UVB radiation on skin: loss of dendricity and membrane ATPase activity in epidermal Langerhans cells, and inhibition of the induction of allergic contact sensitivity (51, 52). On this basis, TNF$\alpha$ has been proposed as a mediator of the immunosuppressive effects of UVB radiation. In fact, susceptibility to the suppressive effects of UVB on allergic contact sensitivity in mice has been reported to be regulated by two genes for TNF$\alpha$ and for the susceptibility to the effects of bacterial lipopolysaccharide (53).

## E. Other Soluble Factors

Cultured mouse keratinocytes have also been shown to secrete distinct immunosuppressive factors following exposure to UVA and to UVB radiation, respectively (54). The factor released after UVA treatment suppressed allergic contact sensitivity to DNFB (but not delayed-type hypersensitivity to subcutaneously injected allogeneic cells). Conversely, the factor released after UVB treatment suppressed delayed-type hypersensitivity to the injected allogeneic cells (but not allergic contact sensitivity to DNFB) (54).

In yet another study, unirradiated cultured human keratinocytes have been reported to secrete a cytokine termed epidermal cell-derived lymphocyte-differentiating factor (ELDIF), which can inhibit the in vitro proliferation of murine splenocytes to various mitogens by inhibiting IL-2 production by T cells (55). On the other hand, effects of UV radiation on the synthesis of ELDIF have not been studied.

UV light has also been shown to induce keratinocytes to produce IL-6; circulating levels of this cytokine were shown to be increased following UV exposure (56). However, it has yet to be shown that IL-6 can mediate an immunosuppressive effect.

## VI. SUMMARY

The birth of photoimmunology as a modern science was heralded by the observation that the immune system plays a key role in preventing the growth of UVB radiation-induced skin cancer. This concept has since spawned numerous studies that have yielded a fuller understanding not only of photocarcinogenesis but also of the effects of UVB radiation on the immunologic functions of epidermal cells, particularly of Langerhans cells and keratinocytes.

Available evidence point to at least three mechanisms that may explain how UVB irradiation leads to T-cell-mediated suppression (Fig. 5). The first mechanism is the abrogation by UVB radiation of the normal antigen-presenting function of Langerhans cells, thus allowing inherently "suppressor" antigen-presenting cells in epidermis to activate "suppressor" T cells. A second mechanism is alteration by UVB radiation of Langerhans cell function from immunogenic to tolerogenic antigen-presenting cells, via the induction of clonal anergy in Th1 and the preferential activation of Th2, which have been proposed to serve as "suppressor" T cells. The third mechanism is the capacity of UVB radiation to induce keratinocytes to secrete soluble mediators that lead to the activation of "suppressor" T cells.

## REFERENCES


1. Scotto J, Fears TR, Fraumeni JF. Incidence of non-melanoma skin cancer in the United States. US Department of Health and Human Services, 1982; NIH publication no. 82-2433.
2. Sunlight, ultraviolet radiation, and the skin, 1989 Bethesda: National Institutes of Health Consensus Development Conference.
3. Roberts LK, Lynch DH, Samlowski WE, Daynes RA. Ultraviolet light and modulation of the immune response. In: Norris DA, ed. Immune Mechanisms in Cutaneous Disease. New York: Marcel Dekker, 1989:167–215.
4. Romerdahl CA, Okamoto H, Kripke ML. Immune surveillance against cutaneous malignancies in experimental animals. In: Norris DA, ed. Immune Mechanisms in Cutaneous Disease. New York: Marcel Dekker 1989:749–767.
5. Graffi A, Horn K-H, Pasternak G. Antigenic properties of tumors induced by different chemical and physical agents. In: Harris RJC, ed. Specific Tumor Antigens. Copenhagen: Munksgaard, 1967:204–209.
6. Kripke ML. Antigenicity of murine skin tumors induced by ultraviolet light. J Natl Cancer Inst 1974;53:1333–1336.
7. Kripke ML. Immunologic mechanisms in UV radiation carcinogenesis. Adv Cancer Res 1981;34:69–106.

8. Boyle J, Briggs JD, Mackie RM, Junor BJR, Aitchison TC. Cancers, warts, and sunshine in renal transplant patients. Lancet 1984;1:702–704.
9. Penn I. The incidence of malignancies in transplant recipients. Transplant Proc 1975;7:323–326.
10. Fisher MS, Kripke ML. Suppressor T lymphocytes control the development of primary skin cancer in ultraviolet-irradiated mice. Science 1982;216:1133–1134.
11. Kripke ML, Lofgreen JS, Beard J, Jessup JM, Fisher MS. In vivo immune respones of mice during carcinogenesis by ultraviolet irradiation. J Natl Cancer Inst 1977;59:1227–1230.
12. Kripke ML, Thorn RM, Lill PH, Civin CI, Pazmino NHY, Fisher MS. Further characterization of immunological unresponsiveness induced in mice by ultraviolet radiation. Transplantation 1979;28:212–217.
13. Bergstresser PR. Sensitization and elicitation of inflammation in contact dermatitis. In: Norris DA, ed. Immune Mechanisms in Cutaneous Disease. New York: Marcel Dekker, 1989:219–245.
14. Jessup JM, Hanna N, Palaszynski E, Kripke ML. Mechanisms of depressed reactivity to dinitrochlorobenzene and ultraviolet-induced tumors during ultraviolet carcinogenesis in BALB/c mice. Cell Immunol 1978;38:105–115.
15. Elmets CA, Bergstresser PR, Tigelaar RE, Wood PJ, Streilein JW. Analysis of the mechanism of unresponsiveness produced by haptens painted on skin exposed to low dose ultraviolet radiation. J Exp Med 1983;158:781–794.
16. Glass MJ, Bergstresser PR, Tigelaar RE, Streilein JW. UVB radiation and DNFB skin painting induced suppressor cells universally in mice. J Invest Dermatol 1990;94:273–278.
17. Streilein JW, Bergstresser PR. Genetic basis of ultraviolet B effects on contact hypersensitivity. Immunogenetics 1988;27:252–258.
18. Yoshikawa T, Rae V, Bruins-Slot W, Van den Berg J-W, Taylor JR, Streilein JW. Susceptibility to effects of UVB radiation on induction of contact hypersensitivity as a risk factor for skin cancer in humans. J Invest Dermatol 1990;95:530–536.
19. Toews GB, Bergstresser PR, Streilein JW. Epidermal Langerhans cell density determines whether contact hypersensitivity or unresponsiveness follows skin painting with DNFB. J Immunol 1980;124:445–453.
20. Stingl G, Gazze-Stingl LA, Aberer W, Wolff K. Antigen presentation by murine epidermal Langerhans cells and its alteration by ultraviolet B light. J Immunol 1981;127:1707–1713.
21. Sullivan S, Bergstresser PR, Tigelaar RE, Streilein JW. Induction and regulation of contact hypersensitivity by resident, bone marrow-derived dendritic epidermal cells: Langerhans cells and Thy-1+ epidermal cells. J Immunol 1986;137:2460–2467.
22. Cruz PD Jr, Nixon-Fulton J, Tigelaar RE, Bergstresser PR. Disparate effects of in vitro low-dose UVB irradiation on intravenous immunization with purified epidermal cell subpopulations for the induction of contact hypersensitivity. J Invest Dermatol 1989;92:160–166.
23. Granstein RD, Lowy A, Greene MI. Epidermal antigen-presenting cells in activation of suppression: identification of a new functional type of ultraviolet radiation-resistant epidermal cell. J Immunol 1984;132:563–565.

24. Granstein RD, Askari M, Whitaker D, Murphy GF. Epidermal cells in activation of suppressor lymphocytes: further characterization. J Immunol 1987;138:4055–4062.
25. Cooper KD, Fox PF, Katz SI. Effects of ultraviolet radiation on human epidermal cells alloantigen presentation: initial depression of Langerhans cell function is followed by the appearance of T6-DR+ cells which enhance epidermal alloantigen presentation. J Immunol 1984;134:129–137.
26. Cooper KD, Neises GR, Katz SI. Antigen-presenting OKM5+ melanophages appear in human epidermis after ultraviolet radiation. J Invest Dermatol 1986;86:363–370.
27. Baadsgaard O, Wulf HC, Wantzin GL, Cooper KD. UVB and UVC, but not UVA, potently induce the appearance of T6-DR+ antigen-presenting cells in human epidermis. J Invest Dermatol 1987;1:113–118.
28. Baadsgaard O, Fox DA, Cooper KD. Human epidermal cells from ultraviolet light-exposed skin preferentially activate autoreactive CD4+2H4+suppressor-inducer lymphocytes and CD8+ suppressor/cytotoxic lymphocytes. J Immunol 1988;140:1738–1744.
29. Simon JC, Cruz PD Jr, Bergstresser PR, Tigelaar RE. Low-dose UVB-irradiated Langerhans cells preferentially activate CD4+ cells of the Th1 subset. J Immunol 1990;145:2087.
30. Simon JC, Tigelaar RE, Bergstresser PR, Edelbaum D, Cruz PD Jr. Ultraviolet B radiation converts Langerhans cells from immunogenic to tolerogenic antigen presenting cells. J Immunol 1991;146:485–491.
31. Cruz PD Jr, Bergstresser PR. Antigen processing and presentation by epidermal Langerhans cells: induction of immunity or unresponsiveness. Dermatol Clin 1990;8:633–647.
32. Simon JC, Tigelaar RE, Mosmann T, Edelbaum D, Bergstresser PR, Schoepf E, Cruz PD Jr. In vivo evidence that UVB induced suppression of contact sensitivity is due in part to functional inactivation of $CD4^+$ Th1 cells. J Invest Dermatol 1991;96:537.
33. Applegate LA, Ley RD, Alcalay J, Kripke ML. Identification of the molecular target for the suppression of contact hypersensitivity by ultraviolet radiation. J Exp Med 1989;170:1117–1131.
34. Alcalay J, Goldberg LH, Wolf JE Jr, Kripke ML. Ultraviolet radiation-induced damage to human Langerhans cells in vivo is not reversed by ultraviolet A or visible light. J Invest Dermatol 1990;95:144–146.
35. Swartz RP. Role of UVB-induced serum factor(s) in suppression of contact hypersensitivity in mice. J Invest Dermatol 1984;83:305–307.
36. Schwarz T, Urbanski A, Gschnait F, Luger TA. Inhibition of the induction of contact hypersensitivity by a UV-mediated epidermal cytokine. J Invest Dermatol 1986;87:289–291.
37. De Fabo EC, Noonan FP. Mechanism of immune suppression by ultraviolet irradiation in vivo: evidence for the existence of a unique photoreceptor in skin and its role in photoimmunology. J Exp Med 1983;157:84–98.
38. Ross JA, Howie SEM, Norval M, Maingay J. Systemic administration of urocanic acid generates suppression of the delayed type hypersensitivity response to herpes simplex virus in a murine model of infection. Photodermatology 1988;5:9–14.

39. Noonan FP, De Fabo EC, Morrison H. Cis-urocanic acid, a product formed by ultraviolet B irradiation of the skin, initiates an antigen presentation defect in splenic dendritic cells in vivo. J Invest Dermatol 1988;90:92–99.
40. Reeve VE, Greenoak GE, Canfield PJ, Boehm-Wilcox C, Gallagher CH. Topical urocanic acid enhances UV-induced tumour yield and malignancy in the hairless mouse. Photochem Photobiol 1989;49:459–464.
41. Dinarello CA. Biology of interleukin 1. FASEB J 1988;2:108–115.
42. Hauser C, Saurat J-H, Schmitt A, Jaunin F, Dayer J-M. Interleukin 1 is present in normal human epidermis. J Immunol 1986;136:3317–3323.
43. Ansel JC, Luger TA, Lowy D, Perry P, Roop DR, Mountz JD. The expression and modulation of IL-1a in murine keratinocytes. J Immunol 1988;140:2274–2278.
44. Kupper TS, Chua AO, Flood P, McGuire J, Gubler U. Interleukin 1 gene expression in cultured human keratinocytes is augmented by ultraviolet irradiation. J Clin Invest 1987;80:430–436.
45. Granstein RD, Sauder DN. Whole body exposure to ultraviolet radiation results in increased serum interleukin-1 activity in humans. Lymphokine Res 1987;6:187–193.
46. Schwarz T, Urbanski A, Gschnait F, Luger TA. UV-irradiated epidermal cells produce a specific inhibitor of interluekin 1 activity. J Immunol 1987;138:1457–1463.
47. Robertson B, Gahring L, Newton R, Daynes R. In vivo administration of interleukin 1 to normal mice depresses their capacity to elicit contact hypersensitivity responses: prostaglandins are involved in this modification of immune function. J Invest Dermatol 1987; 88:380–387.
48. Pentland AP, Mahoney MG. Keratinocyte prostaglandin synthesis is enhanced by IL-1. J Invest Dermatol 1990; 94:43-46.
49. Gilchrest BA, Soter NA, Stoff JS, Mihm MC Jr. The human sunburn reaction: histologic and biochemical studies. J Am Acad Dermatol 1981; 5:411–422.
50. Kock A, Schwarz T, Kirnbauer R, Urbanski A, Perry P, Ansel JC, Luger TA. Human keratinocytes are a source for tumor necrosis factor $\alpha$: evidence for synthesis and release upon stimulation with endotoxin or ultraviolet light. J Exp Med 1990;172:1609–1614.
51. Yoshikawa T, Streilein JW. TNF$\alpha$ and UVB light have similar effects on contact hypersensitivity in mice. Reg Immunol 1990;3:139–144.
52. Vermeer M, Streilein JW. Ultraviolet B light-induced alterations in epidermal Langerhans cells are mediated in part by tumor necrosis factor alpha. Photodermal Photoimmunol Photomed 1990;7:258–265.
53. Yoshikawa T, Streilein JW. Genetic basis of the effect of ultraviolet B light on cutaneous immunity: evidence that polymorphism at the TNF$\alpha$ and LPS loci governs susceptibility. Immunogenetics 1990;32:398–405.
54. Kim T-Y, Kripke ML, Ullrich SE. Immunosuppression by factors released from UV-irradiated epidermal cells: selective effects on the generation of contact and delayed hypersensitivity after exposure to UVA and UVB radiation. J Invest Dermatol 1990;94:26–32.

55. Nicholas JF, Kaiserlian D, Dardenne M, Faure M, Thivolet J. Epidermal cell-derived lymphocyte differentiating factor (ELDIF) inhibits in vitro lymphoproliferation responses and interleukin 2 production. J Invest Dermatol 1987;88:161–166.
56. Urbanski A, Schwarz T, Neuner P, Krutmann J, Kirnbauer R, Kock A, Luger TA. Ultraviolet light induces increased circulating interleukin 6 in humans. J Invest Dermatol 1990; 6:808–811.

# IV

# Photosensitivity Diseases

9

# Evaluation of the Photosensitive Patient

**Thomas Meola, Henry W. Lim, and Nicholas A. Soter**
*New York University School of Medicine, New York, New York*

In 1960 Wilkinson saw the first of several patients with eruptions confined to light-exposed areas. Through extensive history taking he discovered that some of these patients were employed in the same factory. After a visit to this factory, Wilkinson found that 29 workers were similarly affected, all of whom were employed in the same shop, which was the only shop using a certain brand of germicidal soap. The affected workers complained of worsening of their eruptions over the weekends and long holidays when they were outdoors. Investigation into the manufacture of this soap showed that a new germicidal agent, tetrachlorosalicylanilide, had been incorporated a few months prior to the appearance of these cases. Wilkinson performed patch tests with tetrachlorosalicylanilide on affected individuals and found frequent positive reactions, which were enhanced with suberythemogenic dosages of ultraviolet radiation. Skin biopsy specimens from affected individuals were interpreted as showing changes suggestive of a photoallergic dermatitis. In 1961 Wilkinson published his description of 53 cases of photosensitivity to tetrachlorosalicylanilide (1).

The approach employed by Wilkinson 30 years ago, which includes a thorough history and physical examination with the appropriate phototests, photopatch tests, histopathologic examination of skin biopsy specimens, and other laboratory studies, is a paradigm for the evaluation of photodermatoses.

The first objective in the evaluation of the photosensitive patient is to decide if the presenting symptoms and signs actually are induced or exacerbated by exposure to light. Conditions that need to be differentiated from photosensitivity disorders include airborne contact dermatitis and allergic contact dermatitis to topical agents including sunscreens. If the patient is found to be photosensitive,

**Table 1** Classification of Photosensitivity Disorders

| |
|---|
| Idiopathic |
| Polymorphous light eruption |
| Solar urticaria |
| Chronic actinic dermatitis |
| Associated with exogenous photosensitizers |
| Systemic phototoxicity |
| Systemic photoallergy |
| Phototoxic contact dermatitis |
| Photoallergic contact dermatitis |
| Associated with endogenous photosensitizers |
| Porphyrias |

the next objective is to determine the type of photodermatosis and its cause for those processes with known causes. This approach will direct treatment decisions and provide information to the patient regarding the prognosis. Even in those cases for which a cause cannot be ascertained, the photosensitivity evaluation may provide information concerning the action spectrum of a patient's photosensitivity disorder and may also provide an objective means of monitoring the course of the disease and its response to treatment.

There is no widely accepted classification of photosensitivity diseases. We have organized some of the more common entities that will be discussed in subsequent chapters into broad categories based on cause (Table 1). Because not all of the diseases in the spectrum of chronic actinic dermatitis have an identifiable photosensitizer, this group is considered idiopathic (2,3).

## I. HISTORY

A thorough history is the first and possibly most important area of the evaluation. However, not all patients who are photosensitive can directly relate their problem to ultraviolet radiation. History of the distribution of the lesions is helpful; however, patients who are exquisitely photosensitive may describe lesions on covered as well as exposed areas. This distribution is attributable to the inadequate protection from ultraviolet radiation afforded by some textiles (4).

Questions concerning the morphology of the lesions are essential to narrow the diagnostic possibilities. Vesicles and bullae may suggest one of the porphyrias, phototoxicity from an exogenous agent, or a vesicular variant of polymorphous light eruption. Papules and plaques may represent polymorphous light eruption or lupus erythematosus. A history of wheals after sun exposure is suggestive of solar urticaria or erythropoietic protoporphyria. The patient also may note erythematous patches as the initial manifestation of phototoxic and photoallergic eruptions. These areas can become hyperpigmented over time. In the case of chronic actinic dermatitis or photoallergic contact dermatitis, lichenification may develop, which the patient may no longer relate to sun exposure.

A history of burning, stinging, or pruritus prior to the appearance of lesions may not be helpful to the clinician due to their lack of specificity. Polymorphous light eruption, solar urticaria, phototoxicity, photoallergy, and erythropoietic protoporphyria may present with one or more of these symptoms.

The time course of the photosensitivity eruptions is one of the most important identifying characteristics. Of all the photosensitivity eruptions, solar urticaria has the closest temporal relationship to sun exposure, with its onset usually occurring within minutes and resolution within a few hours, although a rare delayed form of solar urticaria has been described. Polymorphous light eruption typically appears between 16 and 48 hr after sun exposure and gradually fades over the next 7–10 days. In contrast to the relatively rapid resolution of these eruptions, chronic actinic dermatitis, photoallergic contact dermatitis, and the porphyrias may persist from months to years.

Some of the photosensitivity diseases have a seasonal variation. This feature is characteristic of polymorphous light eruption, which typically appears in the spring or early summer with subsequent episodes becoming less severe as the summer progresses. This seasonal pattern is also a feature of solar urticaria.

It may be possible to identify the action spectrum of a photosensitivity disorder by the history alone. While ultraviolet A (UVA) and visible light are readily transmitted by window glass, only a minimal amount of ultraviolet B (UVB) is transmitted. Therefore, eruptions that can be produced or exacerbated by exposure through glass have an action spectrum that includes UVA or the visible light range.

The age at onset of the eruption and the patient's sex are also important in the history of photosensitivity disorders. For example, polymorphous light eruption has a predilection for young women while chronic actinic dermatitis is observed more often in elderly men. The onset of photosensitivity in childhood may suggest one of the childhood porphyrias, xeroderma pigmentosum, or hereditary polymorphous light eruption.

Since several dermatologic conditions can be exacerbated by sun exposure and possibly mimic a primary photosensitivity disorder (Chap. 16), special attention must be paid to this aspect of the history. A history of nondermatologic illnesses also must be obtained. Hepatic disease, cholelithiasis, or diabetes mellitus can on occasion be associated with certain porphyrias, and systemic lupus erythematosus can be exacerbated by sun exposure.

When assessing a patient for photosensitivity disorders, a detailed medication history including both systemic and topical preparations should be obtained. This information is most useful when a phototoxic or photoallergic eruption is sus-

**Table 2** Occupational Photodermatoses

| Occupation | Photosensitizer | Photodermatosis |
|---|---|---|
| Bartenders | Lime (furocoumarin) | Phototoxic contact dermatitis |
| Cement workers, lithographers, tanning industry | Chromium | Photoallergic contact dermatitis |
| Dye manufacturing | Anthroquinone dyes | Photoallergic and phototoxic contact dermatitis |
| Handlers of carrots, celery, dill, fennel, figs, parsley, parsnips | Furocoumarins | Phototoxic contact dermatitis |
| Herbicide manufacturing | Chlorinated phenols | Porphyria cutanea tarda |
| Nurses, pharmacists | Phenothiazines | Photoallergic and phototoxic contact dermatitis |
| Roofers | Coal tar pitch | Phototoxic contact dermatitis |
| Ultraviolet-cured ink manufacturing | Dimethyl aminobenzoate | Phototoxic contact dermatitis |

pected. The use of topical products such as sunscreens, soaps, cosmetics, and perfumes should not be overlooked since these historically have been some of the most common causes of photosensitivity disorders. A history of excessive alcohol intake or ingestion of estrogen-containing medications is often obtained in cases of porphyria cutanea tarda.

Questions concerning the occupational history of a photosensitive patient may uncover exposure to chemicals that are known photosensitizers in the workplace (Table 2). The occupational history also should include potential exposure to artificial light sources such as welding arcs, lasers, germicidal lamps, mercury vapor lamps, and sunlamps.

In rare cases, a photodermatosis such as solar urticaria or the photoinduced lesions of lupus erythematosus may be associated with systemic symptoms. Therefore, it is appropriate to question photosensitive patients about symptoms such as shortness of breath or arthralgias, which may be associated with the eruption.

## II. PHYSICAL EXAMINATION

In the absence of a history suggestive of a photosensitivity disorder, the physical examination may serve as the first clue in establishing the diagnosis. In a photosensitive patient, the lesions are usually confined to exposed areas. However, as noted previously, those with severe photosensitivity may develop lesions on covered areas as well. While the potential sites of the photosensitivity eruption are relatively limited, the morphologic lesions are varied (Table 3).

Commonly involved sites include the face, posterior aspect of the neck, ears, "V" area of the chest, dorsa of the hands, and extensor aspects of the forearms. The shins and dorsa of the feet may also be involved in patients who have exposure of the lower legs. The most striking feature of a photosensitivity eruption is the sparing of characteristic areas. These include the upper eyelids, submental area, posterior auricular area (Wilkinson's triangle), upper lip, web spaces of the fingers, and creases within folds of skin. However, in patients exposed to reflected radiation or artificial light sources, the pattern of distribution may not fit these general rules. Areas covered by watches, bra straps, or jewelry also may be uninvolved. In contrast, areas characteristically spared in a photosensitivity eruption would be involved in an airborne contact dermatitis. A photocontact dermatitis secondary to cosmetics or perfumes may follow the pattern of application.

Certain physical findings are suggestive of a systemic process that may be associated with photosensitivity. Examples are hepatic disease that may occur in some types of porphyria or the systemic signs of lupus erythematosus.

## III. PHOTOTESTING

The purposes of phototests are to establish the presence of a photosensitivity disorder, define its action spectrum, and reproduce the clinical lesion when possible. The former two are accomplished by minimal erythema dose (MED) determinations for UVA and UVB and by visible light testing. These results may assist in the diagnosis of a specific photosensitivity disorder (Table 4) and in the selection of a sunscreen if indicated. In the case of polymorphous light eruption, repetitive phototests may be employed to reproduce the clinical lesions.

### A. MED Determinations for UVA and UVB

A variety of artificial light sources are adequate for MED determinations. Two of the more commonly used sources of UVA are fluorescent "black light" tubes and Hanovia hot quartz lamps with a window glass filter to block UVB. Both of these are relatively inexpensive and practical for office use. Practical sources of UVB include fluorescent "sunlamp" tubes and Hanovia hot quartz lamps without a window glass filter. The light sources are discussed in detail in Chapter 3.

Patients undergoing MED determinations should be advised to avoid the application of topical corticosteroids to the test sites, and to avoid taking systemic corticosteroids and nonsteroidal anti-inflammatory agents for several

**Table 3** Common Lesions of Photosensitivity Disorders

| Disorder | Lesion | | | | | | |
|---|---|---|---|---|---|---|---|
| | Eczematous Dermatitis | Erythematous Patches | Lichenification | Papules | Plaques | Vesicles or Bullae | Wheals |
| Polymorphous light eruption | + | | | + | + | + | + |
| Solar urticaria | | | | | | | + |
| Chronic actinic dermatitis | + | | + | + | + | | |
| Phototoxicity | | + | | | | + | |
| Photoallergy | + | + | + | + | | + | |
| Porphyrias | | | | + | + | + | + |

**Table 4** Expected Phototest and Photopatch Test Results in Photosensitivity Disorders

| Disorder | MED-A | MED-B | Visible Light | Photopatch Tests |
|---|---|---|---|---|
| Polymorphous light eruption | nl | nl | nl | – |
| Solar urticaria | nl | nl | nl | – |
| Chronic actinic dermatitis | ↓ or nl | ↓ or nl | ↓ or nl | + or – |
| Systemic phototoxicity | ↓ or nl | ↓ or nl | nl | IR or – |
| Systemic photoallergy | ↓ or nl | ↓ or nl | nl | + or – |
| Phototoxic contact dermatitis | nl | nl | nl | IR |
| Photoallergic contact dermatitis | nl | nl | nl | + |

nl, normal; ↓, decreased; +, positive; –, negative; IR, irritant reaction.

days prior to testing, if possible, since these agents may suppress the erythema response and result in false-negative results. To perform the MED determinations for both UVA and UVB, a commercially available template with a series of approximately 1.5 × 1.5 cm openings with removable flaps is required (Fig. 1). The template is placed on typically covered, uninvolved skin such as the inner arm or buttocks. The template openings are then exposed to incremental increases of UVA or UVB. At our medical center exposures of 3, 6, 12, and 18 J/cm$^2$ are used in determining the MED to UVA (MED-A). The MED to UVB (MED-B) is evaluated at exposures of 6, 12, 24, 36, 48, 72, 90, and 108 mJ/cm$^2$. The test sites are evaluated immediately after exposure and then 30 min later for evidence of solar urticaria. The MED determination is made 24 hr later. Erythema to any of the UVA test doses specified above is an abnormal response, and the patient is considered to be photosensitive to UVA. The clinical relevance of an MED-B result is determined by comparison with a range of accepted values based on skin type (Chap. 17). Photosensitivity disorders associated with abnormal responses to UVA and UVB are listed in Table 4.

## B. Visible Light Testing

The least expensive and most accessible light source for visible light testing is a slide projector with a quartz halogen projector lamp. A water-filled polystyrene flask can be attached to serve as a heat filter. The projector lamp is positioned 15 cm from the inner arm of the patient and exposure times of 15, 30, 45, and 60 min are used with the assistance of a template.

The test sites are evaluated immediately after exposure and then 30 min later

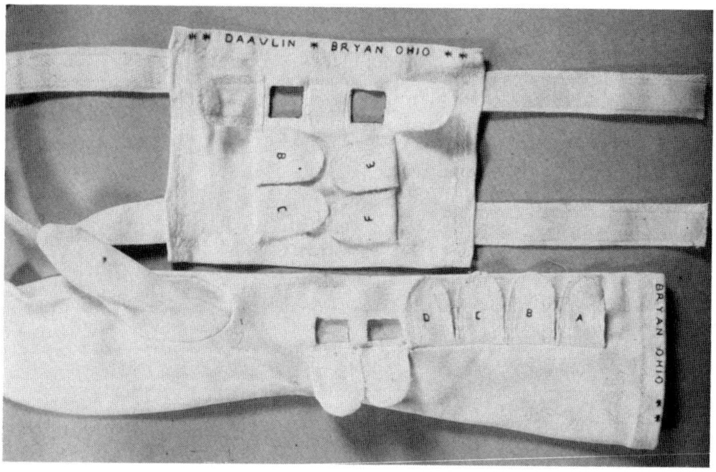


**Figure 1** Templates used for MED determinations. The upper one can be used on the trunk and the lower on the forearm.

for evidence of solar urticaria. A final reading is taken 24 hr later. The development of erythema or a lesion is considered an abnormal response. Photosensitivity disorders characterized by abnormal responses to visible light are listed in Table 4.

## C. Phototesting to Reproduce Clinical Lesions

Repetitive phototesting to reproduce a photosensitivity eruption is used mainly for the diagnosis of polymorphous light eruption. Although the MED-A, MED-B, and visible light tests are typically normal in patients with polymorphous light eruption, the lesions have been reproduced with high cumulative dosages of UVA or UVB.

Two techniques for the reproduction of clinical lesions have been described. In 1962 Epstein introduced a repetitive phototesting technique (5), a modification of which is used at our medical center and described below. The first step is to determine the MED-A and MED-B as described previously. The test sites would ideally be those previously clinically affected. However, involvement of technically difficult testing areas (e.g., face) does not always permit this. In these cases the flexor aspects of the forearms are preferred. Two 4 × 4 cm test sites are used. On the first day of testing, 80% of the MED-A and MED-B is given to separate sites. If no erythema is elicited during the MED-A determination, the initial dosage of UVA should be 10 J/cm². On the following 3 days, if no lesions are evident at the test sites, the dosages are increased by 30% of the previous day's dosage. The second technique, as described by Hölzle et al. (6), involves irradiation of previously affected sites with high dosages of UVA (50–100 J/cm²). The dosage is kept relatively constant and given every day or every other day to 5 × 10 cm test sites for no more than four doses. For each of these techniques, a final examination of the test site should be performed daily for several days after the last exposure, given the typically delayed onset of polymorphous light eruption. Should lesions be induced, a biopsy is performed to confirm the diagnosis. If polymorphous light eruption cannot be induced with artificial light sources, controlled sun exposure may reproduce the lesions. Lesions of lupus erythematosus may also be produced using artificial light sources. Consequently, it is necessary to differentiate these disorders by the serologic, histopathologic, and immunopathologic techniques outlined below.

## IV. PHOTOPATCH TESTS

### A. General Concepts

If the history, physical examination, and MED determinations suggest the possibility of a photoallergic contact dermatitis or one of the disorders included in the spectrum of chronic actinic dermatitis, photopatch tests should be performed. Photopatch testing does not aid in the diagnosis of polymorphous light eruption, solar urticaria, or the porphyrias. Compared to the history, physicial examination, and phototests, photopatch tests are expensive and time-consuming and should be performed only if there is adequate clinical suspicion.

As with any diagnostic procedure, photopatch testing may produce false-negative or false-positive results. In a survey by the German, Austrian, and Swiss Photopatch Test Group, of 2041 positive photopatch test results, only 108 were considered to be secondary to photoallergy and possibly clinically relevant (7). The poor specificity of photopatch tests also was demonstrated in a study from the Mayo Clinic in which only 14 of the 27 patients with photocontact reactions were considered to have relevant reactions (8). Another reason to be selective in deciding when to use photopatch testing is that inadvertent photosensitization to the test materials can occur (9).

There is currently no widely accepted standard photopatch test procedure, but several have been proposed (7,10). Consequently, it is difficult to compare the findings of different techniques: variations in the concentrations of test substances, types of vehicles, and duration of application have been shown to alter results (11). Standardization of the photopatch test procedure is problematic because not all photoallergens react similarly. The photosensitizer 6-methylcoumarin illustrates this difficulty: the brief interval (30–60 min) between application and irradiation that is required to obtain optimal results is considerably shorter than the 24–48 hr interval used for other test substances (12).

Since the action spectrum of almost all photoallergens is in the UVA range, this is the only radiation used in photopatch testing. Many of the patients who undergo photopatch testing, such as those with chronic actinic dermatitis, are sensitive to UVB as well as UVA. Therefore, the light source used in photopatch testing should be free of UVB to avoid false-positive reactions. In addition, light sources used in photopatch testing should be able to deliver a uniform dosage of radiation to a large area. Suitable devices include a panel of fluorescent "black light" tubes or a Hanovia hot quartz lamp with a window glass filter.

## B. Selection of Test Substances

One of the greatest variations in photopatch test procedures in different medical centers is the photoallergens used. Since the photoallergic patient often is unable to implicate a specific substance as the cause of photosensitivity, testing is performed with a variety of photoallergens. When a specific photoallergen is suspected, testing is also performed with this substance. The two most commonly used vehicles are alcohol and petrolatum. Whichever of these is used, it is essential that a low concentration of the test substance be used so as not to cause a phototoxic response that may be misinterpreted as photoallergy.

The photoallergens used at our medical center include the North American Contact Dermatitits Group photopatch test series (Table 5) and the New York University Skin and Cancer Unit photopatch test/sunscreen sensitivity series (Table 6). This combination provides most of the commonly described photoallergens as well as the sunscreen ingredients. Not included in these series are oleoresin extracts from Compositae plants. Frain-Bell and colleagues have shown that allergic contact sensitivity to these substances frequently occurs in patients with chronic actinic dermatitis (13,14). Testing with these additional substances may be helpful if chronic actinic dermatitis is suspected.

## C. Test Procedure

Since there is no standard procedure for photopatch testing, the procedure used at our medical center will be outlined to provide a general framework. It can be modified to meet one's office capabilities and to accommodate the diagnostic needs of individual patients. Four visits are required with this protocol. The first three visits are on consecutive days, and the fourth and final visit is 1 week after the initial visit. As with phototesting, patients are instructed to avoid exposure to

**Table 5** North American Contact Dermatitis Group Photopatch Test Series

| |
|---|
| Chlorpromazine hydrochloride 0.1% in petrolatum |
| Musk ambrette 1% in petrolatum |
| Para-aminobenzoic acid (PABA) 5% in alcohol |
| Petrolatum control |
| Tribromosalicylanilide 1% in petrolatum |
| Octyl dimethyl PABA 5% in alcohol |
| Benzophenone-2 2% in petrolatum |
| Promethazine 1% in petrolatum |
| Oak moss 1% in petrolatum |
| Octyl dimethyl PABA 5% in petrolatum |
| Bithionol 1% in petrolatum |
| PABA 5% in petrolatum |
| Musk ambrette 1% in alcohol |
| Fentichlor 1% in petrolatum |
| Amyl dimethyl PABA 5% in petrolatum |
| Atranorin 0.5% in petrolatum |
| Amyl dimethyl PABA 5% in alcohol |
| Parsol 1780 5% in petrolatum |
| 6-Methylcoumarin 1% in alcohol |
| Benzophenone-4 1% in petrolatum |
| Dichlorophene 1% in petrolatum |
| Sorbitan sesquiolate 5% in petrolatum |

**Table 6** New York University Skin and Cancer Unit Photopatch Test/Sunscreen Sensitivity Series

| |
|---|
| Diphenhydramine hydrochloride 1% in petrolatum |
| Trichlorocarbanilide 1% in petrolatum |
| Tetrachlorocarbanilide 1% in petrolatum |
| 4,5-Dibromosalicylanilide 1% in petrolatum |
| Sulfanilamide 1% in petrolatum |
| Triclosan 2% in petrolatum |
| Hexachlorophene 1% in petrolatum |
| Aqueous chlorhexidine diacetate 0.5% |
| Thiourea 0.1% in petrolatum |
| Sandalwood Oil |
| Methyl anthranilate 10% in petrolatum |
| 3-(4-Methylbenzyliden)–camphor 2% in petrolatum |
| 2-Ethylhexyl-p-dimethyl aminobenzoate 2% in petrolatum |
| Ethoxyethyl-p-methoxycinnamate 1% in petrolatum |
| 2-Ethylhexyl-p-methoxycinnamate 2% in petrolatum |
| 2-Ethoxymethyl-p-methoxycinnamate 2% in petrolatum |
| 1-(4-Isopropylphenyl) 3 phenyl 1,3 propandione 2% in petrolatum |
| Tiscreen #15 |
| Photoplex |

topical and systemic corticosteroids and nonsteroidal anti-inflammatory agents for several days prior to testing. Our UVA light source consists of a bank of high output-UVA bulbs mounted in a phototherapy unit.

On day 1, the MED-A is performed as described previously. Also during this visit, duplicate sets of photoallergens (Tables 5 and 6) are applied to the back and then covered by opaque tape. Patients are instructed to avoid washing this area after the patches are applied.

On day 2, prior to irradiation of one set of photoallergens, the MED-A is quantitated. One set of patches is then removed and the test sites are evaluated for a possible allergic contact or irritant dermatitis. In patients with an MED-A greater than 18 J/cm$^2$, one set of photoallergens is exposed to 10 J/cm$^2$ of UVA while the duplicate set remains covered. If the MED-A is less than 18 J/cm$^2$, the exposure is 50% of the MED-A. After this exposure the unirradiated side is briefly uncovered for evaluation of test sites. Both sets are then covered for an additional 24 hr. It is important to wipe all residual test substances from the sites to be irradiated prior to UVA exposure, since some of these compounds may exert a sunscreen effect and generate false-negative results. In the above protocol, 6-methylcoumarin is applied 30–60 min prior to UVA exposure on the irradiated side for reasons noted previously. However, it is applied simultaneously with the other photoallergens on the nonirradiated side.

On day 3, both the irradiated and nonirradiated test sites are uncovered and evaluated. Responses are graded as described below.

One week after the initial visit, the patch and photopatch test sites are evaluated for delayed reactions.

## D. Grading of Results and Interpretation

The grading of reactions at both the irradiated and nonirradiated patch test sites is based on a scoring system recommended by the American Academy of Dermatology (Table 7). A positive response at the irradiated site in the absence of a response at the nonirradiated site indicates a photoallergic contact dermatitis,

**Table 7** American Academy of Dermatology Grading of Patch and Photopatch Tests

| | |
|---|---|
| + | Weak (nonvesicular) reaction: erythema, infiltration, possibly papules |
| ++ | Strong (edematous or vesicular) reaction |
| +++ | Extreme (spreading, bullous, ulcerative) reaction |
| ? | Doubtful reaction, macular erythema only |
| IR | Irritant reaction |
| – | Negative reaction |
| NT | Not tested |

**Table 8** Interpretation of Photopatch Test Results

| Reaction at Nonirradiated Site | Reaction at Irradiated Site | Interpretation |
|---|---|---|
| – | – | Normal |
| + | Equal to nonirradiated | Allergic contact dermatitis |
| – | + | Photoallergic contact dermatitis |
| + | > Nonirradiated | Photoallergic contact and allergic contact dermatitis |
| Irritant reaction | Irritant reaction | Irritant contact dermatitis |

while positive responses of equal intensity and duration at both the irradiated and nonirradiated sites indicate an allergic contact dermatitis (Table 8). The presence of both allergic contact dermatitis and photocontact dermatitis is indicated by positive responses at both irradiated and nonirradiated sites in which the response at the irradiated site is more pronounced or persists for a longer period than that of the nonirradiated site. Differentiation of irritant and allergic reactions is often not possible. In severe irritant responses, well-defined erythema that resolves promptly may be observed. Allergic responses are more commonly eczematous, with poorly defined borders and slower resolution.

## V. LABORATORY EVALUATION

The two instances in which laboratory studies are necessary in the assessment of photosensitivity disorders are the evaluation of the porphyrias and lupus erythematosus.

The biochemical tests of blood, urine, and feces necessary for the diagnosis of porphyria are reviewed in Chapter 15. When lupus erythematosus is suspected, relevant laboratory studies include a complete blood cell count, erythrocyte sedimentation rate, and measurement of antinuclear antibodies, including anti-Ro, anti-La, anti-Sm, and anti-RNP antibodies. In differentiating lupus erythematosus from polymorphous light eruption, the serologic tests are most useful since they are negative in polymorphous light eruption. These two entities may also be differentiated by histopathologic and immunopathologic techniques.

## VI. HISTOPATHOLOGIC AND IMMUNOPATHOLOGIC EXAMINATIONS

Skin biopsy specimens for histopathologic examination may be taken from lesions present at the time of the initial physical examination or from lesions induced by phototesting. In the differentiation of polymorphous light eruption from lupus erythematosus, direct immunofluorescence tests of lesional and nonlesional skin (lupus band test) show the absence of immunoglobulins or complement proteins at the dermoepidermal junction in polymorphous light eruption, while these immunoreactants are present in lupus erythematosus.

Histopathologic examination also is useful in differentiating actinic reticuloid from the other disorders in the spectrum of chronic actinic dermatitis. This differentiation is based on the presence of atypical mononuclear cells in the dermis and epidermis.

## REFERENCES

1. Wilkinson DS. Photodermatitis due to tetrachlorosalicylanilide. Br J Dermatol 1961; 73:213–219.
2. Hawk JLM, Magnus IA. Chronic actinic dermatitis—an idiopathic photosensitivity syndrome including actinic reticuloid and photosensitive eczema. Br J Dermatol 1979: 101 (Suppl 17): 24.
3. Lim HW, Buchness MR, Ashinoff R, Soter NA. Chronic actinic dermatitis: study of the spectrum of chronic photosensitivity in 12 patients. Arch Dermatol 1990; 126:317–323.
4. Berne B, Fischer T. Protective effects of various types of clothes against UV radiation. Acta Derm Venereol (Stockh) 1980; 60:459–460.
5. Epstein JH. Polymorphous light eruptions: phototest technique studies. Arch Dermatol 1962; 85:502–504.
6. Hölzle E, Plewig G, Hofmann C, Roser-Maass E. Polymorphous light eruption. Experimental reproduction of skin lesions. J Am Acad Dermatol 1982; 7:111–125.
7. Hölzle E, Neumann N, Hausen B, Przybilla B, Schauder S, Hönigsmann H, Bircher A, Plewig G. Photopatch testing: the 5-year experience of the German, Austrian, and Swiss Photopatch Test Group. J Am Acad Dermatol 1991; 25:59–68.
8. Menz J, Muller SA, Connolly SM. Photopatch testing: a six-year experience. J Am Acad Dermatol 1988; 18:1044–1047.
9. Harber LC, Bickers DR. Photosensitivity Diseases: Principles of Diagnosis and Treatment, 2nd ed. Philadelphia: BC Decker, 1989: 198, 404.
10. Jansén CT, Wennersten G, Rystedt I, Thune P, Brodthagen H. The Scandinavian standard photopatch test procedure. Contact Derm 1982; 8:155–158.
11. Suhonen R. Photoepicutaneous testing: influence of the vehicle, occlusion time and concentration of the test substances on the results. Contact Derm 1976; 2:218–226.
12. Jackson RT, Nesbitt LT, DeLeo VA. 6-Methylcoumarin photocontact dermatitis. J Am Acad Dermatol 1980; 2:124–127.
13. Frain-Bell W, Hetherington A, Johnson BE. Contact allergic sensitivity to chrysanthemum and the photosensitivity dermatitis and actinic reticuloid syndrome. Br J Dermatol 1979; 101:491–501.
14. Frain-Bell W, Johnson BE. Contact allergic sensitivity to plants and the photosensitivity dermatitis and actinic reticuloid syndrome. Br J Dermatol 1979; 101:503–512.

# 10

# Polymorphous Light Eruption

**Herbert Hönigsmann**
*University of Vienna, Vienna, Austria*

## I. HISTORICAL ASPECTS

Historical sources let us assume that Robert Willan, at the end of the 18th century, reported for the first time in the medical literature a recurrent sunlight-induced rash for which the term "eczema solare" was coined (1). Nearly a century later, Jonathan Hutchinson (2) used the term "summer prurigo" to define an eruption of red, itchy papules mainly affecting the arms and the face that started after puberty and continued for many years during summer with remissions during winter. It is very likely that he described some forms of polymorphous light eruption and perhaps some other idiopathic photodermatoses such as hydroa aestivale or photosensitive eczema. In 1900, Rasch (3), and later in 1918, Haxthausen (4), grouped eczematous and papular photodermatoses and introduced the term chronic polymorphous light eruption. During subsequent decades a variety of photosensitivity dermatoses of unknown pathogenesis, such as solar urticaria, photoallergic dermatitis, hydroa vacciniforme, erythropoietic protoporphyria, and lupus erythematosus were probably lumped together (5). These diseases are now clearly defined as distinct clinical entities. The term polymorphous light eruption is now restricted to a better delineated group of sun-induced skin rashes. However, it is likely that this term still encompasses several photosensitivity conditions with different pathogenetic mechanisms. The different action spectra and the inter-individual polymorphism of the lesions may support this concept (5–7).

## II. DEFINITION

Polymorphous light eruption is an idiopathic, recurrent, acquired abnormal skin reaction to sunlight of delayed onset. It is characterized by a varied appearance of erythematous papules, papulovesicles, plaques, and lesions resembling erythema multiforme lesions on sun-exposed skin areas. Within each patient the single morphologic feature of the lesions always remains the same. The term "polymorphous" designates the inter-individual variation in the clinical appearance of the disease.

## III. EPIDEMIOLOGY

Polymorphous light eruption is the most common photosensitivity disease, and, according to a survey of apparently healthy individuals (8), it may be even more common than one would assume when considering only the number of patients who seek medical advice. It may perhaps represent a "normal" cutaneous reaction that anyone can develop under adequate irradiation and certain environmental conditions. The observation that some control subjects without a history of polymorphous light eruption developed typical lesions on exposure to artificial ultraviolet radiation could support such an assumption (9).

Polymorphous light eruption most commonly affects females with fair skin color, although it has been reported in persons of all races and skin colors (6,10–13). The age of onset can be any time from childhood to adult life; the majority of patients experience their first attack before the age of 30. A positive family history is occasionally reported in some series (14), although it is possible that the high incidence in some reports may be due to the inclusion of persons with other photodermatoses (15).

The familial photosensitivity rash with autosomal dominant inheritance that is common among Native Americans (16–18) has been considered a unique variant of polymorphous light eruption. However, there are many differences, such as primary eczematous and prurigo lesions, cheilitis, and pterygia formation that help to differentiate this condition as a separate entity (17,19,20). This peculiar rash closely resembles actinic prurigo (21,22), a disease that is clearly different from polymorphous light eruption.

## IV. CAUSES

### A. Pathogenesis

The pathogenetic mechanisms of polymorphous light eruption remain unknown. The only causative factor that reproducibly induces skin lesions is sunlight. However, the question of how solar radiation can mediate the development of the disease is unclear. Attempts to identify endogenous or exogenous photosensitizers that may be involved have so far failed. The presence of perivascular lymphocytic infiltrates in the dermis suggests a delayed (photo)immunologic reaction, but evidence from immunologic studies is conflicting (23–25). Repair of ultraviolet (UV)-induced DNA damage appears to be normal (24,26).

## B. Action Spectrum

Determination of the action spectrum of polymorphous light eruption by experimental reproduction of skin lesions using artificial radiation sources has led to conflicting results. This may be attributed to differences in individual sensitivity, technical equipment, test sites, diagnostic criteria and, in particular, the lack of universally accepted, standardized phototest protocols.

Several investigators claimed that ultraviolet B (UVB) radiation is most effective in reproducing skin lesions (9,11,12,27). However, other studies have shown that UVA and at times visible light can provoke typical reactions (7,9,10,12,27,28). In our series of patients, 56% had their action spectrum in the UVA range, 17% were elicited by UVB only, and 26% had a positive test result in the UVA and UVB range (6). Similar results were reported recently by Hölzle et al. (29). The high rate of UVA-sensitive patients is in accordance with the observation that many patients experience their rash after sun exposure through window glass while riding in cars, trains, or airplanes. In some patients, no positive reaction can be achieved by testing with artificial radiation sources despite an unequivocal history. It is possible that in these patients the action spectrum lies in a waveband range emitted below threshold doses from the radiation source used or, alternatively, additional unknown environmental factors may be lacking. Negative test results also may be due to the fact that some patients are tested in early fall when their skin is perhaps less sensitive due to hardening (see below).

Some authors have assumed that patients who report protection from polymorphous light eruption by being behind window glass, applying sunscreens, or staying in the shade could most likely have their action spectrum in the UVB range (30). A comparison of the history and the phototest results in our series unexpectedly did not confirm this assumption: there were more positive results with UVA than with UVB in this group (6).

An increasing number of patients is now reporting that their rash has become more severe after application of sunscreens with high sun protection factor in the UVB range. An explanation for this phenomenon could be that these patients have their action spectrum in the UVA range and, by effectively filtering the sunburning UVB radiation, they are encouraged to stay much longer in the sun, thereby receiving a much higher UVA dose than they would receive without or with lower UVB protection.

## V. CLINICAL FEATURES

Polymorphous light eruption characteristically occurs in spring or early summer following the first exposure to sunlight after winter. However, outdoor activities in winter may also induce an attack. The lesions develop within hours or days following sun exposure, which may or may not be accompanied by sunburn. In the absence of further irradiation the rash subsides within 1–10 days. Pruritus may be quite severe and frequently precedes the eruption. On rare occasions, patients may develop severe pruritus without visible skin lesions (polymorphous light eruption sine eruptione) (31). Upon further sun exposure during summer the eruptions typically tend to become less severe ("decrescendo"), until tolerance to sunlight is achieved. This phenomenon, known as "hardening," is attributed to increased melanization and stratum corneum thickening, although other mechanisms, such as depletion of hypothetical mediators or immune mechanisms, are also possible. After many years of persistence of this annual disease pattern, during which the disease may become more severe with time ("crescendo"), it may finally resolve spontaneously. Polymorphous light eruption heals without scar formation.

The lesions develop exclusively on sun-exposed areas, but there are certain sites of predilection: the V area of the chest, the dorsa of the hands, the extensor aspects of the arms, and sometimes the lower legs, the cheeks, and the shoulders.

As mentioned above, the skin lesions of polymorphous light eruption are polymorphic in different patients but they maintain the same morphologic pattern in the same subject over the years. Pruritus frequently precedes the appearance of skin lesions. They may range from red papules, to papulovesicles, plaques, lesions resembling erythema multiforme, and those resembling insect-bites (strophulus). Hemorrhage is a very rare event (7). The papular type, frequently incorrectly described by the patients as blisters, is the most common type (Fig. 1, see color plate). After coalescing, the papules may form erythematous, sometimes urticaria-like plaques (Fig. 2) that subsequently, induced by scratching, may become eczematized, lichenified, and excoriated. These latter changes represent a secondary phenomenon. Erythematous plaques may sometimes mimic lupus erythematosus, particularly if confined to the face. The papulovesicular type, a more recently defined subset, appears to be not uncommon among North American women (32) but is rarely detected in Europe (33). Varieties resembling erythema multiforme and insect bites are also relatively rare (7,34)

## VI. HISTOPATHOLOGIC CHANGES

The histopathologic changes vary with the morphologic type of polymorphous light eruption (5,7). They are quite characteristic but not diagnostic. The epidermis shows edema, focal spongiosis, and occasionally small vesicles.

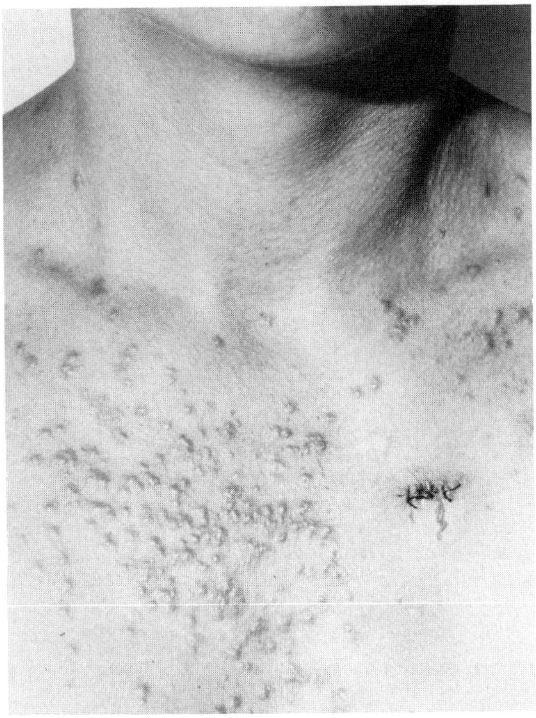

**Figure 2** Coalescent papules forming erythematous plaques resembling urticaria.

Acanthosis, focal parakeratosis, and basal vacuolization can be present. Sunburn cells are not a typical feature. The dermal changes are more characteristic and very similar in all clinical variants. A dense lymphocytic perivascular infiltrate is present in the upper and middermis (Fig. 3). The subepidermal compartment may show edema that can be quite pronounced in the erythema multiforme-type lesions. These changes occur in a number other dermatoses especially in lupus erythematosus. The latter can be ruled out by the absence of basement membrane zone thickening, epidermal atrophy, follicular plugging, and by the negative lupus band test in direct immunofluorescence (35,36).

Using monoclonal antibody techniques, the dermal infiltrate has been shown to consist mostly of T cells of the CD8-positive (suppressor) type (37). In a more recent study, a preponderance of T cells of the CD4-positive (helper) type was described (38). Perivascular deposits of fibrin, complement, and IgM were found inconsistently in some patients (39), but these observations cannot be used to confirm the clinical diagnosis.

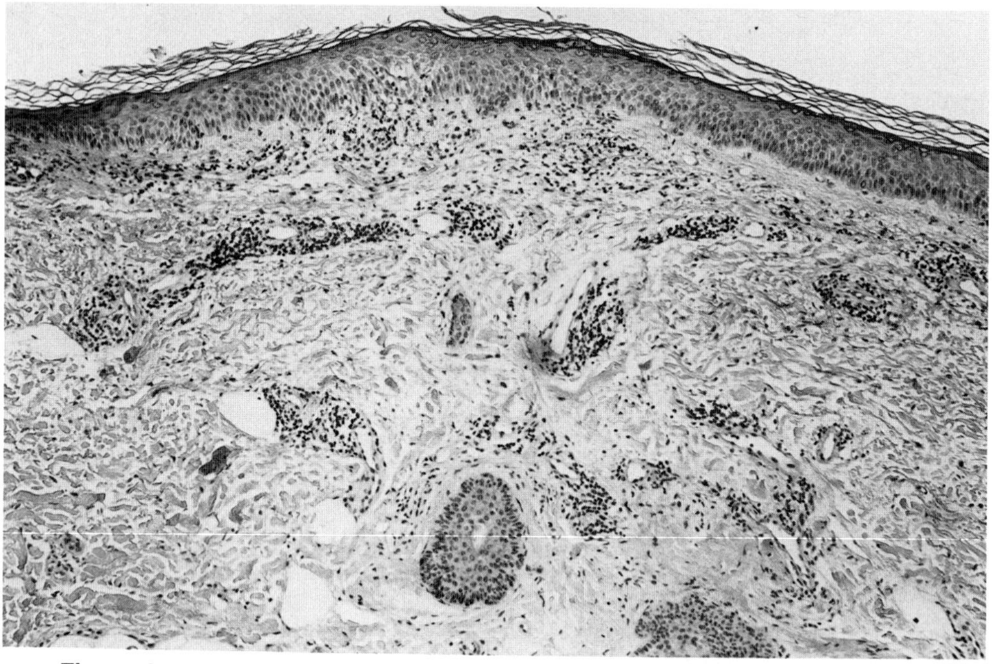

**Figure 3** Polymorphous light eruption. Papular lesion. Note mild epidermal spongiosis, dense perivascular lymphocytic infiltrate in the upper and middermis. Skin adnexae are spared. (Hematoxylin and eosin; original magnification, x 100).

## VII. LABORATORY FINDINGS

There are no diagnostic laboratory tests available for polymorphous light eruption. Laboratory examinations are usually performed to exclude other dermatoses such as erythropoietic protoporphyria and photosensitive lupus erythematosus. In particular, tests for antinuclear antibodies and for Ro (SSA) and La (SSB) antibodies should be done.

## VIII. PHOTOTESTS

Since polymorphous light eruption cannot be defined by any laboratory parameter, phototesting represents an important means of establishing the diagnosis, particularly in patients without clinically manifest disease (5–7,11,13,36). Although controversy about the eliciting wavelengths does not seem to be settled yet, evidence indicates that the action spectrum can be in the UVA, UVB, or in both wavelength ranges (6,9,29,34,40). The role of visible light has not been satisfactorily established with modern light source equipment. Hölzle et al. (7) were unable to elicit specific lesions with monochromatic and polychromatic visible light.

In our experience, the repeat phototest technique as originally outlined by Epstein appears to be the most useful procedure for the reproduction of specific lesions. Prior to the actual test we determine the minimal erythema dose for broad-band UVA and UVB. In this context it should be emphasized that patients with polymorphous light eruption have a normal erythema response to UVB (6,7,41–46) and UVA (6,7). For induction of lesions, we select two symmetrical test areas measuring at least 4 × 4 cm, preferably on previously affected body areas. It is important to note that the use of previously noninvolved skin areas may lead to false-negative results. Also the time of year may be important: after induction of tolerance by sunlight in late summer or fall the skin can be refractory. The test areas are exposed daily for 4–8 days with UVA and UVB. The initial irradiation dose used is approximately 100–150% of the predetermined minimal erythema dose in either waveband range. If neither lesions characteristic for polymorphous light eruption nor erythema develop, the doses are increased by 20–40% at each session. It should be noted that the eruption sometimes may not develop for 2 or 3 days after the last exposure. The test is considered positive when a rash typical for the patient's clinical variant of the disease is elicited (Fig. 4, see color plate). In our experience, positive test results can be obtained with this procedure in the majority of the patients using one or both wavelengths bands.

## IX. DIFFERENTIAL DIAGNOSIS

In the majority of patients who present with the typical history and the classic clinical picture of polymorphous light eruption with papules or papulovesicles, the diagnosis will hardly cause any problem. The patients are usually aware of the association between exposure to sunlight and their disease, which they commonly refer to as "sun allergy". However, if no lesions are present, phototesting is the most useful diagnostic procedure.

Some cases cannot be diagnosed based on clinical features alone. Secondarily eczematized polymorphous light eruption has to be differentiated from photoallergic contact dermatitis, airborne contact dermatitis, and other contact sensitivity to multiple allergens. These can be excluded by appropriate tests, such as photopatch or patch tests. Chronic actinic dermatitis, which includes persistent light reactivity and actinic reticuloid, is a well defined entity and should be easily differentiated (See Chap. 12). These diseases are chronic and often become lichenified due to scratching. In addition, they exhibit a lowered minimal erythema dose in the UVA and/or UVB range.

Solar urticaria shows a monomorphic urticarial wheal arising within a very short time after sun exposure and clearing within 1–2 hr (see Chap. 11). This is in contrast to the delayed reaction of polymorphous light eruption. Solar urticaria can easily be confirmed by phototesting with a broad spectrum of UV radiation and often with visible light.

Erythropoietic protoporphyria has a diagnostic history of life-long light sensitivity, typical waxy thickening and "pseudolichenification" on the exposed skin areas, and can be excluded by red cell fluorescence and porphyrin analysis (see Chap. 15).

The plaque-type lesions must be differentated from lymphocytic infiltrate of Jessner-Kanof, granuloma faciale, other granulomatous infiltrates and, most importantly, from lupus erythematosus. Histopathological examination is the most reliable means of excluding most of these diseases. Some lesions of lupus erythematosus may mimic polymorphous light eruption, particularly because in some patients lesions may also be reproduced by phototesting, although with a much longer latency period. However, lupus lesions usually persist for weeks and even months without further irradiation. Again histopathologic examination, direct immunfluorescence, and serologic tests for antinuclear antibodies together with other criteria of lupus will help to establish the diagnosis. In rare instances the clear differentiation may remain impossible. In our own series of over 200 patients with polymorphous light eruption, a subgroup of 7 female subjects was encountered during phototesting, with anti-Ro antibodies, long-lasting atypical skin lesions, negative results of direct immunofluorescence, and no other criteria of lupus. A follow-up study over several years did not show a progression into true lupus erythematosus (unpublished data).

Hydroa vacciniforme, a rare idiopathic condition, can be differentiated by the typical varioliform vesicles, pustules, crusts and, finally, scars. Lesions can be provoked by UVA only (47–51).

## X. TREATMENT

### A. General Remarks

The treatment of polymorphous light eruption has to be subdivided into therapy for the acute exacerbation and the prophylactic therapy before expected sun exposure.

Patients with fully developed disease require topical corticosteroids, in some cases in the form of wet dressings, for several days. Exceptionally severe cases may be treated with a short course of systemic (oral) corticosteroids, beginning with 60 mg prednisolone equivalent and tapering to 10 mg during a period of 8–10 days.

Since polymorphous light eruption will subside spontaneously and is not a life-threatening condition, all possible risks of therapy should be carefully considered.

Many patients will agree to undergo some sort of preventive measures. Prophylactic treament consists of several approaches: avoidance of sunlight during the summer, the use of sunscreens with appropriate filters, systemic treatment, and preventive phototherapy.

## B. Topical Prevention

Topically applied sunscreens predominantly absorb in the UVB range and thus protect patients who have their action spectrum in this particular waveband range. In our experience, these sunscreens are helpful in less sensitive patients but ineffective in more severe cases. Obviously, the mode of application influences the degree of protection. With a new German brand of sunscreen containing a strong UVA absorber (Contralum® Ultra, Hermal, Reinbek, Germany) with a UVB protection factor of 10 and a UVA protection factor of 7 (52), we were able to suppress the UVA-induced positive test reaction in experimental subjects. Whether this preparation also protects under field conditions has not been sufficiently investigated.

The use of sunblocks containing zinc oxide or titanium dioxide is helpful in some cases, but the color of these preparations is usually not cosmetically acceptable. The prophylactic application of topical corticosteroids, for obvious reasons, cannot be recommended.

## C. Systemic Prevention

Several drugs are in use for systemic prophylactic intervention in polymorphous light eruption. These include beta-carotene, the antimalarials chloroquine and hydroxychloroquine, and nicotinamide. However, their true benefit has remained controversial.

Beta-carotene has been reported to be of value in some patients (53–56), but several other studies have not found the substance to be reliably effective in any photodermatosis except erythropoictic protoporphyria (5,13,34,57–61).

Oral antimalarials, in particular chloroquine, have been widely used and reported to be effective in selected patients (40,62–65). However, its benefit has been questioned by controlled studies (59). In addition, the potentially serious toxicity restricts the use of antimalarials.

A more recent study suggested that oral nicotinamide may be effective in preventing polymorphous light eruption (66). This was based on the assumption that polymorphous light eruption may depend on a disturbance of tryptophan metabolism. High dosages of nicotinamide should reduce the level of the potential photosensitizer kynurenic acid. Unfortunately, this also could not be confirmed in a controlled trial (67).

Recently, azathioprine was reported to be successful in two patients with severe, disabling polymorphous light eruption (68). This treatment is clearly not indicated for the average patient but may be considered for selected cases. This report awaits controlled confirmation in a larger group of patients. Prophylactic systemic corticosteroid treatment certainly represents inappropriate therapy for a benign, self-limited condition.

## D. Phototherapy

Both phototherapy with UVB and photochemotherapy have been reported as being effective in preventing polymorphous light eruption (6,58,69–71). The rationale of phototherapy regimens is based on the observation that many patients become less sensitive after repeated exposure to sun as summer progresses. Increased pigmentation and hyperkeratosis following phototherapy appear to be similar to this natural "hardening" phenomenon.

UVB phototherapy (71) is quite useful in the prophylaxis of the disease. A treatment schedule of 5 exposures a week for 3 weeks in the spring followed by regular exposures to sunlight will protect a considerable number of patients. UVB-sensitive patients may develop the eruption during treatment since the treatment dose and the dose that may elicit the rash can be close together. Sometimes prednisone must be used to suppress the reaction during treatment (72).

The most effective treatment is the prevention by photochemotherapy with oral psoralens (PUVA). Since Gschnait et al. (69) demonstrated the induction of tolerance to sunlight by short-term PUVA, several studies have compared this treatment with other forms of preventive treatment in patients with polymorphous light eruption and have confirmed its effectiveness (6,30,58,70,73,74). PUVA bears the advantage of a rapid and intense pigment induction at relatively low UVA dosages that usually remain well below the threshold dosages for eliciting the rash. Thus, treatment of the very sensitive patient is possible with a lower incidence of papular reactions. Our own treatment schedule consists of 3–4 treatments per week for 3–4 weeks in early spring. PUVA preventive treatment offers only temporary protection, and patients are advised to have regular repeated sun exposure to maintain protection.

The mechanisms by which phototherapy induces tolerance to sunlight has not been clarified. Hyperpigmentation and thickening of the stratum corneum may be important factors for the protective effect. However, other mechanisms may also be involved since polymorphous light eruption does occur in dark-skinned subjects. In this context, modulation of cutaneous immune functions is being discussed, although there is presently no evidence to confirm the possible role of immune mechanisms.

Prophylactic UVB or short-term PUVA is indicated in all patients with severe disease, who cannot avoid sun exposure, and do not respond to appropriate sunscreen use.

# REFERENCES

1. Bateman T. A Practical Synopsis of Cutaneous Diseases According to the Arrangement of Dr. Willan. Philadelphia: Collins & Croft, 1817:251–253.
2. Hutchinson J. Lectures on Clinical Surgery: On Certain Rare Diseases of the Skin. Vol. 1. London: Churchill, 1897:126–136.

3. Rasch C. Om et polymorft (erythematøst, vesikuløst og ekzematoidt) lysudslet. Hospitalstidende 1900; 43:478–480.
4. Haxthausen H. Studier over lysdermatiterne: II. Undersøgelser over pathologiske sera's virkning paa fotokemiske processer. Hospitalstidende 1918; 19–20:577–585.
5. Epstein JH. Polymorphous light eruption. J Am Acad Dermatol 1980; 3:329–343.
6. Ortel B, Tanew A, Wolff K, Hönigsmann H. Polymorphous light eruption: action spectrum and photoprotection. J Am Acad Dermatol 1986; 14:748–753.
7. Hölzle E, Plewig G, Hofmann C, Roser-Maass E. Polymorphous light eruption. Experimental reproduction of skin lesions. J Am Acad Dermatol 1982; 7:111–125.
8. Morison WL, Stern RS. Polymorphous light eruption: a common reaction uncommonly recognized. Acta Derm Venereol (Stockh) 1982; 62:237–240.
9. Jansen CT. The polymorphic phototest reaction. Arch Dermatol 1982; 118:638–642.
10. Wiskemann A, Wulf K. Untersuchungen über den auslösenden Spektralbereich und die direkte Lichtpigmentierung bei chronischen und akuten Lichtausschlägen. Arch Klin Exp Dermatol 1959; 209:443–453.
11. Epstein JH. Polymorphous light eruption. Ann Allergy 1966; 24:397–405.
12. Frain-Bell W, Dickson A, Herd J, Sturrock I. The action spectrum in polymorphic light eruption. Br J Dermatol 1973; 89:243–249.
13. Magnus IA. Polymorphic light eruption and summer prurigo. In: Dermatological Photobiology. Philadelphia: JB Lippincott, 1976:174–188.
14. Jansen CT. Heredity of chronic polymorphic light eruptions. Arch Dermatol 1978; 114:188–190.
15. Johnson BE. Abnormal reactions associated with skin disorders. In: Jarrett A, ed. The Physiology and Pathophysiology of the Skin. Vol 8. London: Academic Press, 1984:2494–2497.
16. Everett MA, Crockett W, Lamb JH, Minor D. Light-sensitive eruptions in American Indians. Arch Dermatol 1961; 83:243–248.
17. Birt AR. Photodermatitis in Indians of Manitoba. Can Med Assoc J 1968; 98:392–397.
18. Birt AR, Davis RA. Photodermatitis in North American Indians: familial actinic prurigo. Int J Dermatol 1971; 10:107–114.
19. Hojyo-Tomoka MT, Domínguez-Soto L. Clinical and epidemiological characteristics of polymorphous light eruption in Mexico. Castellania 1975; 3:21–23.
20. Scheen SR III, Connolly SM, Dicken CH. Actinic prurigo. J Am Acad Dermatol 1980; 5:183–190.
21. Calnan CD, Meara RH. Actinic prurigo (Hutchinson's summer prurigo). Clin Exp Dermatol 1977; 2:365–371.
22. Aoki T, Fujita M. Actinic prurigo: A case report with successful induction of skin lesions. Clin Exp Dermatol 1980; 5:47–52.
23. Raffle EJ, MacLeod TM, Hutchinson F. In vitro lymphocyte studies in chronic polymorphic light eruption. Br J Dermatol 1973; 89:143–148.
24. Jung EG, Bohnert E. Chronische polymorphe Lichtdermatose. Dermatologica 1974; 148:209–212.
25. Horkay I, Bodolay E, Krajczár J, Alföldi G. In situ identification of T lymphocytes in polymorphic light eruption. Photodermatology 1984; 1:250–252.

26. Ichihashi M, Ramsay CA. Excision repair of DNA in some photodermatoses. Br J Dermatol 1976; 95:13–17.
27. Magnus IA. Studies with a monochromator in the common idiopathic photodermatoses. Br J Dermatol 1964; 76:245–264.
28. Levine MJ. Polymorphic light eruption. Exacerbation from photocopier exposure. Arch Dermatol 1981; 117:373–374.
29. Hölzle E, Plewig G, von Kries R, Lehmann P. Polymorphous light eruption. J Invest Dermatol 1987; 88:32s–38s.
30. Gschnait F, Schwarz T, Ladich I. Treatment of polymorphous light eruption. Arch Dermatol Res 1983; 275:379–382.
31. Dover JS, Hawk JLM. Polymorphic light eruption sine eruptione. Br J Dermatol 1988; 118:73–76.
32. Elpern DJ, Morison WL, Hood AF. Papulovesicular light eruption. A defined subset of polymorphous light eruption. Arch Dermatol 1985; 121:1286–1288.
33. Lehmann P, Hölzle E, Plewig G. Vesikulobullöse Form der polymorphen Lichtdermatose. Allergologie 1986; 9:32–53.
34. Frain-Bell W. Cutaneous Photobiology. Oxford: Oxford University Press, 1985.
35. Fisher DA, Epstein JH, Kay DN, Tuffanelli DL. Polymorphous light eruption and lupus erythematosus: differential diagnosis by fluorescent microscopy. Arch Dermatol 1970; 101:458–461.
36. Jansén C, Karvonen J. Polymorphous light eruption: a seven-year follow-up on 114 patients. Arch Dermatol 1984; 120:862–865.
37. Muhlbauer JE, Bhan AK, Harrist TJ, Bernhard JD, Mihm MC: Papular polymorphic light eruption: an immunoperoxidase study using monoclonal antibodies. Br J Dermatol 1983; 108:153–162.
38. Moncada B, González-Amaro R, Baranda ML, Loredo C, Urbina R. Immunopathology of polymorphous light eruption: T lymphocytes in blood and skin. J Am Acad Dermatol 1984; 10:970–972.
39. Muhlbauer JE, Mihm MC, Harrist TJ. Papular polymorphous light eruption. Fibrin, complement, and immunoglobulin deposition. Arch Dermatol 1984; 120:866–886.
40. Epstein JH. Polymorphous light eruption. Dermatol Clin 1986; 4:243–251.
41. Epstein JH. Polymorphous light eruptions: wavelength dependency and energy studies. Arch Dermatol 1962; 85:82–88.
42. Thune P. Chronic polymorphic light eruption. Particular wavebands and the effect of carotene therapy. Acta Derm Venereol (Stockh) 1976; 56:127–133.
43. Rottier PB, Baart de la Faille H. MED-action spectra in polymorphic light eruption and in porphyria compared with model normal spectra. Acta Dermatol Venereol (Stockh) 1977: 57 (suppl 77):3–35.
44. Plewig G, Hofmann C, Hölzle E. Polymorphe Lichtdermatose. In: Braun-Falco O, Wolff HH, eds. Fortschritte der praktischen Dermatologie und Venerologie. Berlin: Springer Verlag, 1979:117–127.
45. Jansén CT. Erythemal and pigmentary phototest reactions in polymorphous light eruptions. Acta Derm Venereol (Stockh) 1979; 59:499–503.
46. Diffey BL, Farr PM. The erythemal response to ultraviolet radiation in subjects with polymorphic light eruption. Br J Dermatol 1986; 114:103–108.
47. Jaschke E, Hönigsmann H. Hydroa vacciniforme—Aktionsspektrum, UV-Toleranz nach Photochemotherapie. Hautarzt 1981; 32:350–353.

48. Goldgeier MH, Nordlund JJ, Lucky AW, Sibrack LA, McCarthy MJ, McGuire J. Hydroa vacciniforme: diagnosis and therapy. Arch Dermatol 1982; 118:588–591.
49. Halasz CL, Leach EE, Walther RR, Poh-Fitzpatrick M. Hydroa vacciniforme: induction of lesions with ultraviolet A. J Am Acad Dermatol 1983; 8:171–176.
50. Sonnex TX, Hawk JLM. Hydroa vacciniforme: a review of ten cases. Br J Dermatol 1988; 118:101–108.
51. Eramo LR, Garden JM, Esterly NB. Hydroa vacciniforme. Diagnosis by repetitive ultraviolet-A phototesting. Arch Dermatol 1988; 122:1310–1313.
52. Hönigsmann H. Wirkung einer therapeutischen Lichtschutzcreme auf die Phototestreaktion bei polymorpher Lichtdermatose. Z Hautkr 1989; 64:1065–1068.
53. Swanbeck G, Wennersten G. Treatment of polymorphous light eruptions with beta-carotene. Acta Derm Venereol (Stockh) 1972; 52:462–466.
54. Nordlund JJ, Klaus SN, Mathews-Roth MM. A new therapy for polymorphous light eruption. Arch Dermatol 1973; 108:710–712.
55. Jansén CT. β-Carotene treatment of polymorphous light eruptions. Dermatologica 1974; 140:363–373.
56. Wennersten G, Swanbeck G. Treatment of light sensitivity with carotenoids. Acta Derm Venereol (Stockh) 1974; 54:491–499.
57. Mathews-Roth MM, Pathak MA, Fitzpatrick TB, Harber LC, Kass EM. Beta-carotene therapy for erythropoietic protoporphyria and other photosensitivity diseases. Arch Dermatol 1977; 113:1229–1232.
58. Parrish JA, Levine MJ, Morison WL, Gonzalez E, Fitzpatrick TB. Comparison of PUVA and beta-carotene in the treatment of polymorphous light eruption. Br J Dermatol 1979; 100:187–198.
59. Corbett MF, Hawk JLM, Herxheimer A, Magnus IA. Controlled therapeutic trials in polymorphic light eruption. Br J Dermatol 1982; 107:571–581.
60. Suhonen R, Plosila M. The effect of beta-carotene in combination with canthaxanthin, Ro 8-8427, in treatment of polymorphous light eruptions. Dermatologica 1981; 163:172–176.
61. Jansén CT. Oral carotenoid treatment in polymorphous light eruption: a cross-over comparison with oxychloroquine and placebo. Photodermatology 1985; 2:166–169.
62. Cahn MM, Levy EJ, Shaffer B. Polymorphous light eruption: the effect of chloroquine phosphate in modifying reactions to ultraviolet light. J Invest Dermatol 1956; 26:201–207.
63. Christiansen JV, Brodthagen H. The treatment of polymorphic light eruptions with chloroquine. Br J Dermatol 1956; 68:204–208.
64. Epstein JH. Synthetic antimalarial drug therapy in lupus erythematosus and polymorphous light eruptions. Calif Med 1960; 92:135–138.
65. Isaacson D, Elgart M, Turner ML. Antimalarials in dermatology. Int J Dermatol 1982; 21:379–395.
66. Neumann R, Rappold E, Pohl-Markl H. Treatment of polymorphous light eruption with nicotinamide: a pilot study. Br J Dermatol 1986; 115:77–80.
67. Ortel B, Wechdorn D, Tanew A, Hönigsmann H. Effect of nicotinamide on the phototest reaction in polymorphous light eruption. Br J Dermatol 1988; 118:669–673.
68. Norris PG, Hawk JLM. Successful treatment of severe polymorphous light eruption with azathioprine. Arch Dermatol 1989; 125:1377–1379.

69. Gschnait F, Hönigsmann H, Brenner W, Fritsch P, Wolff K. Induction of UV light tolerance by PUVA in patients with polymorphous light eruption. Br J Dermatol 1978; 99:293–295.
70. Jansén CT, Karvonen J, Malmiharju T. PUVA therapy for polymorphous light eruptions: comparison of systemic methoxsalen and topical trioxalen regimens and evaluation of local protective mechanisms. Acta Derm Venereol (Stockh) 1982; 317–320.
71. Morison WL, Momtaz, K. Mosher DB, Parrish JA. UV-B phototherapy in the prophylaxis of polymorphous light eruption. Br J Dermatol 1982; 106:231–233.
72. Morison WL. Phototherapy and Photochemotherapy of Skin Disease. Praeger, 1983:80–81.
73. Murphy GM, Logan RA, Lorell CR, Morris RW, Hawk JLM, Magnus IA. Prophylactic PUVA and UVB therapy in polymorphic light eruption—a controlled trial. Br J Dermatol 1987; 116:531–538.
74. Addo HA, Sharma SC. UVB phototherapy and photochemotherapy (PUVA) in the treatment of polymorphic light eruption and solar urticaria. Br J Dermatol 1987; 116:539–547.

# 11

# Solar Urticaria

**Takeshi Horio**
*Kansai Medical University, Osaka, Japan*

## I. INCIDENCE

Solar urticaria is a rare photosensitivity disease. In the past 15 years, the author has seen approximately 30 patients. Ive et al. described 17 patients with solar urticaria over a period of 5 years (1). Patients with solar urticaria have been reported throughout the world. Although it occurs equally in both sexes in the author's own experience, a slight preponderance in women has been described in the literature. The age of onset ranges from the first to the eighth decades. There has been only one instance of siblings with solar urticaria (2).

## II. CLINICAL MANIFESTATIONS

Solar urticaria develops at the site of exposure to sunlight within a few minutes. It begins as an itching or burning sensation along with erythema; subsequent development of the wheal is sometimes associated with a surrounding erythema (Fig. 1). When the exposure time is short, the wheal may be minimal or absent and only erythema develops. Within 1–2 hr, the urticaria disappears completely without any residual skin changes. If large areas of the body are exposed to sunlight for a long period of time, systemic signs may occur, including dizziness, wheezing, and syncope.

Chronically sun-exposed skin, such as the face, the dorsa of the hands, and the extensor aspects of the forearms, are often less sensitive than are covered areas. In summer months the eruption may develop on covered parts of the body as a result of small amounts of light penetrating thin clothing.

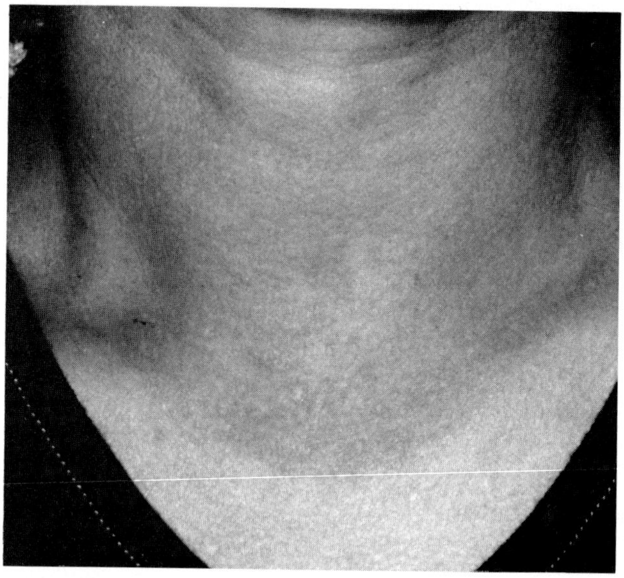

**Figure 1** Solar urticaria developed on the neck after exposure to natural sunlight.

## III. CLASSIFICATION

Although the clinical features of solar urticaria are almost identical in all patients, the eliciting wavelengths and the results of passive transfer tests vary from person to person.

Rajka first showed that solar urticaria could be passively transferred to normal subjects by means of an intradermal injection of serum from an affected patient and subsequent exposure to light in the appropriate action spectrum 24 hr after the serum injection (3). A wheal appears immediately after the irradiation. A positive result suggests the presence of antibody in the patient's serum.

Rajka also tried reverse passive transfer by initially exposing the skin of normal subjects and then injecting the patient's serum into the irradiated site. No response was observed. Epstein reported that both reverse passive transfer and passive transfer techniques could be successfully accomplished in a patient who developed urticaria in response to ultraviolet B (UVB) (4). It has been assumed that a substance produced in normal human skin by solar radiation might be the antigen for solar urticaria.

**Table 1** Classification of Solar Urticaria

| Type | Action spectrum (nm) | Passive transfer | Reverse passive transfer | Mechanism |
|---|---|---|---|---|
| I | 285–320 | Positive | Positive | Allergic |
| II | 320–400 | Negative | Negative | Unknown |
| III | 400–500 | Negative | Negative | Unknown |
| IV | 400–500 | Positive | Negative | Unknown; probably allergic |
| V | 280–600 | Negative | Negative | Unknown |
| VI | 400 | Negative | Negative | Protoporphyrin in skin |

Source: Ref. 5.

Harber et al. have classified solar urticaria into six types, depending on the action spectra and results of passive and reverse passive transfer tests (5) (Table 1). Patients who are sensitive to UVB and in whom both passive and reverse passive transfer tests are positive belong to type I. Patients in whom the action spectrum lies in the UVA range and in whom neither passive nor reverse passive transfer can be demonstrated are classified as type II. In type III and type IV solar urticaria, the action spectrum is in the visible light spectrum ranging from 400 nm to 500 nm. The passive transfer test is negative in type III but is positive in type IV. The reverse passive transfer test is negative in both type III and type IV. Type V patients show a broad action spectrum ranging from UVB to visible light (280–600 nm). In this type, neither passive nor reverse passive transfer can be demonstrated. In type VI, solar urticaria is associated with erythropoietic protoporphyria, as reported by Magnus et al. (6). There are some variations and overlappings in this classification. Other investigators have also attempted to classify solar urticaria. The patients can be divided, depending only on the action spectrum, into those sensitive to UVB, UVA, or visible light or combinations (7). In addition, solar urticaria may occur in patients with systemic lupus erythematosus.

## IV. ACTION SPECTRUM STUDIES

For the determination of the action spectrum in solar urticaria, the following artificial light sources are useful, easily available, and inexpensive: a slide projector lamp for a visible light source, a fluorescent black light for a UVA source, and a fluorescent sunlamp for a UVB source. However, it should be kept in mind that the black light and sunlamp also emit a small amount of UVB and

UVA, respectively. When necessary, various colored glass filters are used in combination with these light sources to eliminate shorter wavelengths. A monochromator, which can emit selected narrow bands of wavelengths, is useful for more detailed and sophisticated studies.

Usually a wheal appears within 5–10 min after termination of the irradiation. In some patients, however, the urticaria develops after a longer latent time. Therefore, the exposed area should be observed at least 30 min after the irradiation. Ihm reported an unusual patient with a delayed type of solar urticaria (8). Monfrecola et al. also reported a similar patient who developed a wheal 5 hr after light exposure (9). Although the latter patient showed normal porphyrin metabolism, delayed edematous reactions after light exposure should be differentiated from porphyrias. It is possible that the delayed onset of solar urticaria is due to the existence of an inhibition spectrum, which will be discussed later.

In 14 of the 20 (70%) Japanese patients with solar urticaria in our study, the wavelengths responsible for the urticaria were found to lie within the visible light range. Two patients reacted only to UVA, three to a broad spectrum of light ranging from UVB to visible light, and one patient to UVB alone. In contrast, Frain-Bell found that the majority of their patients (20 of 26) in Scotland reacted to a wide spectrum of light (7). Therefore, a difference in the predominant type of solar urticaria may exist among races.

The wavelengths involved may vary from time to time with repeated phototesting in a few patients (7). Murphy and Hawk described a patient who showed a marked expansion of the action spectrum (10). At the first examination, this patient developed urticaria only to 380 nm light from a monochromator. Later, all wavelengths from 300 nm to 600 nm induced a wheal-and-flare reaction.

## V. ETIOLOGY AND PATHOGENESIS

Differences in the action spectrum among cases indicate that the chromophore (photosensitizer) may differ from patient to patient. The inconstant results of passive transfer studies suggest that the mechanisms involved may not be uniform in all cases.

The passive transfer test in solar urticaria is a modification of the Prausnitz-Küstner technique used in immediate hypersensitivity reactions. Therefore, a positive result suggests that a specific IgE antibody may exist in the patient's serum. In the passive transfer of this type, an incubation period of at least a few hours is needed between the injection of the patient's serum and antigen challenge. Kojima et al. reported that positive results were obtained from the passive transfer test at 2–6 hr after the injection of serum, whereas the results were negative when the injection sites were irradiated at 30 min (11). This observation indicates that the positive passive transfer test could not be due to a phototoxic substance in the patient's serum. An intradermal injection of a phototoxic

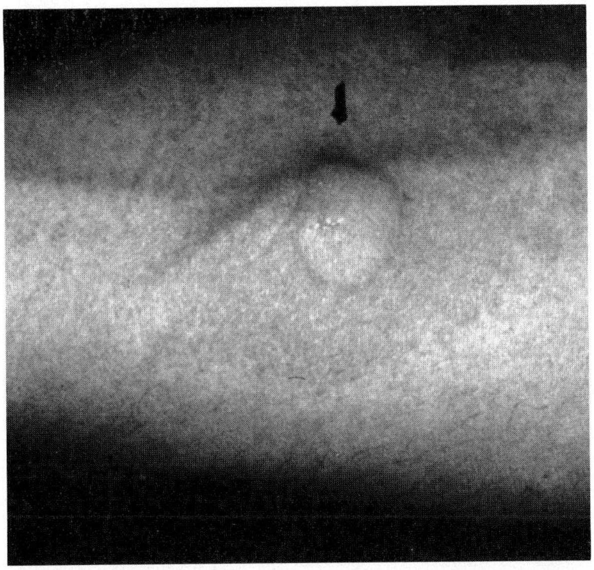

**Figure 2** Photosensitizer in patient's serum. A patient developed a wheal at the site of injection of his own serum, which had been previously exposed to light in vitro.

substance can produce a wheal even if light exposure is administered immediately after the injection.

The reverse passive transfer test is performed by first exposing the skin of normal persons (antigen formation) and then injecting the patient's serum (antibody). The mechanism of a positive reaction cannot be explained by IgE-mediated hypersensitivity, in which the antibody must be injected in the normal subject at least a few hours before the antigen challenge.

In some cases of solar urticaria, a circulating photoallergen in the patients' sera seems to be an etiological factor (12). These patients developed a wheal at the site of injection of their own serum, which had been previously exposed to light in vitro (Fig. 2). The results indicate that the wheal-forming factor is a substance produced by light energy in the patient's serum. Normal control subjects do not produce such a reaction to their own irradiated sera. The patients' sera irradiated in vitro do not produce the urticarial response in the normal subjects. Therefore, the photoresponse is specific for the patients. Furthermore, the passive transfer test was positive in normal persons, when the skin sites injected with the patients' sera were irradiated in vivo, indicating that the antibody also exists in the patients' serum. Some patients develop the urticaria to their own in vitro–irradiated serum but not to normal irradiated serum. This observation suggests that the photoallergen could be an abnormal endogenous substance made only in vivo. However, other patients react not only to their own serum but also to normal serum irradiated in vitro. This suggests that a normal serum factor could also be the photoallergen of solar urticaria. Kojima et al. reported a patient who reacted to in vitro–irradiated plasma from both the patient and a normal subject but not to in vitro–irradiated serum (11).

These observations indicate that an autoimmune type of solar urticaria might exist. However, a responsible photosensitizer and photoallergen have not been identified. Using ultrafiltration techniques, we suggested that the photosensitizer (a precursor of a photoallergen) might be a serum factor whose molecular weight is more than 100 kDa (13).

Although the photoallergen may be produced from an endogenous serum factor in most patients, solar urticaria can be induced also by exogenous factors such as chemical substances. One of our patients with solar urticaria showed an immediate type of photoallergy to chlorpromazine (14). This patient developed a wheal at the photopatch test site with chlorpromazine immediately after the exposure to UVA (Fig. 3). Since the peak reaction occurred within 2 hr, this was

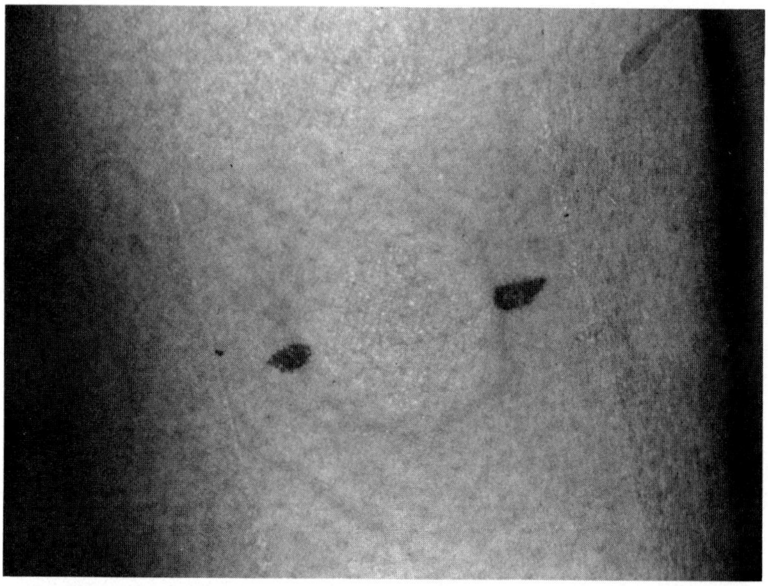

**Figure 3** A wheal appeared immediately after the exposure to UVA at a photopatch test site with chlorpromazine.

not a usual delayed type of positive photopatch test. The passive transfer study was negative. Photocontact urticaria was found in a patient who also showed delayed photosensitive eczema (15).

Magnus et al. reported a patient with erythropoietic protoporphyria associated with solar urticaria (6). This patient developed a wheal immediately after sunlight exposure, although the typical skin changes are delayed in porphyrias. The authors claimed that porphyrins were responsible for the solar urticaria. However, protoporphyrin is not known to be an antigenic substance. This type of photosensitivity may develop on a nonimmunological, possibly phototoxic basis.

The exact nature of the antibody in solar urticaria has not been identified. Sams examined two cases of solar urticaria in which both the passive and reverse passive transfer studies were positive and demonstrated that the antibody was neither IgG nor IgM but possibly was IgE (16). The association with an atopic history is low in solar urticaria. The IgE levels are within the normal range.

## VII. CHEMICAL MEDIATORS

Some investigators have found elevated histamine levels in venous blood draining the skin in which urticaria had been induced (17, 18). Chemoattractants for neutrophils and eosinophils have been detected in the venous blood after experimental challenge (18). On the other hand, other investigators were unable to detect histamine and kinins in dermal perfusates (19). A report that some patients respond to a kinin inhibitor but not to antihistamines suggests the involvement of kinins as a chemical mediator (20).

## VII. INHIBITION SPECTRUM

In 1982, an inhibition spectrum was first demonstrated in solar urticaria (20). Wheal formation induced by visible light was inhibited by irradiation with longer wavelengths than those of the action spectrum (Fig. 4). In some patients with solar urticaria, monochromatic light but not a broad-band spectrum can induce urticaria (10). It is possible that the broad-band spectrum may include an inhibition spectrum. The mechanism of inhibition has not been clarified. Suppression of mast cell degranulation or inactivation of released chemical mediators is not likely, since the inhibition spectrum does not suppresses the wheal induced by the intradermal injection of the mast cell degranulator, compound 48/80. Wheal formation by histamine injection is not affected by exposure to the inhibition spectrum. This observation may exclude an effect on the vascular response of chemical mediators. Although we assumed that the photoallergen produced by the action spectrum might be inactivated by the subsequent exposure to the inhibition spectrum (21), this explanation is unlikely in the case in which preirradiation of the inhibition spectrum suppresses the urticaria (22).

We have demonstrated that the inhibition spectrum in four of seven patients

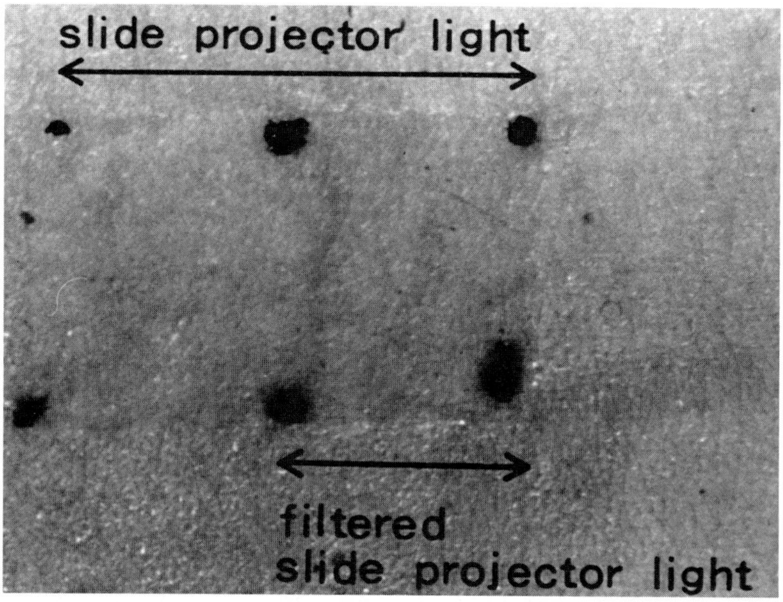


**Figure 4** Inhibition spectrum. A patient's skin was exposed to slide projector light. The right half of the area was additionally irradiated with the filtered slide projector light from which the action spectrum was excluded.

was always longer than the action spectrum. Recently, however, it was reported that UVA inhibited solar urticaria induced by visible light in one patient (23).

## VIII. AUGMENTATION SPECTRUM

An augmentation spectrum was also found in a patient with solar urticaria (24). The irradiation with visible light (450–500 nm) before exposure to activating wavelengths (320–420 nm) enhanced the urticaria. Postirradiation with visible light had no effect. Therefore, this spectrum appeared to have an augmentative but not an additive effect on the action spectrum. This theory was supported by the observation that a monochromatic light of 450 nm showed a stronger enhancing effect than 360 nm, the peak of the action spectrum.

Based on the fact that only preirradiation with the augmentation spectrum enhanced urticaria, one can speculate that a precursor of photosensitizer that absorbs light energy in the augmentation spectrum is altered to a state that can easily absorb the urticaria-eliciting energy (24).

## IX. HISTOPATHOLOGY

Histopathologic findings of solar urticaria are identical to those of nonphotosensitive urticaria. There is edema in the upper and middle dermis with separation of the connective tissue. In addition, a minimal-to-moderate perivascular infiltrate, consisting of lymphocytes, monocytes, and sometimes a few eosinophils, is present.

## X. TREATMENT

Systemic administration of H1 antihistamines is somewhat beneficial only in selected patients. Chlorpheniramine (25), homochlorcyclizine hydrochloride (22), and terfenadine (26) have been reported to be effective.

Antimalarial drugs have been beneficially used in some patients. Willis and Epstein reported control in two patients with the use of chloroquine (27).

The topical application of suncscreens can increase the minimal urticarial dose in experimental induction and may be helpful in UVB-sensitive patients.

Plasmapheresis has been used with a beneficial effect for the treatment of patients with solar urticaria in whom a photoallergen can be detected in the serum or plasma (28, 29).

So-called "desensitization" or "hardening" can occur in patients with solar urticaria. Skin constantly exposed to the natural sunlight and also an area in which urticaria has recently been produced are tolerant to subsequent irradiation (Fig. 5). Patients with solar urticaria usually avoid sunlight exposure. However, repeated exposures to natural sunlight or to artificial light sources often show beneficial therapeutic effects (30), although the treatment must be performed repeatedly at an interval of a few days. The mechanism by which tolerance is induced remains unclear. However, it is possible that the amount of chemical mediators is reduced by repeated exposures (13).

Psoralen plus ultraviolet A (PUVA) photochemotherapy has also been used with beneficial effect (31, 32). Remission time is more long-lasting than that obtained by UVB irradiation. PUVA is effective not only in UVA-sensitive patients, but also in those in whom the action spectrum does not include the UVA range. Therefore, the mode of action of PUVA appears to be different from that of the "desensitization" produced by natural sunlight. The mechanism by which PUVA exerts a therapeutic effect has not been established. PUVA induces melanogenesis and thickening of the stratum corneum, which could diminish the penetration of sunlight. In experimental systems, it was shown that PUVA irradiation inhibits mast cell degranulation and the release of chemical mediators (33, 34).

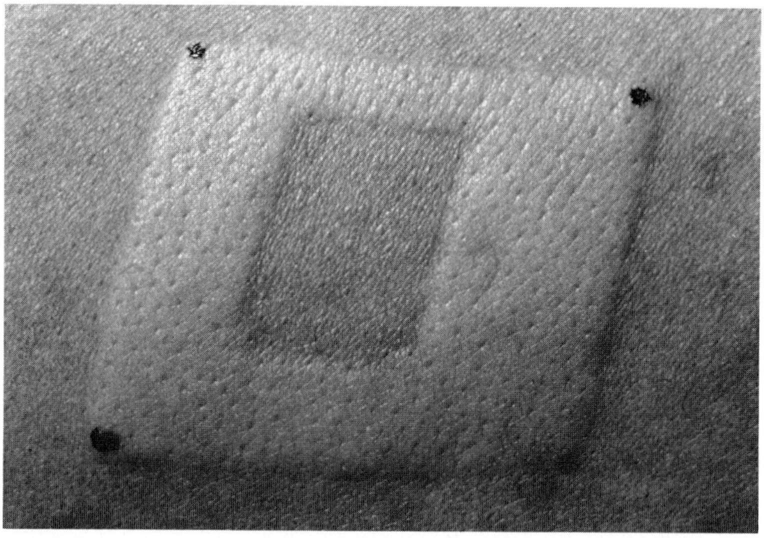

**Figure 5** At the site of repeated exposure (small square), a wheal did not develop on irradiation; test performed 24 hr after initial exposure.

## XI. COURSE

Solar urticaria tends to persist indefinitely in most patients. There is very little information available on how long the photosensitivity state continues. In Frain-Bell's series, 2 of 30 patients cleared completely, three showed some degree of improvement, and the remainder were unchanged or worse (7). Harber's group also reviewed the clinical course in 13 of 15 patients. Five of these experienced spontaneous remission after an average of 4.5 years. In three of them the action spectrum was in the visible light range, one in the UVA and visible light range, and one only in the UVA range. Patients with an action spectrum predominantly in the visible light range may have the best prognosis (35). The present author has never observed a cure in our experience with more than 25 patients.

## REFERENCES

1. Ive H, Lloyd J, Magnus IA. Action spectra in idiopathic solar urticaria. Br J Dermatol 1965; 77:229–243.
2. Samter M. Immunologic Diseases. Boston: Little-Brown, 1965.

3. Rajka E. Passive transfer in light urticaria. Clin Immunol 1942; 13:327–345.
4. Epstein S. Urticaria photogenica. Ann Allergy 1949; 7:443–457.
5. Harber LC, Holloway RM, Wheatley VR, Baer RL. Immunologic and biophysical studies in solar urticaria. J Invest Dermatol 1963; 41:439–443.
6. Magnus IA, Jarrett A, Prankerd TAJ, Rimington C. Erythropoietic protoporphyria: a new porphyria syndrome with solar urticaria due to protoporphyria. Lancet 1961; 2:448–451.
7. Frain-Bell W. Cutaneous Photobiology. Oxford: Oxford University Press, 1985.
8. Ihm CW. Solar urticaria: report of an unusual case. Cutis 1979; 784–786.
9. Monfrecola G, Nappa P, Pini D. Solar urticaria with delayed onset: a case report. Photodermatology 1988; 5:103–104.
10. Murphy GM, Hawk JLM. Broadening of action spectrum in a patient with solar urticaria. Clin Exp Dermatol 1987; 12:455–456.
11. Kojima M, Horiko T, Nakamura Y, Aoki T. Solar urticaria: the relationship of photoallergen and action spectrum. Arch Dermatol 1986; 122:550–555.
12. Horio T, Minami K. Solar urticaria: photoallergen in a patient's serum. Arch Dermatol 1977; 113:157–160.
13. Horio T. Photoallergic urticaria induced by visible light: additional cases and further studies. Arch Dermatol 1978; 114:1761–1764.
14. Horio T. Chlorpromazine photoallergy: coexistence of immediate and delayed type. Arch Dermatol 1975; 111:1469–1471.
15. Lovell CR, Cronin E, Rhodes EL. Photocontact urticaria from chlorpromazine. Contact Dermatitis 1986; 14:290–291.
16. Sams WM Jr. Solar urticaria: studies of the active serum factor. J Allergy Clin Immunol 1970; 45:295–301.
17. Hawk JLM, Eady RAJ, Challoner AVJ, Kobza-Black A, Keahey TM, Greaves MW. Elevated blood histamine levels and mast cell degranulation in solar urticaria. Br J Clin Pharmacol 1980; 9:183–186.
18. Soter NA, Wasserman SI, Pathak MA, Parrish JA, Austen KF. Solar urticaria: release of mast cell mediators into the circulation after experimental challenge (abstract). J Invest Dermatol 1979; 72:282.
19. Sams WM Jr, Epstein JH, Winkelmann RK. Solar urticaria: investigation of pathogenetic mechanisms. Arch Dermatol 1969; 99:390–397.
20. Hasei K, Ichihashi M. Solar urticaria: determinations of action and inhibition spectra. Arch Dermatol 1982; 188:346–350.
21. Horio T, Yoshioka A, Okamoto H. Production and inhibition of solar urticaria by visible light exposure. J Am Acad Dermatol 1984; 11:1094–1099.
22. Torinuki W, Kumai N, Miura T. Solar urticaria inhibited by visible light. Dermatologica 1983; 166:151–155.
23. Leenutaphong V, von Kries R, Hölzle E, Plewig G. Solar urticaria induced by visible light and inhibited by UVA. Photodermatology 1988; 5:170–174.
24. Horio T, Fujigaki K. Augmentation spectrum in solar urticaria. J Am Acad Dermatol 1988; 18:1189–1193.
25. Michell P, Hawk JLM, Shafrir A, Corbett MF, Magnus IA. Assessing the treatment of solar urticaria: the dose-response as a quantifying approach. Dermatologica 1980; 160:198–207.

26. Diffey BL, Farr PM. Treatment of solar urticaria with terfenadine. Photodermatology 1988; 5:25–29.
27. Willis I, Epstein JH. Solar- vs heat-induced urticaria. Arch Dermatol 1974; 110:389–392.
28. Duschet P, Leyen P, Schwarz T, Hocker P, Greiter J, Gschnait F. Solar urticaria: treatment by plasmapheresis. J Am Acad Dermatol 1986; 15:712–713.
29. Leenutaphong V, Hölzle E, Plewig G, Grabensee B, Kutkuhn B. Plasmapheresis in solar urticaria. Photodermatology 1987; 4:308–309.
30. Bernhard JD, Jaenicke KF, Momtaz K, Parrish JA. Ultraviolet A phototherapy in the prophylaxis of solar urticaria. J Am Acad Dermatol 1984; 10:29–33.
31. Parrish JA, Jaenicke KF, Morison WL, Momtaz K, Shea C. Solar urticaria: treatment with PUVA and mediator inhibitors. Br J Dermatol 1982; 106:575–580.
32. Plewig G, Hölzle E, Lehmann P. Phototherapy for photodermatoses. Curr Probl Dermatol 1986; 15:254–264.
33. Danno K, Toda K, Horio T. The effect of 8-methoxypsoralen plus long-wave ultraviolet (PUVA) radiation on mast cells: PUVA suppresses degranulation of mouse skin mast cells induced by compound 48/80 or concanavalin A. J Invest Dermatol 1985; 85:110–114.
34. Toda K, Danno K, Tachibana T, Horio T. Effect of 8-methoxypsoralen plus long-wave ultraviolet (PUVA) radiation on mast cells. II. In vitro PUVA inhibits degranulation of rat peritoneal mast cells induced by compound 48/80. J Invest Dermatol 1986; 87:113–116.
35. Harber LC, Bickers DR. Photosensitivity Diseases. Philadelphia: BC Decker, 1989.

12

# Chronic Actinic Dermatitis

**John Hawk**
*St. John's Institute of Dermatology, St. Thomas's Hospital, London, England*

**Wai-Kwong Cheong**
*National Skin Centre, Singapore*

## I. INTRODUCTION

Chronic actinic dermatitis is an uncommon, sometimes disabling eczematous photosensitivity eruption affecting predominantly elderly men. It arises spontaneously or in association with other pre-existing photosensitive or eczematous disorders and is induced by ultraviolet (UV) and occasionally also visible irradiation. Sun-exposed and often covered skin is affected by eczematous or on occasion pseudolymphomatous changes. Concomitant contact or, less commonly, photocontact sensitivity to airborne or other allergens may be present. The disease generally persists throughout the year, although exacerbations tend to occur during summer. Most patients retain the disorder indefinitely in the absence of treatment, although remission may rarely occur.

## II. TERMINOLOGY

The term chronic actinic dermatitis was originally proposed by Hawk and Magnus (1) to embrace the conditions photosensitive eczema and actinic reticuloid. Actinic reticuloid, described by Ive et al. (2) in 1969, is a chronic photodermatosis affecting predominantly elderly men and characterized by infiltrated erythematous plaques of sun-exposed skin on an eczematous background. The histopathologic features in all cases are suggestive of cutaneous T-cell lymphoma, while there are abnormal morphologic responses to lower than normal minimal erythema dosages of monochromatic or broad-spectrum ul-

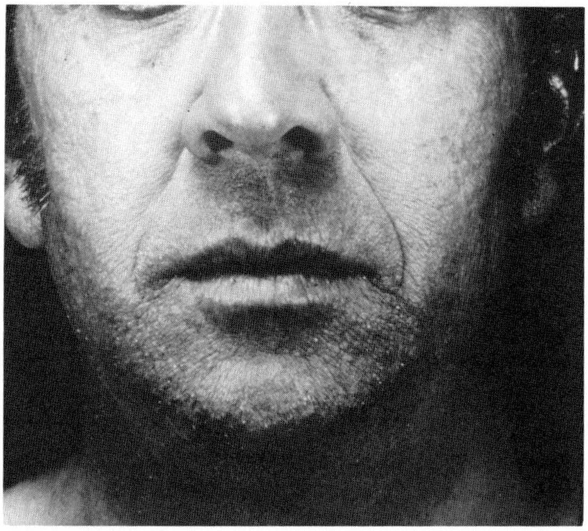

**Figure 1** Patient with photosensitive eczema variant of chronic actinic dermatitis. (From Hawk JLM, Cheong WK. Eczematous photodermatoses. In: Marks R, ed. Eczema. Martin Dunitz, 1991.)

traviolet B (UVB), ultraviolet A (UVA), and sometimes also visible irradiation. Results of limited photopatch tests were negative in the original report. Ramsay and Kobza Black (3) in 1973 described photosensitive eczema, a similar but milder skin disorder occurring in men with longstanding eczema of differing causes which slowly progressed to widespread eczematous photosensitivity (Fig. 1). Phototests showed sensitivity to UVB and occasionally also short-wavelength UVA. Photopatch tests were negative. Transitions, however, were shown to occur between these two conditions (4) and there were patients with the clinical and histological features of photosensitive eczema and the photobiological abnormalities of actinic reticuloid, and vice versa (1). These findings led Hawk and Magnus to propose that the term chronic actinic dermatitis be used to include photosensitive eczema, actinic reticuloid, and the continuum of intermediate states (1). The concept that the two conditions may not be separate entities was also independently suggested by Frain-Bell et al., who coined the term photosensitivity dermatitis and actinic reticuloid syndrome (5). The postulate has since been reinforced by more recent studies showing that the action spectrum of photosensitivity does not correlate with the severity of the lymphocytic infiltrate (6).

Persistent light reactivity or reaction refers to persistence of eczematous photosensitivity following acute photocontact dermatitis and despite avoidance of the photocontact allergen, along with spread of the dermatitis to affect all sun-exposed areas and sometimes also covered areas, not just those previously in contact with the relevant allergen. In addition, the nonexposed skin also becomes sensitive to UV exposure, responding with an eczematous eruption similar to that on exposed areas. Wilkinson (7) was the first to report this phenomenon; he found that a proportion of patients with photocontact dermatitis to tetrachlorosalicylanilide continued to display photosensitivity despite avoidance of the photoallergen, calling such patients persistent light reactors. Subsequently, various other chemical agents including bithionol (8), quinoxaline dioxide (9), and musk ambrette (10) have also been reported as photoallergens able to induce this condition. The initial photocontact dermatitis in these patients is usually UVA induced, but as persistent light reactivity develops, the action spectrum alters to include UVB wavelengths as well and the syndrome becomes clinically indistinguishable from chronic actinic dermatitis.

More recently, reports of persistence of eczematous photosensitivity following exposure to systemic photosensitizers such as thiazides or quinidine, with features similar to persistent light reactivity except for negative photopatch test results, suggest that the term persistent light reactivity might now be broadened also to include these events (11,12).

It is therefore apparent that marked similarities exist between persistent light reactivity and chronic actinic dermatitis as originally described. Furthermore, progression from photoallergic contact dermatitis to persistent light reactivity and to actinic reticuloid has also been reported (12,13). It therefore seems appropriate to include persistent light reactivity within the definition of chronic actinic dermatitis (12,14), thereby also considerably simplifying the terminology associated with this form of eczematous photosensitivity.

## III. PATHOGENESIS

The exact mechanism underlying chronic actinic dermatitis remains unknown. Any hypothesis must explain the abnormal sensitivity found on light testing of covered areas not previously in contact with relevant chemical agents. Baer and Kopf proposed that the carrier protein originally bound to photoallergen may become changed during the process of photocontact dermatitis into a neoantigen, thus subsequently stimulating the immune system (15). This idea was later supported by Kochevar and Harber, who demonstrated photo-oxidation of the histidine moiety of albumin in the presence of tetrachlorosalicylanilide, which binds to albumin; the carrier protein was thereby modified to weak antigen (16). Subsequent absorption of UV radiation by the carrier protein alone in the absence of the initiating exogenous allergen might then produce sufficient oxidized protein to bring about a delayed hypersensitivity-type immune response at all skin sites. In cases not preceded by photocontact dermatitis or systemic drug photosensitivity, a similar process of neoantigen formation might presumably occur at least in some patients, either in the presence of endogenous photosensitizer alone, for example kynurenic acid, which is perhaps increased in concentraion in chronic actinic dermatitis skin (17), or in the absence of any photosensitizer. In many other patients, however, contact allergy to various ubiquitous chemical agents, namely oleoresins from Compositae plants, fragrance materials, lichens, chromates, and rubber chemicals, has been noted (18–21). In one report, for example, 52 of 69 patients with chronic actinic dermatitis reacted to one or more Compositae oleoresin extracts on patch testing (18). Although there is no firm evidence of a photocontact reaction to Compositae oleoresins, it has been demonstrated that these compounds (22) as well as a number of fragrance materials (21) have phototoxic potential in vitro including the ability to photooxidize histidine. It is therefore conceivable that relapses and persistence of chronic actinic dematitis might be induced or assisted in these patients by continual exposure to such photoactive chemical agents. Moreover, avoidance of these substances combined with other appropriate treatment has been shown to help in gradual resolution of the disorder (23).

A further causal factor in chronic actinic dermatitis may be that any preceding allergic or photoallergic contact dermatitis, both delayed-type hypersensitivity responses, may conceivably predispose to the apparently similar chronic actinic dermatitis reaction purely by enhancing cutaneous immune function sufficiently to recognize endogenous photoallergen at exposed sites. Similarly, chronic actinic dermatitis also has been reported to develop in long-standing endogenous eczema (3) and, rarely, long-standing polymorphous light eruption (24), both also postulated to be delayed-type hypersensitivity responses.

The histologic features in the pseudolymphomatous lesions of chronic actinic dermatitis of a dense epidermotropic infiltrate of lymphocytes and macrophages containing giant cells and hyperchromatic nuclei (2,6), and the immunohistochemical finding of CD8+ cytotoxic/suppressor T-cell predominance over CD4+ helper/inducer T-cells further support the concept of a delayed-type hypersensitivity response as described above. They resemble those in persistent forms of allergic contact dermatitis (25–27), a known reaction of this type.

Chronic actinic dermatitis affects mainly elderly men of all races (12,28), especially outdoor workers and leisure enthusiasts (29). This may be because they are exposed at length to both sunlight and environmental allergens, thus perhaps simultaneously affecting both the presumed endogenous photosensitizer and cutaneous immunologic activity. Furthermore, the aged skin of most affected persons also may have compromised barrier function and allergen removal capability (30), perhaps further predisposing to the disease.

A separate theory of causation proposing an abnormal sensitivity of fibroblasts in actinic reticuloid patients to UVA irradiation (31) has not been supported by other workers (32).

## IV. CLINICAL FEATURES

More than 90% of patients are men, although women may also be affected, and many follow outdoor pursuits. The age of onset is usually in the fifth decade or later, but most commonly in the seventh or eighth decades. Familial incidence is not a feature. There is frequently a history of preceding long-standing, usually endogenous, often nonspecific eczema. The eruption gradually becomes more persistent and localized to light-exposed areas as chronic actinic dermatitis supervenes. The clinical picture is then most frequently one of chronic or subacute eczematous changes, often with lichenification, affecting predominantly light-exposed areas, namely the face (Fig. 2, see color plate), back and sides of the neck, upper chest, and the backs of the hands, with cut-off at the lines of clothing. The upper eyelids, submental area, skin behind the earlobes, skin creases, skin folds, and fingerwebs are characteristically spared, although such sparing is often not apparent if the eczema is patchy or if airborne contact dermatitis is also present. On the other hand, covered areas are often affected, particularly in more severely afflicted subjects, either with patchy eczema or confluent erythroderma. In other patients, erythematous, shiny infiltrated papules or confluent plaques on a background of erythematous, eczematous, or normal skin may develop (Fig. 3, see color plate). The plaques occasionally resemble cutaneous T-cell lymphoma, particularly if arising on an area of normal skin. Palmar and plantar eczema may also be present in some cases (Fig. 4, see color plate), while eyebrows, eyelashes, and even scalp hair may be absent or irregularly stubbly, probably because of rubbing or scratching. Well-demarcated areas of moderate or marked, often widespread, geographically irregular hyper- or hypopigmentation may also occur, not uncommonly persisting after clearing of the disease and occasionally forming white patches resembling vitiligo. The condition is generally worse in summer, and particularly after solar or artificial ultraviolet exposure, but is frequently present to some degree throughout the year. Once established, it usually persists indefinitely, although gradual resolution may rarely occur over months to years (4). Lymphoreticular malignant conditions have been reported to occur in a few patients with chronic actinic dermatitis (33–35), but DNA flow cytometric studies of lymphocytes in the infiltrate have shown no aneuploidy (36). There is also a predominance of CD8+

(6) rather than CD4+ (37,38) lymphocytes in the cutaneous infiltrate, suggesting that the condition is unlikely to predispose to such malignant conditions.

## V. HISTOPATHOLOGIC FEATURES

The histologic features (29) of chronic actinic dermatitis are generally those of epidermal spongiosis with irregular acanthosis and dermal perivascular lymphocytic infiltration (Fig. 5). In pseudolymphomatous chronic actinic dermatitis, however, this infiltration is dense and deep and often contains atypical lymphocytes with large, hyperchromatic, convoluted nuclei or mitotic figures, along with macrophages and occasional eosinophils and plasma cells. Marked exocytosis of lymphocytes into the epidermis, often with formation of scattered spongiotic Pautrier-like microabscesses, is frequent in such patients, sometimes simulating the features of cutaneous T-cell lymphoma. The lymphocytes are of mixed types, being either CD4+ or CD8+ with a tendency to CD8+ predominance in the pseudolymphomatous lesions (6).

## VI. DIAGNOSIS

The diagnosis of chronic actinic dermatitis is suggested by the clinical findings, supported if necessary by the histologic features, and confirmed by broad-

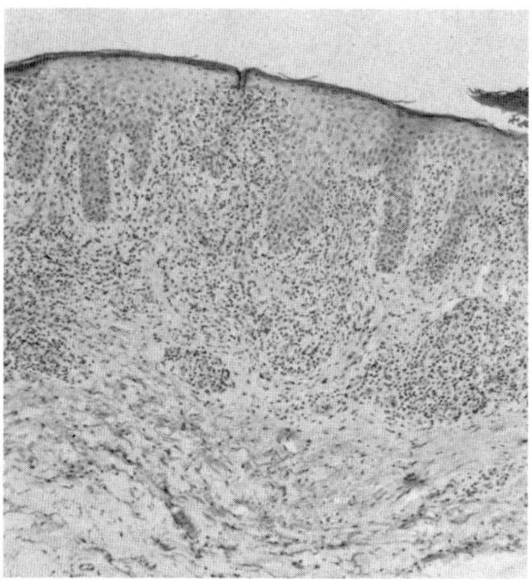

**Figure 5** Histopathologic appearance of chronic actinic dermatitis showing a dense, deep lymphocytic infiltrate and exocytosis of cells into the epidermis (hematoxylin and eosin).

band or monochromatic irradiation testing of normal skin. This last investigation always leads to abnormal erythematous or eczematous responses, usually at dosages lower than the minimal erythema dose and is an essential investigation in the diagnosis of chronic actinic dermatitis. Very rarely, however, onset of this abnormality appears to be delayed by a few weeks to months early in the course of the disease and irradiation tests should be repeated if clinical suspicion of chronic actinic dermatitis persists. The action spectrum for induction of chronic actinic dermatitis includes the UVB wavelengths in all patients, the UVA as well in many and also the visible in some.

Patch and photopatch tests should also be undertaken in all patients with chronic actinic dermatitis, since contact and photocontact dermatitis to exacerbating allergens, particularly Compositae plant oleoresins, colophony, fragrances such as musk ambrette, and sunscreen constituents not uncommonly coexist.

Antinuclear antibody, anti-SSA(Ro) and anti-SSB(La) titers, and blood, urine, and fecal porphyrin concentrations are normal.

Other eczematous light-sensitive conditions are usually readily distinguishable from chronic actinic dermatitis. Photocontact dermatitis affects only exposed sites also in contact with allergen, and phototests of normal skin in the absence of allergen are always normal, while phototests with allergen present generally demonstrate abnormalities only in the UVA range. Oral drug photosensitivity rarely resembles chronic actinic dermatitis, and irradiation tests are generally normal or positive only to UVA wavelengths. Light-exacerbated atopic eczema and seborrheic dermatitis may resemble chronic actinic dermatitis, but photosensitivity is generally less severe in these conditions and, more importantly, phototests are generally normal. Cutaneous T-cell lymphoma rarely also resembles chronic actinic dermatitis, especially the pseudolymphomatous variant, but patients with the former disorder only infrequently suffer clinical photosensitivity, even then minimal in nature, and phototests are only slightly abnormal in such patients, principally to UVA wavelengths alone (39). Furthermore, the dermal infiltrate in chronic actinic dermatitis contains greater numbers of CD8+ cytotoxic/suppressor T cells (6,24) than in cutaneous T-cell lymphoma, in which CD4+ helper T cells characteristically predominate (37,38). Erythrodermic chronic actinic dermatitis must be differentiated from other erythrodermas, generally by phototests after the eruption clears by confinement of the patient to a dark room. If clearance does not occur within days, the erythroderma is unlikely to be due to chronic actinic dermatitis. Large numbers of circulating Sézary cells, up to 20% or more of total circulating lymphocytes, rarely may be present in erythrodermic chronic actinic dermatitis (40) and suggest the malignant Sézary syndrome, but the CD4+/CD8+ T lymphocyte ratio in chronic actinic dermatitis is much lower (41).

## VII. TREATMENT

Treatment (Table 1) of chronic actinic dermatitis requires careful restriction of exposure to UV and if necessary visible radiation from sunlight and fluorescent lamps, but not from tungsten lamps and television screens, while exacerbating contact allergens and photoallergens must also be avoided. Appropriate clothing cover and nonirritating, broad-spectrum, high-protection sunscreens of low allergenic potential such as the micronized titanium dioxide preparations are generally necessary. Sunscreens containing chemical absorbers also can be used, but allergic contact or photocontact dermatitis may rarely occur as a result, thus causing patients to worsen in spite of treatment (42,43). Emollients, topical corticosteroids, and sometimes also systemic corticosteroids may be required for symptomatic relief, while hydroxychloroquine also has occasionally seemed effective (12); the use of UV protective films (44) applied to automobile and house windows may be helpful in addition when response is slow or incomplete. With this approach, some patients gradually and significantly improve, while a few recover completely.

Patients who do not respond to these measures may sometimes improve with psoralens plus UVA (PUVA) photochemotherapy (45,46). Hindson et al. (46) successfully used this treatment twice weekly over the whole body or, in some cases, just the exposed areas, starting with a UVA dosage of 0.25 $J/cm^2$ and increasing by 0.25 to 1 $J/cm^2$ at each subsequent treatment to a maximum of 10

**Table 1** Principles of Management of Chronic Actinic Dermatitis

Prophylactic measures
- Avoidance of ultraviolet and, where necessary, visible radiation exposure from sun and artificial sources
- Protection of skin with adequate clothing cover and broad-spectrum, nonirritating, high protection factor sunscreens with low allergenic potential, particularly micronized titanium dioxide preparations
- Avoidance of exacerbating contact and photocontact allergens

Therapeutic measures
- Emollients, topical corticosteroids, and, for severe outbreaks, intermittent courses of systemic corticosteroids
- Psoralen plus UVA (PUVA) photochemotherapy or UVB phototherapy, given with care under topical or systemic corticosteroid cover to avoid exacerbation of the disease
- Azathioprine 50–150 mg daily
  - Monitor blood count and liver function monthly
  - Intermittent courses preferred
- Cyclosporin A 150–400 mg daily if azathioprine is ineffective or unsuitable
  - Monitor blood pressure and renal function regularly

$J/cm^2$, this being then given fortnightly as maintenance. During the initial phase of therapy, patients were confined to UV-protected rooms and received applications of betamethasone valerate ointment to all exposed areas immediately after each UVA treatment. Other workers, however, including the present authors, have found it difficult to prevent exacerbating the disease with this treatment. Moreover, long-term maintenance treatment may sometimes be required to prevent relapse. Ultraviolet B phototherapy also has been claimed to be effective in inducing light tolerance (24). It was given three times weekly, starting at 1/10 of the predetermined minimal erythema dose and followed by cautious increments leading to an eventual maintenance dosage of about 200 $mJ/cm^2$ of erythemally effective radiation. However, flares of disease are also likely with this approach and relapse is common once treatment is withdrawn.

In cases of chronic actinic dermatitis that do not respond to the aforementioned forms of treatment oral azathioprine in a dosage of 1.5–2.5 mg/kg daily, or about 50–150 mg, is often effective (46,47). Complete remission or marked improvement has occurred over a few months in more than two-thirds of patients who tolerate the drug; however, gastrointestinal intolerance necessitating discontinuation of treatment occurs in about one in seven of those treated. Other adverse effects are rare, but include bone marrow depression and hepatotoxicity, usually during the first few weeks of treatment, although bone marrow toxicity may also develop later, sometimes insidiously. Close monitoring of blood hemoglobin, white-cell and platelet counts, and liver function is thus necessary throughout treatment. An increased incidence of malignant skin tumors, principally on sun-exposed skin, and of lymphoreticular malignancy has also been reported in patients receiving prolonged azathioprine treatment even without organ transplant (49), but not yet, however, in patients with chronic actinic dermatitis. Nevertheless, treatment should preferably be stopped once remission is achieved or after a maximum trial period of about 6 months, therapy being reinstituted only if relapse occurs. With this approach, some patients go into permanent remission and others need therapy only in the summer. The exact mode of action of azathioprine in chronic actinic dermatitis is unclear, although it may exert an immunosuppressive effect by replacing naturally occurring purine DNA bases in immunocompetent cells with its active intracellular metabolite, 6-thioguanine nucleotide (50,51). Such cells are also abnormally sensitive to UV radiation (52), especially between 320 and 360 nm (53), and sun-exposed skin may thus be more susceptible to the action of azathioprine than covered skin. Also, bone marrow-derived Langerhans cells may be reduced in number through the bone marrow suppressive effects of azathioprine, thereby further affecting the immune response considered important in chronic actinic dermatitis.

Patients in whom azathioprine is ineffective or unsuitable may instead respond over a few weeks to oral cyclosporin A in a dosage of 3–5 mg/kg daily or about 150–400 mg (54,55). However, impairment of renal function and elevation of blood pressure may occur, necessitating careful, regular monitoring. In particular, serum creatinine levels should not be allowed to increase to more than 30% above pretreatment values, although some authorities believe that the glomerular filtration rate or renal blood flow also should be measured regularly throughout treatment. Other possible adverse effects of cyclosporin include gastrointestinal upset, muscle tremor, hypertrichosis, and gingival hypertrophy, all sometimes necessitating reduction in dosage or cessation of treatment. However, relapse tends to occur promptly after drug withdrawal and reduction in dosage for other reasons should therefore always be gradual. The exact mode of action of cyclosporin in chronic actinic dermatitis is not known, but its primary target is known to be the CD4+ T cells, CD8+ T cells being little affected. Thus, its beneficial effect in chronic actinic dermatitis appears somewhat paradoxical, although CD4+ T-cell activity in the pathogenesis of the disease is very likely also important.

Thus chronic actinic dermatitis, a condition once impossible to treat without long-term confinement of patients in the dark, now generally responds satisfactorily to careful therapy, thereby enabling most patients to follow an acceptable lifestyle.

## REFERENCES

1. Hawk JLM, Magnus IA. Chronic actinic dermatitis—an idiopathic photosensitivity syndrome including actinic reticuloid and photosensitive eczema. Br J Dermatol 1979; 101(suppl 17): 24.
2. Ive FA, Magnus IA, Warin RP, Wilson Jones E. "Actinic reticuloid": a chronic dermatosis associated with severe photosensitivity and the histological resemblance to lymphoma. Br J Dermatol 1969; 81: 469–485.
3. Ramsay CA, Kobza Black A. Photosensitive eczema. Trans St Johns Hosp Dermatol Soc. 1973; 59: 152–158.
4. Hawk JLM, Magnus IA. Resolution of actinic reticuloid with transition to photosensitive eczema. J R Soc Med 1978; 71: 608–609.
5. Frain-Bell W, Lakshmipathi T, Rogers J, Willock J. The syndrome of chronic photosensitivity dermatitis and actinic reticuloid. Br J Dermatol 1974; 91: 617–634.
6. Norris PG, Morris JM, Smith NP, Chu AC, Hawk JLM. Chronic actinic dermatitis: an immunohistological and photobiological study. J Am Acad Dermatol 1989; 21: 966–971.
7. Wilkinson DS. Patch test reactions to certain halogenated salicylanilides. Br J Dermatol 1962; 74: 302–306.
8. Jillson OF, Baughman RD. Contact photodermatitis from bithionol. Arch Dermatol 1963; 88: 409–418.
9. Zaynoun ST, Johnson BE, Frain-Bell W. The investigation of quindoxin photosensitivity. Contact Derm 1976; 2: 343–352.
10. Ramsay CA. Transient and persistent photosensitivity due to musk ambrette: clinical and photobiological studies. Br J Dermatol 1984; 111:423–429.

11. Robinson HN, Morison WL, Hood AF. Thiazide diuretic therapy and chronic photosensitivity. Arch Dermatol 1985; 121: 522–524.
12. Lim HW, Buchness MR, Ashinoff R, Soter NA. Chronic actinic dermatitis: study of the spectrum of chronic photosensitivity in 12 patients. Arch Dermatol 1990; 126: 317–323.
13. Horio T. Actinic reticuloid via persistent light reaction from photoallergic contact dermatitis. Arch Dermatol 1982; 118: 339–342.
14. Norris PG, Hawk JLM. Chronic actinic dermatitis: a unifying concept. Arch Dermatol 1990; 126: 376–378.
15. Baer RL, Kopf AW. Editor's comment. In: Baer RL, Kopf AW, eds. Yearbook of Dermatology, 1963–64 Series. Chicago: Year Book Medical Publishers, 1964: 133.
16. Kochevar IE, Harber LC. Photoreactions of 3,3',4',5-tetrachlorosalicylanilide with proteins. J Invest Dermatol 1977; 68: 151–156.
17. Swanbeck G, Wennersten G. Evidence for kynurenic acid as a possible photosensitizer in actinic reticuloid. Acta Derm Venereol (Stockh) 1973; 53:109–113.
18. Frain-Bell W, Hetherington A, Johnson BE. Contact allergic sensitivity to chrysanthemum and the photosensitivity dermatitis and actinic reticuloid syndrome. Br J Dermatol 1979; 101: 491–501.
19. Frain-Bell W, Johnson BE. Contact allergic sensitivity to plants and the photosensitivity dermatitis and actinic reticuloid syndrome. Br J Dermatol 1979; 101: 503–515.
20. Thune PO, Solberg YJ. Photosensitivity and allergy to aromatic lichen acids, Compositae oleoresins and other plant substances. Contact Derm 1980; 6: 64–71.
21. Addo HA, Ferguson J, Johnson BE, Frain-Bell W. The relationship between exposure to fragrance materials and persistent light reaction in the photosensitivity dermatitis and actinic reticuloid syndrome. Br J Dermatol 1982; 107: 261–274.
22. Addo HA, Sharma SC, Ferguson J, Johnson BE, Frain-Bell W. A study of Compositae plant extract reactions in photosensitivity dermatitis. Photodermatology 1985; 2: 68–79.
23. Addo HA, Frain-Bell W. Persistence of allergic contact sensitivity in subjects with photosensitivity dermatitis and actinic reticuloid syndrome. Br J Dermatol 1987; 117:555–559.
24. Toonstra J, Henquet CJM, van Weelden H, van der Putte SCJ, van Vloten WA. Actinic reticuloid: a clinical, photobiologic, histopathologic and follow-up study of 16 patients. J Am Acad Dermatol 1989; 21: 205–214.
25. Ecker RI, Winkelmann RK. Lymphomatoid contact dermatitis. Contact Derm 1981; 7: 84–93.
26. Orbaneja JG, Diez LI, Lozano JLS, Salazar LC. Lymphomatoid contact dermatitis: a syndrome produced by epicutaneous hypersensitivity with clinical features and a histopathologic picture similar to that of mycosis fungoides. Contact Derm 1976; 2: 139–143.
27. Kanerva L, Estlander T, Jolanki R. Immunohistochemistry of lymphocytes and Langerhans cells in long lasting allergic patch tests. Acta Derm Venereol (Stockh) 1988; 68: 116–122.
28. Kaidbey KH, Messenger JL. The clinical spectrum of the persistent light reactor. Arch Dermatol 1984; 120: 1441–1448.

29. Ferguson J. Photosensitivity dermatitis and actinic reticuloid syndrome (chronic actinic dermatitis). Semin Dermatol 1990; 9: 47–54.
30. Christopher E, Kligman AM. Percutaneous absorption in aged skin. In: Montagna W, ed. Advances in the Biology of Skin Ageing. Vol 6. Oxford: Pergamon Press, 1965: 163–169.
31. Botcherby PK, Magnus IA, Marimo B, Gianelli F. Actinic reticuloid-an idiopathic photodermatosis with cellular sensitivity to near ultraviolet radiation. Photochem Photobiol 1984; 39: 641–649.
32. Johnson BE, Walker EM, Ferguson J, Frain-Bell W. Cellular sensitivity to UVA in photosensitivity dermatitis/actinic reticuloid (PD/AR). Br J Dermatol 1988; 118: 286.
33. Jenson NE, Sneddon IB. Actinic reticuloid with lymphoma. Br J Dermatol 1970; 82: 287–291.
34. Thomsen K. The development of Hodgkin's disease in a patient with actinic reticuloid. Clin Exp Dermatol 1977; 21: 109–113.
35. Ashinoff R, Buchness MR, Lim HW. Lymphoma in a black patient with actinic reticuloid treated with PUVA: possible etiologic consideration. J Am Acad Dermatol 1989; 21: 1134–1137.
36. Norris PG, Newton JA, Camplejohn RS, Hawk JLM. A flow cytometric study of actinic reticuloid. Clin Exp Dermatol 1989; 14: 128–131.
37. McMillan EM, Wasik, R, Beeman K, Everett MA. In situ immunophenotyping of mycosis fungoides. J Am Acad Dermatol 1982; 6: 888–897.
38. Chu AC, Patterson J, Berger C, Vonderheid E, Edelson R. In situ study of T-cell subpopulations in cutaneous T-cell lymphoma: diagnostic criteria. Cancer 1984; 54: 2414–2422.
39. Volden G, Thune PO. Light sensitivity in mycosis fungoides. Br J Dermatol 1977; 97: 279–284.
40. Neild VS, Hawk JLM, Eady RAJ, Cream JJ. Actinic reticuloid with Sézary cells. Clin Exp Dermatol 1982; 7: 143–146.
41. Chu AC, Robinson D, Hawk JLM, Meacham R, Spittle M, Smith NP. Immunologic differentiation of the Sézary syndrome due to cutaneous T-cell lymphoma and chronic actinic dermatitis. J Invest Dermatol 1986; 86: 134–137.
42. Murphy G, White IR. Photoallergic and allergic contact dermatitis from dibenzoylmethanes. Photodermatology 1986; 3: 140–147.
43. Green C, Catterall M, Hawk JLM. Chronic actinic dermatitis and sunscreen allergy. Clin Exp Dermatol 1991; 16: 70–71.
44. Herd J, Sturrock I, Frain-Bell W. The use of plastic material for the protection of patients with severe photodermatoses. Br J Dermatol 1973; 88: 283–285.
45. Morison WL, White HAD, Gonzalez E, Parrish JA, Fitzpatrick TB. Oral methoxsalen photochemotherapy of uncommon photodermatoses. Acta Derm Venereol (Stockh) 1979; 59: 366–368.
46. Hindson C, Spiro J, Downey A. PUVA therapy of chronic actinic dermatitis. Br J Dermatol 1985; 113: 157–160.
47. Leigh IM, Hawk JLM. Treatment of chronic actinic dermatitis with azathioprine. Br J Dermatol 1984; 110: 691–695.

48. Murphy GM, Maurice PDL, Norris PG, Morris RW, Hawk JLM. Azathioprine in the treatment of chronic actinic dermatitis: a double-blind controlled trial with monitoring of exposure to ultraviolet radiation. Br J Dermatol 1989; 121: 639–646.
49. Kinlen LJ. Immunosuppressive therapy and cancer. Cancer Surveys 1982; 1: 565–583.
50. Tidd DM, Patterson ARP. A biochemical mechanism for the delayed cytotoxic reaction of 6-mercaptopurine. Cancer Res 1974; 34: 738–740.
51. Lennard L, Rees C, Lilleyman JS, Maddocks JL. Childhood leukemia: a relationship between intracellular 6-mercaptopurine metabolites and neutropenia. Br J Clin Pharmacol 1983; 16: 359–364.
52. Kaplan HS, Smith KC, Tomlin P. Radiosensitization of *E coli* by purine and pyrimidine analogues incorporated into deoxyribonucleic acid. Nature 1961; 190: 794–796.
53. Lennard L, Thomas S, Harrington CI, Maddocks JL. Skin cancer in renal transplant recipients is associated with increased concentration of 6-thioguanine nucleotide in red blood cells. Br J Dermatol 1985; 113: 723–729.
54. Norris PG, Camp RDR, Hawk JLM. Actinic reticuloid: response to cyclosporin. J Am Acad Dermatol 1989; 21: 307–309.
55. Duschet P, Schwarz T, Oppolzer G, Gschnait F. Persistent light reaction: successful treatment with cyclosporin A. Acta Derm Venereol (Stockh) 1988; 68: 176–178.

# 13

# Cutaneous Phototoxicity

**Craig A. Elmets**
*Case Western Reserve University, Cleveland, Ohio*

Cutaneous phototoxicity is a process that develops when selected drugs and chemicals, capable of absorbing radiant energy, are present in the skin and are activated to produce an inflammatory response by the appropriate wavelengths of ultraviolet and/or visible radiation. The inflammation that results is the direct consequence of cellular damage caused by the light-activated drug.

Phototoxic reactions are common. With the exception of sunburn, they occur more frequently than any other type of acute photosensitivity disease. Cutaneous phototoxicity encompasses photosensitivity caused by the presence of both endogenous and exogenous photosensitizers within the skin (1). For example, the photosensitivity that develops as a result of the accumulation of porphyrins due to inherited enzymatic defects in the heme biosynthetic pathway is an example of cutaneous phototoxicity caused by endogenous photosensitizers (Chap. 15), whereas that developing in patients taking demethylchlortetracycline or chlorpromazine is an example of cutaneous phototoxicity caused by exogenous photosensitizers. This chapter focuses on examples of phototoxicity caused by exogenous agents.

Exogenous agents capable of causing a phototoxic reaction are, in most instances, systemically administered drugs whose distribution extends to the skin. Phototoxic responses can also be produced by drugs and cosmetics applied directly to cutaneous surfaces. In rare instances, industrial compounds, food additives, and dyes in clothing materials have been found to provoke phototoxic reactions. An unusual form of phototoxicity, termed phytophotodermatitis, occurs in individuals who come in contact with plants capable of synthesizing photosensitizing compounds and who are then exposed to solar radiation.

## I. SPECTRUM OF DRUG-INDUCED PHOTOSENSITIVITY DISEASES

Drugs and other exogenous agents may produce two other types of photosensitivity disease besides phototoxicity (2,3). First, they may cause or exacerbate a disease in which photosensitivity is a prominent manifestation. Examples of this form of photosensitivity include hydrochlorothiazide-induced subacute cutaneous lupus erythematosus (4), isoniazid-induced pellagra, and alcohol- or estrogen-exacerbated porphyria cutanea tarda. In most instances, the clinical manifestations or associated laboratory abnormalities of these diseases are sufficiently different from phototoxic reactions that differentiating the former from the latter is not difficult.

Second, drugs and chemicals may initiate an immunologically mediated form of photosensitivity called photoallergic contact dermatitis (5). Photoallergic reactions (Chap. 14) represent a specific type of allergic contact dermatitis in which ultraviolet radiation serves to convert a potential allergen from an inactive to an active form. In susceptible individuals, the light-activated chemical elicts a cell-mediated immune response with the clinical characteristics of allergic contact dermatitis. The main clinical difference between allergic contact dermatitis and photoallergic contact dermatitis is that in the latter the eczematous process only occurs on sun-exposed areas of skin.

Differentiation of phototoxic from photoallergic reactions is often difficult (2,3,5). This is due not only to similarities in the clinical and histologic features of the two types of photosensitivity but also to the fact that most drugs capable of causing photoallergic reactions are able, at least to some extent, to produce a phototoxic response as well. Despite the similarities, several clinical and histologic features serve to differentiate these two types of photosensitivity (Table 1). In particular, photoallergic reactions produce an acute, subacute, or chronic dermatitis, and, with the exception of early lesions, contain histopathologic features identical to other forms of allergic contact dermatitis. Phototoxic reactions, in situations in which excoriation has not obscured the typical appearance, produce erythema that blanches with pressure (i.e., an exaggerated sunburn response that histopathologically exhibits epidermal cell degeneration and a sparse dermal mononuclear infiltrate). Not infrequently phototoxic reactions resolve with residual hyperpigmentation. Although some patients, especially those with skin types 5 or 6, may develop areas of hyperpigmentation with photoallergy, it is distinctly less common than in phototoxic reactions. The time course of these two types of photosensitivity responses is also different. Phototoxic reactions occur in minutes to hours after exposure to ultraviolet radiation, whereas photoallergic responses require at least 24 hr to develop. Other features that serve to differentiate these two types of cutaneous photosensitivity include the dosage of chemical required for elicitation of photosensitivity (small in photoallergy, large in phototoxicity) and the likelihood

**Table 1** Comparison of Cutaneous Phototoxic and Photoallergic Reactions

| Variables | Phototoxicity | Photoallergy |
|---|---|---|
| Distribution of lesions | Exposed areas of skin only | Exposed areas of skin may "bleedover" to unexposed skin |
| Clinical characteristics | Blanching erythema (i.e., exaggerated sunburn reaction) | Acute, subacute, or chronic dermatitis |
| Changes in pigmentation | Frequent, once the acute reaction fades | Unusual |
| Histopathologic findings | Epidermal cell degeneration; dermal edema and vasodilatation; sparse dermal mononuclear infiltrate | Epidermal spongiosis and exocytosis of mononuclear cells; dermal mononuclear cell infiltrate |
| Onset after exposure to ultraviolet radiation | Minutes to hours | At least 24 hr |
| Dosage of drug required for photosensitivity | Large | Small |
| Incidence of photosensitivity | High | Low |

that an individual will develop such a reaction following exposure to sufficient dosages of chemical and light (low in photoallergy, high in phototoxicity).

## II. MECHANISMS OF PHOTOTOXICITY

The mechanism by which drugs and chemicals produce cutaneous photosensitivity is complex and has not been fully elucidated (Table 2). However, it is known that a diverse group of chemicals (Table 3) can cause such a reaction (6). In general, these compounds have a low molecular weight and a resonating, often tricyclic, structure.

Compounds that produce cutaneous phototoxic responses must efficiently absorb radiant energy (2,3,7). The specific wavelengths of light absorbed by a given phototoxic chemical depend on the physicochemical characteristics of the phototoxic agent. In most instances, the wavelengths lie within the ultraviolet A (UVA) range (8). However, a few photosensitizing compounds, such as sulfanilamide, absorb energy within the ultraviolet B (UVB) range, while others, such as the porphyrins, absorb energy from the visible spectrum.

Absorption of radiant energy by a photosensitizing compound transiently elevates electrons within the molecule from a stable ground state to an unstable

**Table 2** Pathogenetic Mechanisms of Drug- and Chemical-Induced Phototoxic Reactions

| | |
|---|---|
| Photosensitizing chemical | |
| ↓ | Ultraviolet or visible radiation |
| Elevation of molecules from a stable ground state to an unstable excited state | |
| ↓ | |
| Production of reactive oxygen intermediates | |
| ↓ | |
| Damage to subcellular organelles | |
| ↓ | |
| Tissue damage and inflammation | |

excited state (3,6). In the excited state, molecules contain a higher amount of energy, which is then released as electrons rapidly return to the more stable ground state. In most cutaneous phototoxic reactions, a photodynamic process occurs in which the released energy is transferred to molecular oxygen, leading to the generation of reactive oxygen intermediates, such as superoxide anion,

**Table 3** Phototoxic Drugs and Chemicals

| Drug or Chemical | Action Spectrum |
|---|---|
| Psoralens | UVA |
| Coal tar and tar pitch | UVA |
| Tetracyclines | UVA |
| Amiodarone | UVA |
| Chlorpromazine and other phenothiazines | UVA |
| Piroxicam | UVA |
| Thiazide diuretics | UVA |
| Nalidixic acid | UVA |
| Furosemide | Unknown |
| Quinidine | UVA |
| Quinine | UVA |
| Sulfanilamide | UVB |
| Hematoporphyrin derivative | Visible |
| Tolbutamide and other sulfonylurea antidiabetic agents | Unknown |
| Vinblastine | UVB |
| Dacarbazine | Unknown |
| 5-Fluorouracil | Unknown |

singlet oxygen, hydroxyl radical, and hydrogen peroxide (1,3,9). This occurs either by direct photosensitizer-mediated transfer of energy to oxygen or by activation of cellular enzymes responsible for the production of reactive oxygen intermediates (10). Reactive oxygen intermediates generated through this process are then capable of damaging subcellular organelles, which in turn results in tissue injury and inflammation. An exception to this process is the photosensitivity caused by psoralens, which is largely not photodynamic in nature (11,12). The pathogenesis of psoralen photosensitivity is discussed in more detail later in this chapter.

The specific organelle that is the target for damage during the course of a phototoxic reaction varies depending on the subcellular localization of the photosensitizing compound. The subcellular localization of a given photosensitizer depends on its physicochemical characteristics, particularly its lipid solubility. Hydrophilic substances lead to destruction of the cell membrane, whereas more hydrophobic substances diffuse within the cell where they are capable of causing cytoplasmic or nuclear damage (7). It should be noted that although effects on one organelle may predominate, most photosensitizers produce alterations in more than one structure. Photosensitizer-mediated destruction of cell membranes, lysosomes, mitochondria, and nuclear material has been described.

Another important feature in the pathogenesis of phototoxic reactions is the cellular target of the photosensitizing chemical. Keratinocytes, mast cells, and endothelial cells have all been implicated as targets in phototoxic reactions (1). The route of administration of the photosensitizer plays an important role in determining the cell type that is targeted (1). Topically applied chemicals produce the greatest damage to keratinocytes. Orally or parenterally administered phototoxic compounds preferentially cause damage to mast cells or endothelial cells.

Damage to cells results in the release of soluble mediators that eventuate in the inflammatory response. Eicosanoids, proteases, histamine, and complement have all been implicated in the generation of photosensitizer-induced inflammatory responses (1). Cytokines such as interleukin-1, interleukin-6, and tumor necrosis factor-alpha, which have been detected in the UVB-induced erythema response, may be involved in phototoxicity caused by drugs and chemicals; however, experimental evidence supporting this concept is lacking for most agents.

## III. CLINICAL AND HISTOLOGIC MANIFESTATIONS OF PHOTOTOXICITY

### A. Acute Phototoxic Erythema

The clinical manifestations of drug-induced phototoxicity are varied (13). Like most other photosensitivity diseases they are confined to exposed areas of skin,

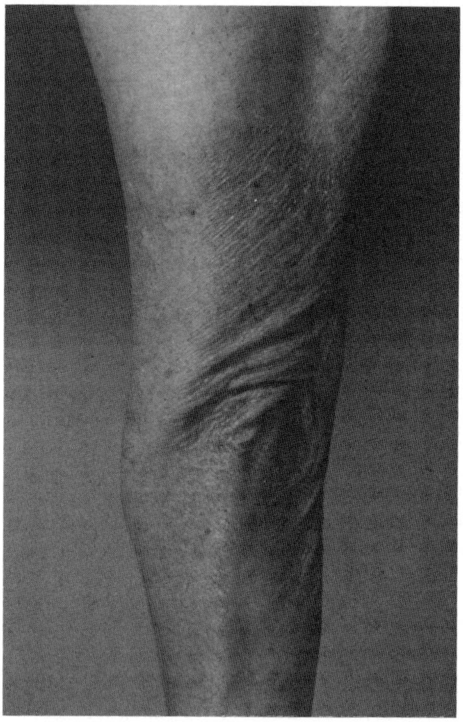

**Figure 1** Phototoxic erythema. (Courtesy of Department of Dermatology, New York University School of Medicine).

which include the forehead, cheeks, nose, chin, ears, V of the neck, and dorsal hands. The scalp, periorbital, postauricular, and submental areas are spared. In the past, the most likely source of ultraviolet or visible light responsible for initiating phototoxic reactions was the sun. However, the popularity of suntanning parlors that use high-output UVA emission sources has made them an increasingly common cause of phototoxic reactions. In most instances, patients develop a tender erythema and edema that has the clinical appearance of an exaggerated sunburn reaction (Fig. 1). In severe cases, blisters may develop. Onset of the reaction occurs minutes to hours after light exposure. Symptoms include pain, tenderness, burning, or prickling. In some cases, the initial discomfort on exposure to light is so severe that erythema does not develop. This situation results because the pain and burning sensation begin almost immediately and are so intolerable that such individuals do not receive sufficient dosages of ultraviolet or visible radiation to produce erythema. This clinical symptom complex is particularly common in individuals exposed to the photosensitizing polyaromatic hydrocarbons in coal tar and pitch and is known as "tar smarts." It also occurs in response to protoporphyrin in patients with erythropoietic protoporphyria (8). In those situations in which the phototoxic reaction does cause erythema, it and the accompanying tenderness resolve within days and often are accompanied by desquamation and hyperpigmentation. Hyperpigmentation is occasionally the most prominent feature of a phototoxic response. In this situation, it is assumed either that the phototoxic chemical produced an erythema response so mild that it was not detected by the patient or that the phototoxic chemical was a direct stimulus for melanogenesis. The hyperpigmentation may persist for weeks to months after initial exposure and may become a cosmetic problem for some patients.

When the skin has repeatedly undergone phototoxic injury, the physical findings may be different. In this situation, pruritus is a more prominent feature, and the skin has a lichenified appearance with erythema and scale.

Phototoxic reactions may manifest in the nails as onycholysis, a finding that may occur in association with, or independent of, other cutaneous manifestations of phototoxicity. The tetracyclines (14) and benoxaprofen (8,13) are particularly common causes of photo-onycholysis.

The histologic appearance of drug- and chemical-induced phototoxic reactions varies depending on the location in the skin where the photosensitizing chemical is concentrated (1). When the photosensitizer has been applied topically, and thus the predominant site of photosensitizer deposition is in the epidermis, necrotic keratinocytes are found, combined with an inflammatory infiltrate in the dermis. The inflammatory infiltrate is initially comprised of neutrophils followed later by an infiltrate of mononuclear cells. Systemic administration of exogenous photosensitizers results in the accumulation of photosensitizing chemicals within the dermis. The histopathologic appearance in this situation is that of a dermal mononuclear infiltrate with little epidermal change.

## B. Uncommon Clinical Manifestations of Drug-Induced Phototoxicity

Although an exaggerated sunburn response is the most common cutaneous manifestation of phototoxicity, other less frequent responses include a blistering eruption similar to that seen in porphyria cutanea tarda (called pseudoporphyria), a photodermatitis resembling lichen planus, and a photodistributed abnormality in skin pigmentation.

### *1. Drug-Induced Pseudoporphyria*

A rare complication of nalidixic acid (15,16), furosemide (17), tetracycline (18), naproxen (19,20), and amiodarone (21) therapy is the development of a photosensitive bullous eruption whose clinical and histologic features bear a striking resemblance to porphyria cutanea tarda. Patients develop blisters and have increased fragility of the skin after exposure to small amounts of sunlight. On histologic appearance, there is a subepidermal blister with a sparse inflammatory infiltrate comprised predominantly of lymphocytes. In some cases, IgM, IgG, and C3 and periodic acid-Schiff-positive, diastase-resistant material have been observed at the basement membrane and around blood vessel walls. Although the clinical and histologic features resemble porphyria cutanea tarda, variegate porphyria, and hereditary coproporphyria, abnormalities in porphyrin metabolism have not been detected. This type of photosensitivity response resolves when use of the implicated drug has been discontinued; however, resolution may take weeks to months.

### *2. Lichen Planus-Like Photosensitivity*

A number of orally administered drugs including demethylchlortetracycline (14,22,23), hydrochlorothiazide (24), quinine (25), quinidine (26,27), chloroquine (13), and hydroxychloroquine (13) can cause a photodistributed lichenoid eruption. This type of reaction is seen more commonly in areas of the world that receive large amounts of sunlight (28). Clinically, the eruption may range from a violaceous erythema with scale to one in which coalescent violaceous papules with Wickham's striae predominate; in all cases, its distribution is largely restricted to exposed areas of skin. Oral involvement has not been described. This feature, while not absolute, has been used to differentiate drug-induced lichen planus-like photosensitivity from idiopathic lichen planus and photoaggravated lichen planus (13). The histopathologic features of lichen planus-like phototoxic reactions are identical to those of idiopathic lichen planus. A lymphocytic infiltrate hugs the epidermis, and hypergranulosis and basal layer degeneration are present.

An impediment to defining the pathogenetic features of this peculiar type of phototoxic response, and to establishing the diagnosis formally in questionable cases, has been the inability to reproduce the skin disease by phototesting. The reasons for this are unclear. Like other types of drug-induced phototoxicity, the lichen planus eruption eventually resolves upon withdrawal of the drug producing it.

### *3. Phototoxic Hyperpigmentation*

For some drugs, the major manifestation of the phototoxic response is the development of hyperpigmentation in sun-exposed areas. Depending on the drug initiating the response, skin discoloration may be caused either by increased amounts of melanin in the skin or by accumulation of the drug or one of its photoproducts. When melanin is responsible for the increased pigmentation, its deposition usually occurs as a consequence of the inflammatory response that precedes it. Almost every drug capable of causing phototoxic erythema has been implicated in producing this type of pigmentary response. Some phototoxic agents also can directly stimulate melanocytes to produce melanin. Phototoxic hyperpigmentation is a common finding in psoralen-induced phytophotodermatitis.

Discoloration of the skin produced by a drug or its photoproduct is particularly common following chlorpromazine and amiodarone (29,30) therapy and in argyria. In individuals taking chlorpromazine on a chronic basis, repeated exposure to natural or artificial light containing UVA radiation produces a purplish–gray discoloration of the skin. Ultrastructural studies have shown that the pigmentary abnormality is caused by a combination of increased amounts of melanin within the dermis, a photoproduct of the chlorpromazine, and a melanin–photoproduct complex (13).

Amiodarone causes a brown-to-blue to slate-gray coloration in sun-exposed areas in 1–10% of patients in whom the drug is administered for prolonged periods. Electron microscopic studies have shown that the pigmentary abnormality is associated with lysosomal inclusions comprised of a phospholipid–amiodarone (or a metabolite) conjugate (30). Although these conjugates are found both in sun-exposed and sun-protected areas, they are much more common in sun-exposed skin, suggesting that ultraviolet radiation acts to increase the rate of conjugate formation (29). In most, but not all, cases, the hyperpigmentation produced by amiodarone resolves after discontinuation of the drug (29).

Argyria is caused by deposition of silver within tissues. It was more common when various forms of silver were used as medicinal agents but is still occasionally seen after occupational exposure (silver reclamation workers, silversmiths, silver mining, silver nitrate manufacturing) (31). The blue-to-slate gray apearance of the sun-exposed skin and lunulae of the nails is the most obvious clinical manifestation. Several studies have demonstrated that equivalent amounts of silver are deposited in both sun-exposed and non-sun-exposed skin sites. However, upon exposure to light, silver compounds undergo a photochemical conversion similar to that occurring during exposure of photographic film, thus accentuating the pigmentary abnormalities in sun-exposed sites (32). Silver also has a stimulatory effect on melanocytes, which is an additional mechanism by which hyperpigmentation occurs.

## IV. FACTORS THAT INFLUENCE THE PHOTOTOXICITY RESPONSE

A number of factors are known to influence drug- and chemical-induced phototoxicity reactions (1). Hair, melanin and other pigments, and the stratum corneum absorb significant amounts of ultraviolet and visible radiation, and therefore limit the degree of radiant energy impinging on the photosensitizing chemical. Heterogeneity in the metabolic capabilities of the host is another determining factor in phototoxic responses. Metabolites of a number of compounds, including chlorpromazine and benoxaprofen, are more potent at eliciting phototoxicity than are the parent compounds (23). Thus, individual differences in the phototoxic response may reflect variation in the extent to which a compound is transformed into a phototoxic metabolite. The photosensitizing properties of topically applied chemicals also depend on their percutaneous absorption. Systemically administered compounds are influenced by their absorption from the gastrointestinal tract and by their distribution within tissues. For unknown reasons, the heat generated by infrared radiation inhibits the development of phototoxicity and the concomitant or subsequent administration of UVB radiation augments the UVA-induced phototoxic response.

## V. DIAGNOSIS OF DRUG- AND CHEMICAL-INDUCED PHOTOTOXIC REACTIONS

The most important clinical feature necessary for the diagnosis of a drug-induced phototoxic reaction is recognition that the eruption has a photodistribution. While other photosensitivity diseases have a similar distribution pattern, phototoxic reactions are noted for their sharp demarcation between sun-exposed and non-sun-exposed skin. This is in contrast to other photosensitivitty diseases in which there is a "bleedover" into non-sun-exposed areas. History of exposures to photosensitizers is also of benefit in establishing the diagnosis. It should be remembered that drugs capable of producing phototoxicity also are able to cause other types of photosensitivity. For example, chlorpromazine can cause both phototoxic and photoallergic reactions and hydrochlorothiazide also can elicit subacute cutaneous lupus erythematosus and chronic actinic dermatitis.

The diagnosis of drug-induced phototoxic reactions is often facilitated by phototesting. In this procedure, non-sun-exposed skin (buttocks or inner aspect of the forearms) is irradiated with gradually increasing amounts of UVB or UVA radiation to determine the minimal erythema dose (MED) for each wavelength band. In most instances, erythema is elicited by smaller amounts of UVA radiation than would be expected while the MED for UVB is normal, since most forms of drug-induced phototoxicity are caused by ultraviolet energy within the UVA range.

Greater precision in the diagnosis of phototoxic reactions can be achieved by obtaining skin biopsy specimens for histopathologic examination because other photosensitivity diseases, such as chronic actinic dermatitis (see Chap. 12), can also have abnormal results on phototesting. The histopathologic apearance of each of these other diseases is usually distinctive, however. If possible, it is preferable to biopsy the phototest site or a relatively new lesion since secondary changes from rubbing and scratching of more well-developed lesions may obfuscate the histopathologic interpretation.

In situations in which the photosensitivity has resolved and in which it is ethically feasible, it may be possible to challenge the patient with the implicated drug to establish the diagnosis definitively.

## VI. TREATMENT

Of utmost importance in the treatment of drug-induced phototoxic eruptions is removal of the offending agent. However, even after this has been achieved, the photosensitivity may not resolve immediately since many exogenous agents persist in the skin for some time after their use has been discontinued. It is often necessary to provide symptomatic therapy for several weeks until the photosensitivity subsides. Cool compresses, emollients, and topical corticosteroids are used. Systemic corticosteroids are reserved for severe cases.

Individuals suspected of experiencing drug-induced photosensitivity should be counseled to wear protective clothing when outdoors and to avoid sunlight as much as possible. Because most agents capable of eliciting a phototoxic response are activated by UVA and visible radiation, patients with severe photosensitivity should be informed that window glass will provide little protection against the development of photosensitivity. Furthermore, fluorescent lamps emit small amounts of UVA radiation, which may exacerbate photosensitivity reactions. It may therefore be important for some patients with severe photosensitivity to use incandescent, rather than fluorescent, lamps for illumination.

When it is impossible to avoid sun exposure completely, sunscreens should be applied. The choice of a sunscreen depends on the chemical producing the photosensitivity response. In most instances the sunscreen should provide photoprotection within the UVA range. For those chemicals, such as the porphyrins, that produce photosensitivity after exposure to visible light, a physical sunblocking agent should be employed.

## VII. SPECIFIC AGENTS CAPABLE OF CAUSING CUTANEOUS PHOTOTOXICITY

### A. Psoralens

The psoralens are a family of naturally occurring tricyclic hydrocarbon compounds whose structure contains a furan ring coupled either linearly or angularly to benzopyrone (12). Unlike most other photosensitizing compounds, psoralens mediate their phototoxic effects primarily through non-oxygen-dependent mechanisms (11). DNA appears to be the major target for damage, although destruction of other structures has also been described. Initially psoralens interact with DNA by intercalating between DNA base pairs. This is a non-light-dependent process. Subsequent exposure to UVA radiation produces monofunctional adducts and bifunctional cross-links with pyrimidine bases. This results both in inhibition of DNA synthesis and activation of DNA repair processes, which, in a manner that has not been precisely determined, translates into a cutaneous inflammatory reaction.

Psoralen phototoxic reactions are known to occur in three clinical situations. The large majority of cases develop in individuals who are taking oral 8-methoxypsoralen or 4,5,8-trimethylpsoralen as photochemotherapeutic agents for psoriasis or other skin diseases. The reaction typically manifests as erythema that develops 24–36 hr after UVA exposure. In mild to moderate cases, the reaction peaks at 48 –72 hr and gradually declines thereafter. In severe cases, the erythema progresses to blisters with systemic symptoms that include fever, nausea, vomiting, and loss of fluids and electrolytes. In this situation, the time course may be more prolonged. It is important to emphasize that the delayed onset of the psoralen-induced erythema is somewhat unusual for drug-induced photosensitivity. The reason why psoralens differ from other phototoxic compounds is unclear.

A second situation in which psoralen photosensitivity reactions develop is through contact with psoralen-producing plants or fruits. This type of photosensitivity reaction has been termed phytophotodermatitis (33,34). Plants capable of causing such a reaction synthesize psoralen compounds and belong to the Umbelliferae, Leguminosae, and Rutaceae families, which include parsnip, meadow grass, the gas plant, fraxinella, burning bush, dittany, Persian limes, and bergamot oranges. Although celery, which belongs to the Umbelliferae family, produces limited amounts of psoralens, its synthesis is greatly accentuated when infected by the fungus *Sclerotinia sclerotiorum*. Individuals at increased risk for developing phytophotodermatitis include children and adults who inadvertently come in contact with these agents during outdoor recreational activities, gardeners in whom it is an occupational hazard, bartenders who are exposed to Persian limes while mixing drinks, and farm workers, grocers, and produce inspectors who handle diseased celery.

In phytophotodermatitis, the psoralen phototoxic reaction manifests initially as a burning erythema that eventually resolves leaving areas of hyperpigmentation. When the reaction is particularly severe, the erythema may progress to form bullae. In other instances, the erythema is mild and may even go unnoticed. These patients often present with a pigmentary disorder. The hyperpigmented areas may persist for several months. A unique feature of phytophotodermatitis is its distribution. In those individuals who brush up against psoralen-containing grasses and plants, this often takes on a linear distribution (Fig. 2). In celery handlers and in bartenders, the eruption is frequently localized to the hands.

Berloque dermatitis (Fig. 3, see color plate) begins as a photodermatitis in areas to which selected perfumes have been applied and resolves with areas of hyperpigmentation (33). All perfumes capable of causing such a reaction contain extracts from the rind of the bergamot orange. Psoralens are present in the rind of the fruit and in its extracts. The incidence of berloque dematitis has declined

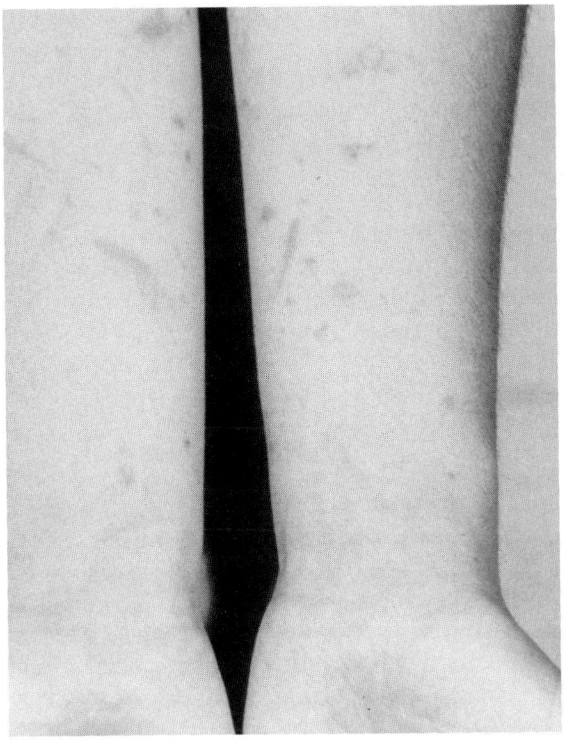

**Figure 2** Phytophotodermatitis.

substantially in recent years due to refinements in the extraction process that allow for greater purification of the aromatic components and to regulations that limit the concentration of bergamot oil in perfumes and colognes.

## B. Crude Coal Tar and Tar Pitch

Crude coal tar is formed through the incomplete combustion of fossil fuels and is widely used by dermatologists in the treatment of psoriasis. It contains large amounts of anthracene, pyrene, fluoranthrene, benzo(a)pyrene, acridine, and phenanthrene, all of which are phototoxic polyaromatic hydrocarbons (35). Tar phototoxicity is also an occupational problem for roofers who are exposed to tar pitch (2). Individuals exposed to tar pitch first describe an intense burning and stinging of the skin that occurs within minutes of sun exposure ("tar smarts") (36). These symptoms alone may be of sufficient severity to preclude further sun exposure, and therefore objective signs of erythema may not develop. After exposure to larger ultraviolet doses, objective signs of phototoxicity occur that follow a characteristic biphasic course (37). An immediate wheal-and-flare reaction develops that fades within an hour. The erythema again increases in intensity 24–48 hr later without further exposure to ultraviolet light and gradually resolves thereafter.

Relatively little is known about the mechanism by which tar-containing compounds produce photosensitivity. It is known that UVA is necessary for photosensitization to occur and that the photosensitivity response is an oxygen-dependent process (38). The subcellular target for tar phototoxic reactions has not been definitively identified, although anthracene, one of the photosensitizing components of tars, is known to be phototoxic to lysosomes (39). It has also been shown that treatment with crude coal tar and ultraviolet light induces the biosynthesis and release of eicosanoids (40).

## C. Nalidixic Acid

Occasionally nalidixic acid causes a photosensitivity syndrome that has many of the clinical features of porphyria cutanea tarda (15,16). Affected individuals develop skin fragility and blisters on the dorsal aspects of the hands, feet, and lower legs following exposure to UVA radiation (16). Biochemical abnormalities of porphyrin metabolism have not been detected in these patients, and the deposition of complement components at the dermal–epidermal junction and periodic acid-Schiff positive material around the dermal microvasculature, have not been observed. After the drug has been discontinued, it may take several months for the blistering process to resolve (15).

## D. Furosemide

Although the chemical structure of furosemide differs substantially from nalidixic acid, it produces an eruption resembling porphyria cutanea tarda that has clinical and histologic features similar to the photosensitivity reaction caused by that agent (17). The furosemide-induced eruption is particularly common in individuals with chronic renal failure who are taking large dosages of the drug. Although the process was at one time considered to be photoinduced, recent studies have questioned whether light is necessary for the disease to occur.

## E. Tetracyclines

Although photosensitivity is a well-recognized complication of therapy with the tetracycline family of antibiotics, the different tetracyclines vary considerably in their phototoxic capacity (14,22,23,41). There is a good correlation between the activity of the different tetracyclines in in vitro assays of phototoxicity and their capacity to cause cutaneous photosensitivity responses in vivo. Demethylchlortetracycline and doxycycline are particularly strong photosensitizers and frequently cause problems with sun sensitivity. Tetracycline and oxytetracycline are relatively weaker phototoxic agents. Minocycline is devoid of phototoxic activity and does not cause photosensitivity.

Cutaneous photosensitivity caused by the tetracyclines occurs as an exaggerated sunburn reaction after exposure to small amounts of sunlight. UVA is responsible for this effect. The onset of the reaction is more rapid than that of a UVB-induced sunburn and may begin with burning and tingling. However, a few patients taking demethylchlortetracycline and doxycycline have been described in whom the onset of delayed erythema occurred 1 week after sun exposure. The erythema resolves within a few days, similar to that of a UVB-induced sunburn.

A photosensitivity reaction resembling porphyria cutanea tarda but without the biochemical abnormalities of porphyria has been described in some patients taking tetracycline hydrochloride and oxytetracycline (18). Photo-onycholysis may occur in patients taking demethylchlortetracycline (14).

The tetracyclines produce their phototoxic effects through processes that result in the generation of singlet oxygen (18). Mitochondria are thought to be the initial target in tetracycline phototoxic reactions (41). Complement components (42) and eicosanoids (40) may also be involved.

## F. Phenothiazines

A phototoxic erythema occurring after exposure to UVA radiation develops in some patients taking phenothiazine tranquilizers (43). One of the phenothiazines, chlorpromazine, is also known to produce a photoallergic contact dermatitis (43). The mechanism by which the phenothiazines produce their phototoxic effects has been investigated using chlorpromazine as a prototype. Two stable metabolites are the moieties that mediate cellular damage (4). The subcellular target of chlorpromazine phototoxic reactions is unknown although damage to DNA, protein, and cell membrane lipids has been demonstrated. Chlorpromazine phototoxicity occurs through both oxidative and nonoxidative means. Complement (45), proteases (46), and histamine (52) have all been implicated in the development of chlorpromazine phototoxicity.

## G. Amiodarone

The antiarrhythmic drug amiodarone causes a phototoxic erythema in up to 50% of patients (29). The eruption begins as a burning and stinging sensation that resolves by 72 hr. Although an individual episode of amiodarone photosensitivity may resolve within days, phototoxic reactions may persist for several months after the drug has been discontinued, which is due to the long half-life of amiodarone (35–105 days) (29). Amiodarone also causes a slate gray hyperpigmentation of the skin that is accentuated in areas of photosensitivity (30,47). Although the reason for this is unclear, increased concentrations of amiodarone-containing complexes have been detected in sun-exposed areas of skin. An amiodarone-induced porphyria cutanea tarda-like eruption (21) has also been described.

The wavelengths of light responsible for amiodarone photosensitivity lie primarily within the UVA spectrum, and the minimal erythema dose response for UVB is normal (29). This is somewhat surprising since in in vitro assays UVB radiation mediates amiodarone phototoxicity more efficiently than does UVA radiation. The disparity between the in vivo and in vitro results may be explained by the fact that amiodarone is deposited primarily within dermal endothelial and perivascular smooth muscle cells that lie beyond the depth of penetration of UVB radiation.

Studies suggest that the cell membrane is the major target of amiodarone phototoxicity and that it occurs through an oxygen-dependent process (29). Both amiodarone and its metabolite desethylamiodarone, which is a more potent photosensitizer than the parent compound, appear to be involved (29).

## H. Nonsteroidal Anti-Inflammatory Agents

Piroxicam, a nonsteroidal anti-inflammatory agent has been associated with cutaneous photosensitivity after exposure to UVA radiation (48–50). Both photoallergic and phototoxic mechanisms have been proposed (43). Photosensitivity develops within a few days of first taking the drug, an observation that would favor a phototoxic mechanism (23). However, the response occurs in only a small percentage of patients and the clinical and histologic features are that of an eczematous dermatitis rather than an exaggerated sunburn response (23,48,49). These features would support a photoallergic mechanism for piroxicam photosensitivity. Moreover, efforts to demonstrate piroxicam phototoxicity in in vivo animal models and in in vitro assays have been unsuccessful (51). However, studies have demonstrated that at least two metabolites of piroxicam are phototoxic (51). It has thus been proposed that piroxicam photosensitivity develops in a small percentage of individuals who are able to convert the drug to one of its phototoxic metabolites.

Phototoxic reactions were also particularly common among patients taking benoxaprofen (23,53–55). After exposure to UVA radiation, individuals experienced an urticarial reaction attributed to photosensitized destruction of mast cell membranes (23). Benoxaprofen photosensitivity is no longer a significant clinical problem because the drug has been removed from the market.

There are also reports of photosensitivity reactions produced by a number of other nonsteroidal anti-inflammatory agents including sulindac, meclofenamate, naproxen, and tiaprofenic acid (48,50).

## I. Thiazide Diuretics

Fewer than 1% of individuals taking thiazide diuretics experience an exaggerated sunburn reaction in response to UVA radiation. In most individuals the photosensitivity resolves shortly after use of the drug has been stopped (24), although a limited number of patients progress to chronic actinic dermatitis. The phototoxic reaction caused by hydrochlorothiazide should be differentiated from subacute lupus erythematosus, which is known to be exacerbated by hydrochlorothiazide (32). The mechanism by which chlorothiazide and hydrochlorothiazide produce photosensitivity has received little attention.

## J. Quinidine

In addition to the lichenoid eruption described earlier in this chapter, quinidine photosensitivity may produce both an eczematous reaction and an exaggerated sunburn reaction in selected individuals (26,27,56). The pathogenesis of quinidine photosensitivity is not understood, although it is known that UVA is responsible for the effect.

## K. Quinine

Phototoxic reactions occasionally develop in patients taking quinine for muscle cramps or malaria and in individuals who drink large quantities of tonic water, which contains quinine (25). Most cases of quinine photosensitivity present either as a burning or pruritic erythema or as a lichen planus-type eruption in sun-exposed areas of skin. UVA wavelengths are responsible for the effect. The reaction resolves once the quinine has been stopped, although it may take several weeks or months.

## L. Other Photosensitizers

Other drugs noted to cause photosensitivity reactions include sulfanilamides, tolbutamide and other sulfonylurea antidiabetic agents, hematoporphyrin derivative, vinblastine, dacarbazine, 5-fluorouracil, and sodium valproate (57).

# REFERENCES


1. Epstein JH. Phototoxicity and photoallergy in man. In: Daynes RA, Krueger G, eds. Experimental and Clinical Photoimmunology. Vol 2. Boca Raton: CRC Press, 1983:107–114.
2. Emmett EA. Phototoxicity from exogenous agents. Photochem Photobiol 1979; 30:429–436.
3. Harber LC, Baer L. Pathogenic mechanisms of drug-induced photosensitivity. J Invest Dermatol 1972; 58:327–342.
4. Reed BR, Huff JC, Jones SK, Orton PW, Lee LA, Norris DA. Subacute cutaneous lupus erythematosus associated with hydrochlorothiazide therapy. Ann Intern Med 1985; 103:49–51.
5. Elmets CA. Drug-induced photoallergy. Dermatol Clin 1985; 4:231–241.
6. Toback AC, Anders JE. Phototoxicity from systemic agents. Dermatol Clin 1986; 4:223–230.

7. Spikes JD. Photosensitized reactions in mammals. In: Daynes RA, Spikes JD, ed. Experimental and Clinical Photoimmunology. Vol 1. Boca Raton: CRC Press, 1983: 69–80.
8. Johnson BE, Ferguson J. Drug and chemical photosensitivity. Semin Dermatol 1990; 9:39–48.
9. Athar M, Mukhtar H, Elmets CA, Zaim MT, Lloyd JR, Bickers DR. In situ evidence for the involvement of superoxide anions in cutaneous porphyrin photosensitization. Biochem Biophys Res Commun 1988; 151:1054–1059.
10. Athar M, Elmets CA, Bickers DR, Mukhtar H. A novel mechanism for the generation of superoxide anions in hematoporphyrin derivative-mediated cutaneous photosensitization: activation of the xanthine oxidase pathway. J Clin Invest 1989; 83:1137–1143.
11. Pathak MA, Kramer DM, Fitzpatrick TB. Photochemistry and photobiology of furocoumarins (psoralens). In: Pathak MA, Harber LC, Seiji M, Kukita A, Fitzpatrick TB, eds. Sunlight and Man. Tokyo: University of Tokyo Press, 1974:335–368.
12. Song P-S, Tapley KJ. Photochemistry and photobiology of psoralens. Photochem Photobiol 1979; 29:1177–1197.
13. Frain-Bell WC. Drug-induced photosensitivity. In: Cutaneous Photobiology. Frain-Bell WC, ed. Oxford: Oxford University Press, 1985:125–152.
14. Orentreich N, Harber LC, Tromovitch TA. Photosensitivity and photo-onycholysis due to demethylchlortetracycline. Arch Dermatol 1961; 83:730–737.
15. Brauner GJ. Bullous photoreaction to nalidixic acid. Am J Med 1975; 58:576–580.
16. Ramsay CA, Obreshkova E. Photosensitivity from nalidixic acid. Br J Dermatol 1974; 91:523–528.
17. Burry JN, Lawrence JR. Phototoxic blisters from high frusemide dosage. Br J Dermatol 1976; 94:495–499.
18. Epstein JH, Tuffanelli DL, Seibert JS, Epstein WL. Porphyria-like cutaneous changes induced by tetracycline hydrochloride. Arch Dermatol 1976; 112:661–666.
19. Howard AM, Dowling J, Varigos G. Pseudoporphyria due to naproxen. Lancet 1985; 1:819–820.
20. Judd LE, Henderson DW, Hill DC. Naproxen-induced pseudoporphyria. Arch Dermatol 1986; 122:451–454.
21. Parodi A, Guarrera M, Rebora A. Amiodarone-induced pseudoporphyria. Photodermatology 1988; 5:146–147.
22. Hasan T, Khan AU. Phototoxicity of the tetracyclines: photosensitized emission of singlet delta dioxygen. Proc Natl Acad Sci USA 1986; 83:4604–4606.
23. Kochevar IE. Mechanisms of drug photosensitization. Photochem Photobiol 1987; 45:891–895.
24. Harber LC, Lashinsky AM, Baer RL. Skin manifestations of photosensitivity due to chlorothiazide and hydrochlorothiazide. J Invest Dermatol 1959; 33:83–84.
25. Ferguson J, Addo HA, Johnson BE, Frain-Bell WC. Quinine induced photosensitivity: clinical and experimental studies. Br J Dermatol 1987; 117:631–640.
26. Marx JL, Eisenstat BA, Gladstein AR. Quinidine photosensitivity. Arch Dermatol 1983; 119:39–43.
27. Wolf R, Dorfman B, Krakowski A. Quinidine-induced lichenoid and eczematous photodermatitis. Dermatologica 1987; 174:285–289.

28. Bhutani LK. The photodermatoses as seen in tropical countries. Semin Dermatol 1982; 1:175–181.
29. Ferguson J, Addo HA, Jones S, Johnson BE, Frain-Bell W. A study of cutaneous photosensitivity induced by amiodarone. Br J Dermatol 1985; 113:537–549.
30. Waitzer S, Butany J, From L, Hanna W, Ramsay C, Downar E. Cutaneous ultrastructural changes and photosensitivity associated with amiodarone therapy. J Am Acad Dermatol 1987; 16:779–787.
31. Bleehen SS, Gould DJ, Harrington CI, Durrant TE, Slater DN, Underwood JCE. Occupational argyria; light and electron microscopic studies and x-ray microanalysis. Br J Dermatol 1981; 104:19–26.
32. Shelley WB, Shelley ED, Burmeister V. Argyria: the intradermal "photograph," a manifestation of passive photosensitivity. J Am Acad Dermatol 1987; 16:211–217.
33. Pathak MA. Phytophotodermatitis. Clin Dermatol 1986; 4:102–121.
34. Kavli G, Volden G. Phytophotodermatitis. Photodermatology 1984; 1:65–75.
35. Kochevar IE, Armstrong RB, Einbinder J. Coal tar phototoxicity: active components and action spectra. Photochem Photobiol 1982; 36:65–69.
36. Crow KD, Alexander E, Buck WHL, Johnson BE, Magnus IA, Porter AD. Photosensitivity due to pitch. Br J Dermatol 1961; 73:220–232.
37. Kaidbey KH, Kligman AM. Clinical and histological study of coal tar phototoxicity in humans. Arch Dermatol 1977; 113:592–595.
38. Tanenbaum L, Parrish, HA, Pathak MA, Anderson R, Fitzpatrick TB. Tar phototoxicity and phototherapy for psoriasis. Arch Dermatol 1975; 111:467–470.
39. Allison AC, Magnus IA, Young MR. Role of lysosomes and of cell membrane in photosensitization. Nature 1966; 209:874–878.
40. Lord JT, Ziboh VA, Poitier J, Legget G, Penneys NS. The effects of photosensitizers and ultraviolet irradiation on the biosynthesis and metabolism of prostaglandins. Br J Dermatol 1976; 95:397–406.
41. Hasan T, Kochevar IE, McAuliffe DJ, Cooperman BS, Abdulah D. Mechanism of tetracycline phototoxicity. J Invest Dermatol 1984; 83:179–183.
42. Lim HW, Novotny H, Gigli I. Role of complement and polymorphonuclear cells in demethylchlortetracycline-induced phototoxicity in guinea pigs. J Clin Invest 1983; 72:1326–1335.
43. Epstein S. Chlorpromazine photosensitivity: phototoxic and photoallergic reactions. Arch Dermatol 1968; 98:354–363.
44. Ljunggren B, Möller H. Phenothiazine phototoxicity: an experimental study on chlorpromazine and its metabolites. J Invest Dermatol 1977; 68:313–317.
45. Torinuki W, Tagami H. Role of complement in chlorpromazine-induced phototoxicity. J Invest Dermatol 1986; 86:142–144.
46. Hunter JA, Bhutani LK, Magnus IA: Chlorpromazine photosensitivity in mice: its action spectrum and the effect of anti-inflammatory agents. Br J Dermatol 1970; 82:157–168.
47. Trimble JW, Mendelson DS, Fetter BF, Ingram P, Gallagher JJ, Shelburne JD. Cutaneous pigmentation secondary to amiodarone therapy. Arch Dermatol 1983; 119:914–918.
48. Diffey BL, Daymond TJ, Fairgreaves H. Phototoxic reactions to piroxicam, naproxen and tiaprofenic acid. Br J Rheumatol 1983; 22:239–242.

49. Kaidbey KH, Mitchell FN. Photosensitizing potential of certain nonsteroidal anti-inflammatory agents. Arch Dermatol 1989; 125:783–786.
50. Stern RS, Bigby M. An expanded profile of cutaneous reactions to nonsteroidal anti-inflammatory drugs. JAMA 1984; 252:1433–1437.
51. Kochevar IE, Morison WL, Lamm JL, McAuliffe DJ, Western A, Hood AF. Possible mechanism of piroxicam-induced photosensitivity. Arch Dermatol 1986; 122:1283–1287.
52. Lam SK, Tomlinson DR. Chlorpromazine-induced histamine release from guinea pig skin in vitro: a photosensitive reaction. Arch Dermatol Res 1976; 255:219–233.
53. DeLeo VA, Hanson D, Schiede S. Benoxaprofen photosensitization of phospholipase activation in mammalian cells in culture. Toxicol Lett 1986; 32:215–220.
54. Ferguson J, Addo HA, McGill PE, Woodcock KR, Johnson BE, Frain-Bell W. A study of benoxaprofen-induced photosensitivity. Br J Dermatol 1982; 107:429–444.
55. Kochevar IE, Hoover KW, Gawienowski M. Benoxaprofen photosensitization of cell membrane disruption. J Invest Dermatol 1984; 82:214–218.
56. Pariser DM, Taylor JR. Quinidine photosensitivity. Arch Dermatol 1975; 111:1440–1443.
57. Levine JI. Medications that increase sensitivity to light: A 1990 listing. Washington D.C: U.S. Department of Health and Human Services 1991; HHS Publication FDA 91-8280.

14

# Photoallergy

**Vincent A. DeLeo**
*Columbia–Presbyterian Medical Center, New York, New York*

Photosensitivity to exogenous chemicals depends on a chemical gaining access to the skin and absorption of light by the chemical. This absorption is followed by photochemical changes and results ultimately in cutaneous pathologic change. The cascade of events that lead from photochemistry to biological alterations and disease are complex and varied. Phototoxicity implies direct photodamage to the skin while the term photoallergy implies an immune mechanism.

Photoallergic reactions can be produced by chemicals applied to the skin or taken systemically. These two different routes of exposure to the exogenous photosensitizer result in reactions termed photoallergic contact dermatitis and systemic photoallergy or photoallergy to a systemic agent, respectively. Since most agents that produce photoallergy after systemic administration are therapeutic agents, this reaction is sometimes called a photoallergic drug reaction.

Although there is some variability of response, most photoallergic reactions are due to classic T-cell-mediated immune mechanisms: Gell and Coombs type-IV reactions (1). The photochemistry occurring on absorption of radiation by the chemical results in production of a complete photoantigen by a poorly defined mechanism that is likely to involve processing by macrophages and complexing to proteins, possibly the HLA-DR molecule. From that point on the process involved in sensitization and elicitation of photoallergy closely resembles allergic contact dermatitis and, as in the latter, the response will only occur in a genetically predisposed part of the population. The reaction will not occur on first exposure since sensitization is required (Table 1).

Similar to phototoxicity, both a chemical and radiation are necessary for photoallergic reactions to occur but the reactions are delayed, usually occurring in 24–48 hr after the combined exposure. The clinical and histologic response is

**Table 1** Differences Between Photoallergic and Phototoxic Reactions

| Variables | Photoallergy | Phototoxicity |
|---|---|---|
| Incidence | Low | High |
| Occurrence on first exposure | No | Yes |
| Onset after UV exposure | 24–48 hr | Minutes to hours |
| Dose dependence | | |
| Chemical | Not crucial | Important |
| Radiation | Not crucial | Important |
| Clinical morphology | Eczematous (erythroderma) | Erythema and edema, bullae, hyperpigmentation |
| Histologic appearance | Eczema | Necrotic keratinocytes |
| Action spectrum | UVA | UVA |
| Route of exposure | Usually topical | Systemic or topical |
| Diagnosis: topical agents | Photopatch tests | Clinical |
| systemic agents | Clinical + phototests | Clinical + phototests |

usually eczematous and does not vary appreciably among agents. Although no accurate epidemiologic data are available, photoallergy is certainly much less common than are phototoxic responses.

Almost all exogenous chemical photosensitivity is due to photons in the ultraviolet A (UVA) and occasionally the visible range, although occasionally ultraviolet B (UVB) may be causative. This is of clinical significance since UVA penetrates window glass and therefore reactions can be triggered by exposure to sunshine in an automobile or even indoors.

## I. PHOTOALLERGY TO SYSTEMIC AGENTS

Although there are many case reports of patients who developed a photodistributed eruption associated with suspected systemic photoallergens, many of these case reports are descriptive only, lack controls, and do not decisively demonstrate that systemically administered drugs can induce photoallergy. Among the most commonly reported drugs capable of causing photoallergic reactions are the sulfonamides, chlorpropamide, chlorpromazine, promethazine, hydrochlorothiazide, chlorothiazide, chlordiazepoxide, calcium cyclamate, tolbutamide, and griseofulvin (2).

Photoallergic responses are delayed in nature and are usually eczematous, but lichenoid-like reactions and, rarely, "wheal and flare" responses may occur. In the acute phase of the photoallergic response, an erythematous, pruritic eruption with vesiculation may occur, whereas patients with chronic photoallergy present with marked lichenification of sun-exposed areas. Although the reactions are generally short-term in duration, persistent light reactivity may evolve from photoallergic reactions.

COLOR PLATES

## CHAPTER 6

**Figure 2** Severe photoaging with multiple prominent actinic keratoses, in addition to telangiectases and irregular epidermal pigmentation.
**Figure 4** Laxity and fine wrinkling on the sun-exposed extensor forearm. Elastic properties of the skin are greatly diminished.
**Figure 5** Severe photoaging in a dark-skinned man (Hispanic, skin type V). The predominant change is elastosis with coarse wrinkling; epidermal changes are virtually absent.

## CHAPTER 10

**Figure 1** Papular lesions of polymorphous light eruption in the V area of a female patient.
**Figure 4** Positive phototest reaction. Papular lesions of polymorphous light eruption are induced by artificial UVA irradiation 24 hr after the fourth exposure.

## CHAPTER 12

**Figure 2** Chronic actinic dermatitis of the face.
**Figure 3** Chronic actinic dermatitis of the pseudolymphomatous (actinic reticuloid) type.
**Figure 4** Palmar eczema in chronic actinic dermatitis.

## CHAPTER 13

**Figure 3** Berloque dermatitis.

## CHAPTER 14

**Figure 1** Photoallergic reactions to piroxicam. Notice the sparing of periorbital areas and nasolabial folds. (Courtesy of C. Halasz, MD.)

## CHAPTER 15

**Figure 4** Increased skin fragility on the dorsum of hand of a patient with porphyria cutanea tarda, manifested as vesicles, erosions, and postinflammatory hyperpigmentation.
**Figure 5** Sclerodermoid skin changes on the preauricular area of a patient with porphyria cutanea tarda.
**Figure 6** Erythema and fine scales on sun-exposed area of a patient with erythropoietic protoporphyria.

## CHAPTER 16

**Figure 1** Systemic lupus erythematosus. Abnormal papular response to sunlight exposure. Note the sparing of eyelids and nasolabial folds.
**Figure 2** Subacute cutaneous lupus erythematosus. Annular erythematous plaques with collarettes of scale.
**Figure 4** Bloom's syndrome. Erythema across the butterfly area of the face and characteristic facies. Note the similarity of the eruption to that of systemic lupus erythematosus.
**Figure 5** Xeroderma pigmentosum. Extensive sun damage in a boy with multiple solar lentigines, actinic keratoses, and basal cell carcinomas.
**Figure 7** Hydroa vacciniforme. Tense vesicles on the dorsal aspects of the hands.

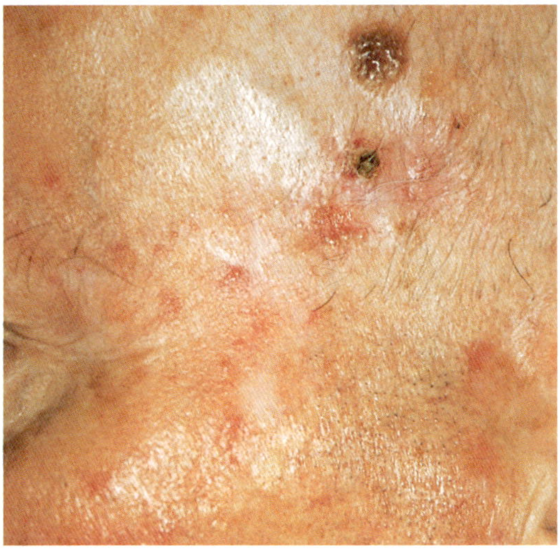
6.2

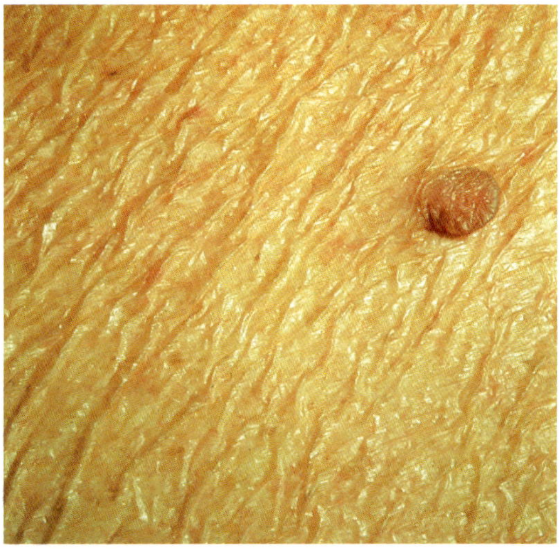
6.4

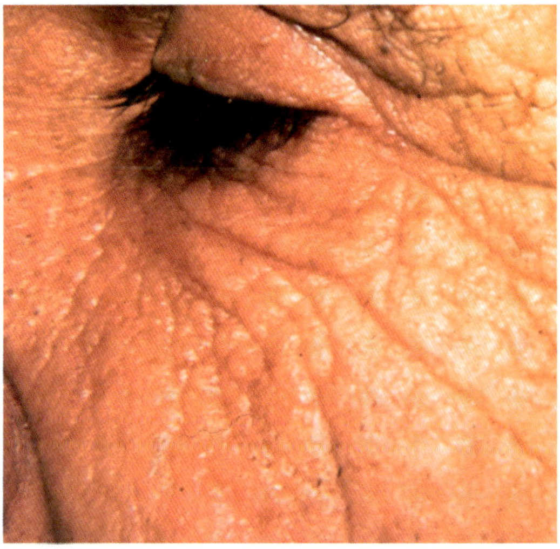
6.5

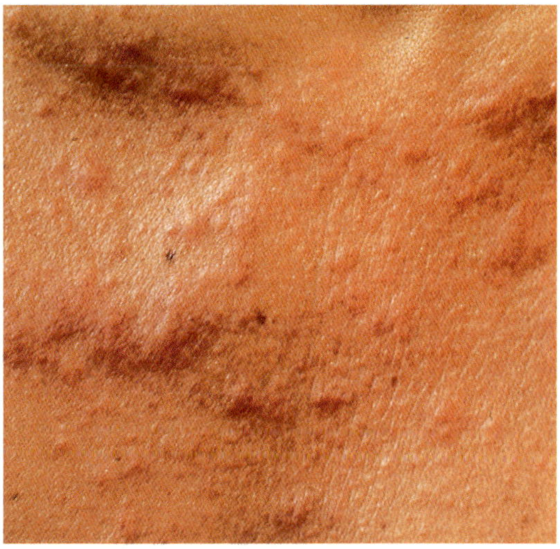
10.1

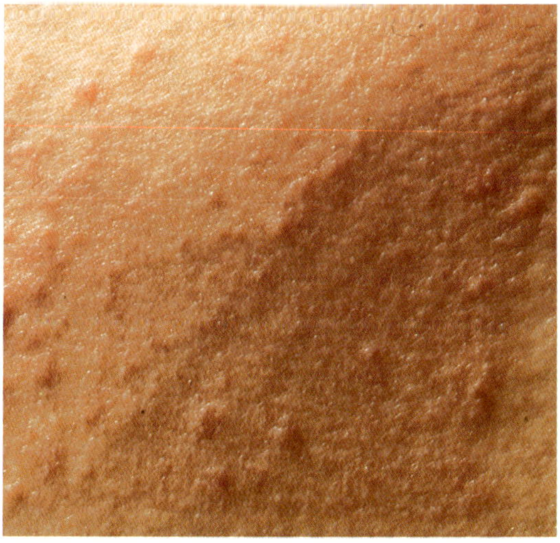
10.4

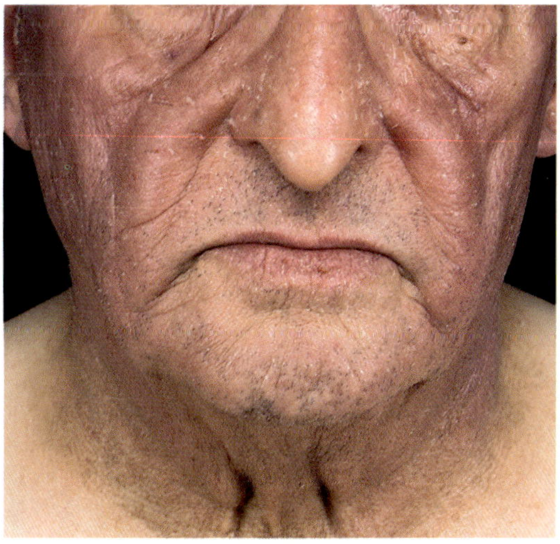
12.2

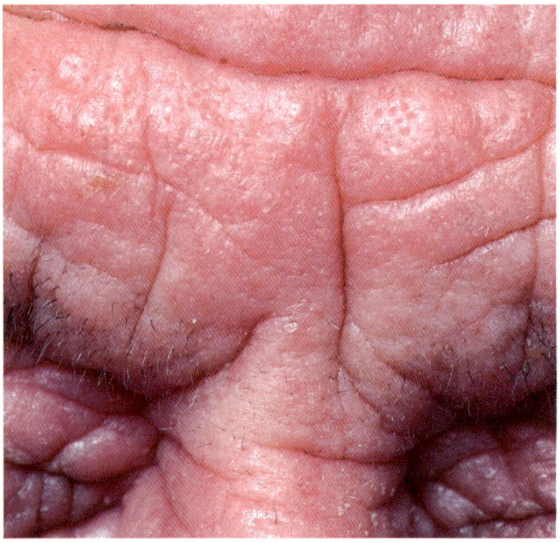
12.3

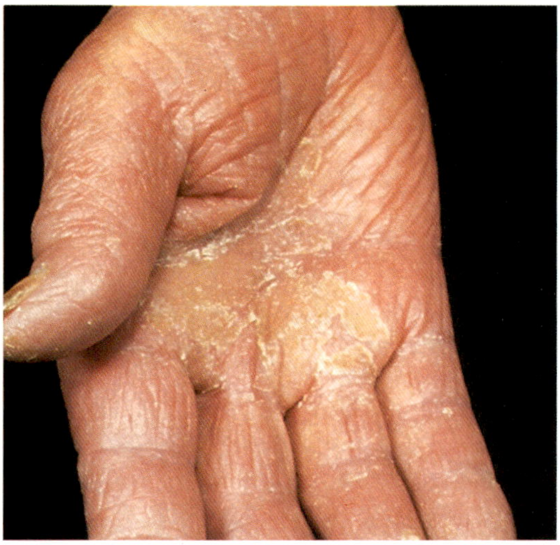
12.4

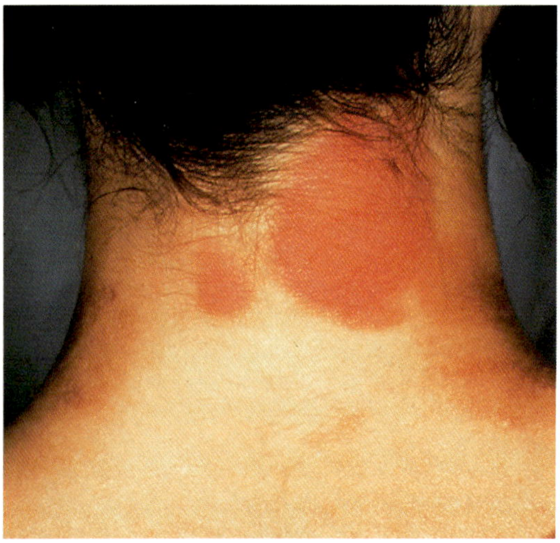
13.3

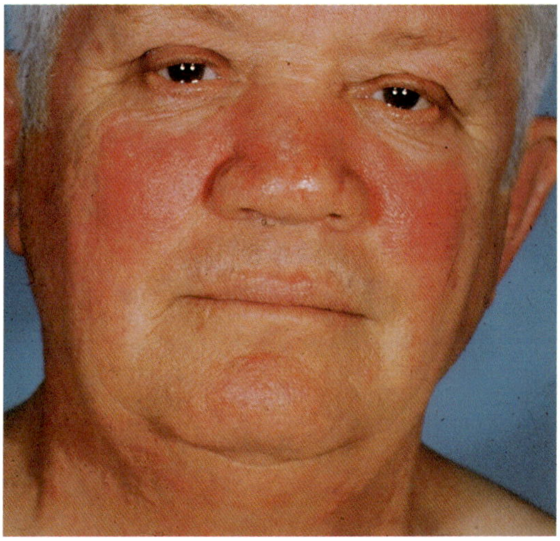
14.1

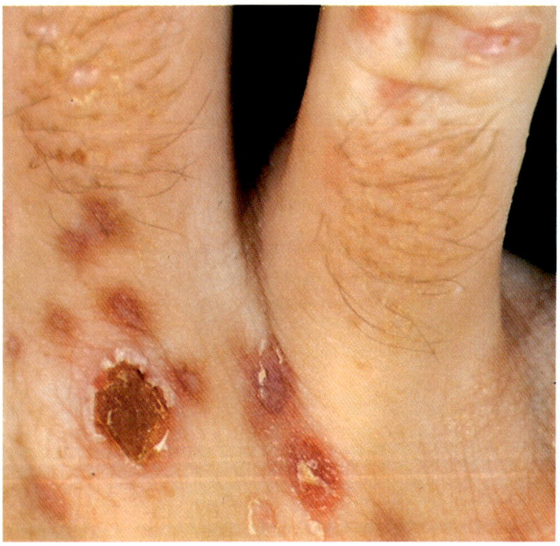
15.4

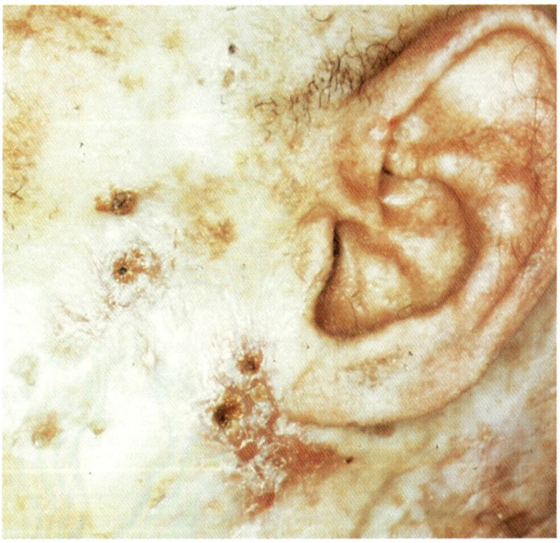
15.5

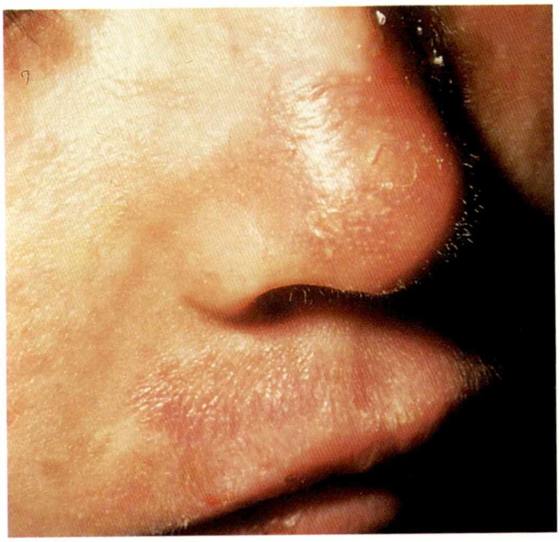

15.6

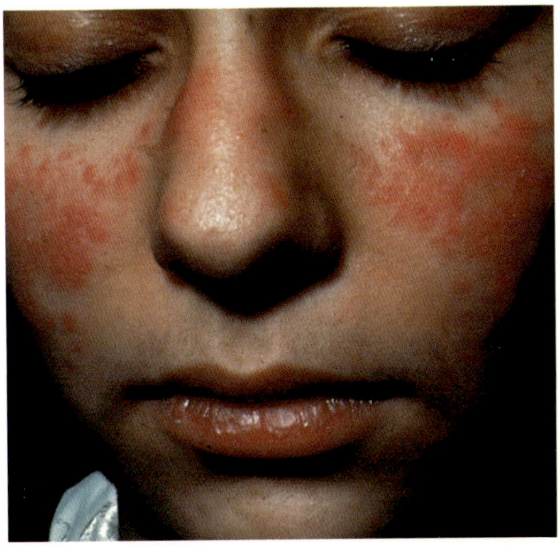

16.1

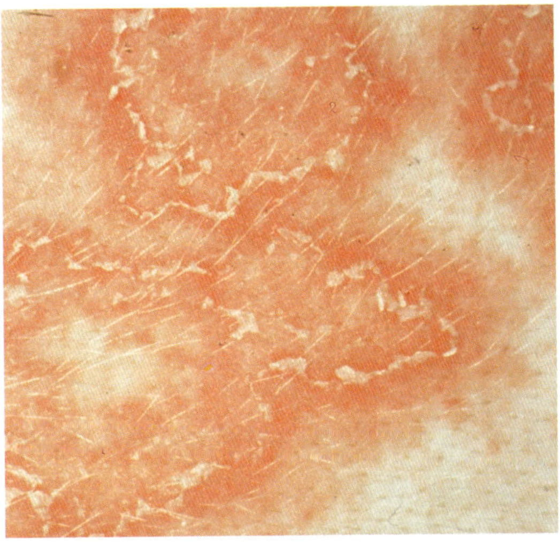

16.2

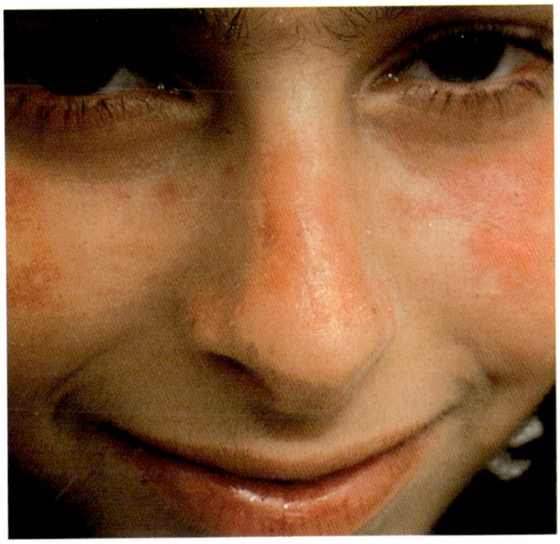

16.4

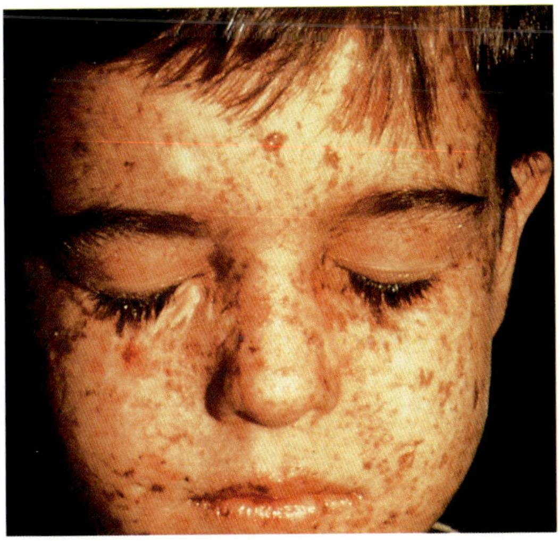

16.5

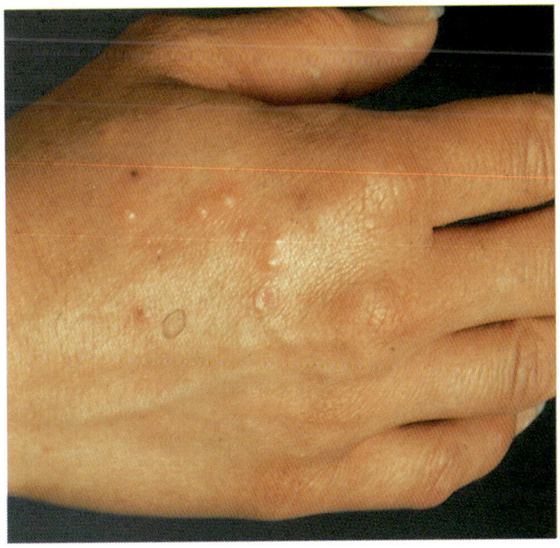

16.7

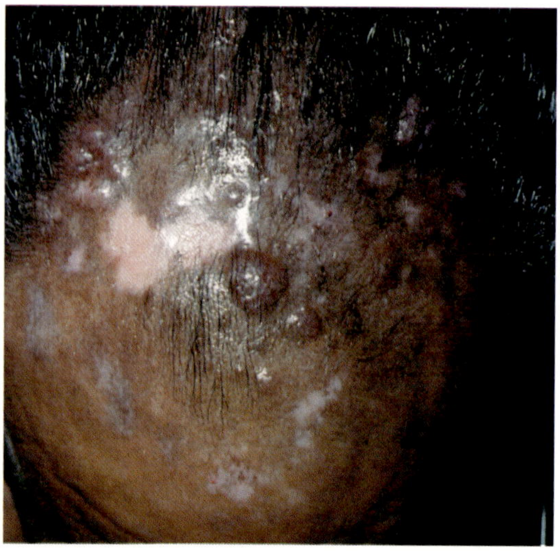
20.1A

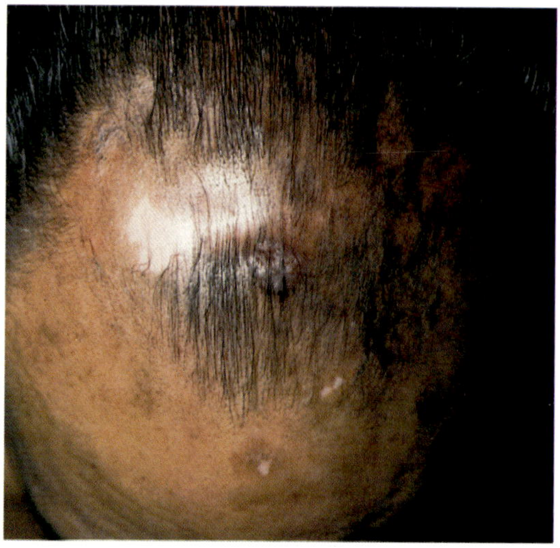
20.1B

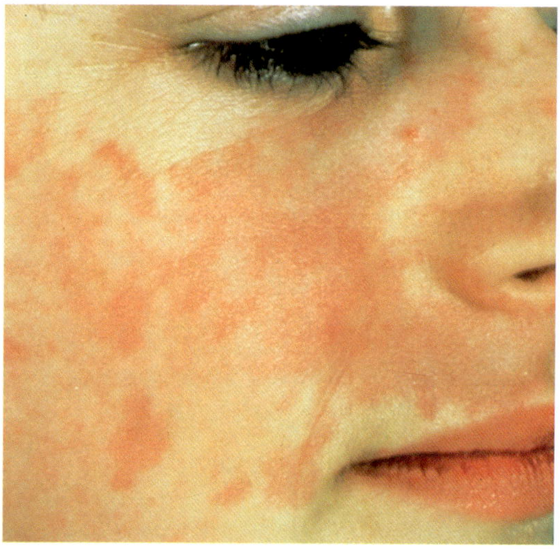
21.2A

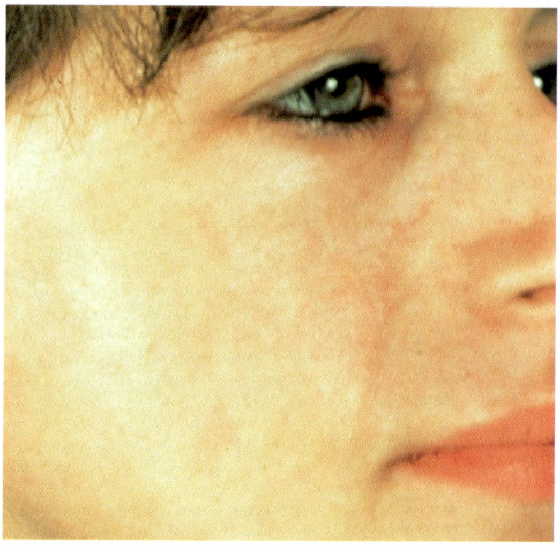
21.2B

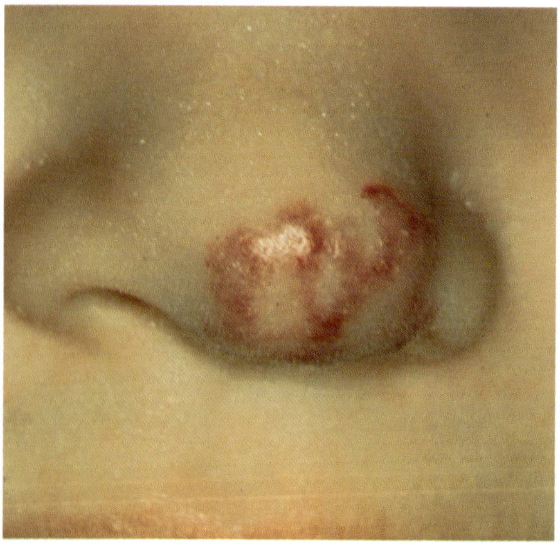
21.4A

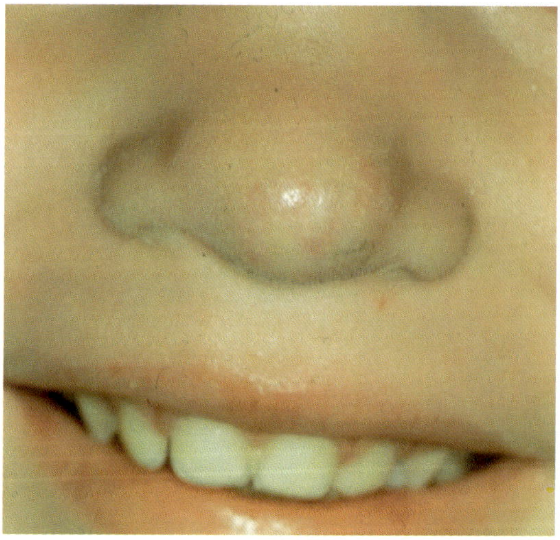
21.4B

CHAPTER 20

**Figure 1** Improvement in the cutaneous manifestations of CTCL in a patient receiving combined therapy with photopheresis and low dosage interferon alpha. An erythrodermic patient with a white blood cell count of 15,500, 46% of which were Sézary cells, and a Southern blot analysis demonstrating evidence of a peripheral blood T-cell clone, had return of the white blood cell count to a normal level after 4 months of photopheresis. At this time, numerous skin tumors developed (A), which contained the identical malignant T-cell clone. After continuation of photopheresis for an additional 16 weeks, in combination with interferon alpha at a dosage of 5 million units subcutaneously every other day, the cutaneous tumors had cleared (B), and a repeat Southern blot analysis of the peripheral blood no longer detected the malignant T-cell clone.

CHAPTER 21

**Figure 2** A. A large red port wine stain involving the right side of the face. B. Eight months later after 5 treatment sessions using a 585 nm pulsed dye laser and 6.5 $J/cm^2$, the majority of the lesion has resolved with a slight residual reticulate pattern.

**Figure 4** A. A capillary hemangioma on the nasal tip of a young child prior to treatment. B. There is a marked decrease in the volume of the lesion, as well as significant lightening of the color after a total of 4 treatments, using a 585 nm pulsed dye laser and an energy density of 7.0 over a period of 8 months.

While diuretics hydrochlorothiazide and chlorothiazide, the antidiabetic medications tolbutamide and chlorpropamide, and the tranquilizers chlorpromazine and promethazine commonly produce phototoxicity, occasionally they have been reported to cause reactions thought to be photoallergic (3). This distinction was usually based on an eczematous morphology and positive photopatch tests. Such testing is fraught with problems of interpretation since all of these agents are phototoxins. A positive response may therefore not differentiate between phototoxic and photoallergic mechanisms. Only with large numbers of normal controls can such testing be properly evaluated. In addition to transient photoallergic responses, hydrochlorothiazide was the first systemically administered drug reported to induce a persistent light reactivity (4). The eruption lasted for months and years after discontinuation of the agent. Such patients were severely debilitated and, on photobiological evaluation, were found to be sensitive to the UVA range.

The frequently used antihistamine diphenhydramine has rarely been reported to cause photoallergy. The responsible wavelengths were found to be in the UVB range (5, 6). Griseofulvin is well known to induce phototoxicity. Photoallergy to this agent was suggested by passive humoral transfer of sensitivity in one report (7). Piroxicam, a nonsteroidal anti-inflammatory drug, is a fairly common photosensitizer (2–3% of patients) (8). It frequently produces an eczematous response and has been associated with positive photopatch tests, suggesting photoallergy (Fig. 1, see color plate) (9–11). Quinidine has been reported to produce both transient and persistent photosensitivity (12). The morphology has been variable, including lichenoid lesions and many features suggesting that at least some of the responses are immunologically mediated.

## II. PHOTOALLERGIC CONTACT DERMATITIS

The first reported cases of photoallergic contact dermatitis occurred during World War II, when sulfanilamides were used topically to treat skin wounds. Photoallergic contact dermatitis reactions in the population at large occurred in the early 1960s when a number of halogenated salicylanilides were incorporated into bar soaps for their antibacterial effects. An "epidemic" of more than 10,000 cases of photoallergic contact dermatitis was thought to have occurred in the United States, Europe, and Japan. Tetrachlorosalicylanilide and tribromosalicylanilide, once recognized as potent photoallergens, were removed from consumer products. In the 1970s and 1980s two fragrance ingredients, musk ambrette and 6-methylcoumarin, were reported as new causes of photoallergic contact dermatitis. These have been removed from marketed products or have been reduced greatly in concentration in such products to supposedly "safe" levels (13–16). More recently, ingredients contained in sunscreens in both para-aminobenzoic acid (PABA) and PABA-free agents have been reported as causes of photoallergic contact dermatitis (Table 2) (17–19).

**Table 2** Agents Inducing Photoallergic Dermatitis

| Systemic Agents | Topical Agents |
|---|---|
| *Diuretics* | *Antibacterials* |
| Hydrochlorothiazide | Sulfanilamide |
| Chlorothiazide | Tetrachlorosalicylanilide |
| | Dibromosalicylanilide |
| *Antidiabetics* | Tribromosalicylanilide |
| Tolbutamide | Dichlorophene |
| Chlorpropamide | Fenticlor |
| | Triclosan |
| *Tranquilizers* | Chlorhexidine |
| Chlorpromazine | Hexachlorophene |
| Promethazine | Bithionol |
| | Jadit |
| *Anti-infectives* | Multifungin |
| Griseofulvin | |
| Sulfonamides | *Fragrances* |
| | Musk ambrette |
| *Miscellaneous* | 6-Methylcoumarin |
| Calcium cyclamate | Sandalwood oil |
| Piroxicam | |
| Diphenhydramine | *Sunscreens* |
| Antimalarials | PABA, PABA esters |
| Quinidine | Benzophenones |
| | Digalloyl trioleate |
| | Dibenzoylmethanes |
| | Cinnamates |
| | |
| | *Therapeutic agents* |
| | Diphenhydramine |
| | Psoralens |
| | Thiazides |
| | Sulfonylureas |
| | Chlorpromazine |
| | Promethiazine |

The treatment of photoallergic contact dermatitis is dependent, like that of allergic contact dermatitis, on avoidance of contact with the allergen. In the majority of patients, once this is achieved the photosensitivity disappears. In rare cases, however, patients with photoallergic contact dermatitis continue to exhibit photosensitivity months and even years after the last exposure to antigen. Such individuals have been called persistent light reactors. In distinction, patients who experience resolution of photosensitivity with antigen avoidance are referred to as transient light reactors. The mechanism of the persistent response is poorly understood. It is not due simply to persistence of antigen in skin, although some cases of localized persistent light reaction have been reported and could result from such persistence. In most patients, however, the photosensitivity involves the entire skin surface, even areas never exposed to the antigen. In addition, the action spectrum of photosensitivity in the presence of antigen both clinically and in photopatch testing, is in the UVA range, while the action spectrum of the photosensitivity in the absence of antigen in persistent light reactivity is in the UVB range, although in some cases the latter may extend into UVA and even visible ranges. The persistent reaction is somewhat agent specific. It has primarily been reported with certain salicylanilides and musk ambrette. Such patients' conditions are more correctly classified under the diagnosis of chronic actinic dermatitis (see Chap. 12).

The diagnosis of photoallergic contact dermatitis is made based on photopatch testing. Such testing should be performed in all patients in whom photosensitivity is suspected. Photoallergic contact dermatitis has been shown to be the final diagnosis in 10–20% of patients referred for evaluation of photosensitivity (19–21). It is difficult to differentiate clinically between photoallergic contact dermatitis and idiopathic photosensitivity disorders, such as polymorphous light eruption. In fact, photoallergic contact dermatitis, especially to sunscreens, can complicate the clinical course in patients with other types of photosensitivity.

## A. Antibacterial Agents

Tetrachlorosalicylanilide and tribromosalicylanilide, the most potent of the photosensitizers in this group, are no longer used in consumer cleansers (i.e., bar soaps and shampoos) in the United States. Testing with one of these is recommended, however, since such agents may be purchased out of the United States. In addition, industrial cleaners containing these agents may still be in use in this country. Patients with long-standing cases of persistent light reactivity may have been exposed to these agents in the past.

Triclosan (Irgasan DP 300) is a widely used antibacterial agent in bar soaps and deodorants. It appears to be a very low level photosensitizer and few cases have been reported despite its widespread usage.

Dichlorophene (G-4) is widely used in this country and in Europe in shampoos, dentrifices, antiperspirants, and "athlete's foot" powder. It is a rarely reported photosensitizer.

Bithionol is a chlorinated phenol used in the 1960s in the United States and more extensively in Japan. It was a potent sensitizer in bar soaps and has been banned from such products in those two countries. It may still be used in personal care products sold in other countries and in industrial cleansers and agricultural and veterinary products marketed in the United States.

Fenticlor is a chlorinated phenol used as an antibacterial and antiseborrheic agent in hair care products made primarily in Canada, the British Isles, and Australia. It was never used extensively in the United States. It appears to be a moderately potent photoallergen. It may produce false-positive phototoxic responses in photopatch testing.

Hexachlorophene was a widely used antibacterial in over-the-counter skin cleansers in the United States; it can only be dispensed by prescription because of reports of neurotoxicity. *PHisoHex* is still used in the United States today, but with much lower frequency. It is a rarely reported photoallergen.

Chlorohexidine is used as an antibacterial in hospital cleansers for both skin and mucosae. It is also used as a dental rinse. It is a rare photoallergen.

Cross-reactivity among many of the above antibacterial agents and other related chemicals has been reported. A list of salicylanilide-related antibacterials is provided in Table 3.

## B. Fragrances

Three fragrance ingredients have been associated with photoallergic contact dermatitis: musk ambrette, 6-methylcoumarin, and sandalwood oil.

Musk ambrette is a synthetic fragrance fixative used primarily in men's cosmetics because of its potent floral odor. Related chemicals extracted from the scent glands of animals and from some plants have been used for years as fixatives and enhancers in perfumes. In the 1970s and 1980s, huge quantities were used in the United States in various cosmetics, primarily men's after shave lotions and colognes. Concentrations of musk ambrette as high as 15% were used in such products. In the late 1970s reports of photoallergy began to appear in the literature. By the 1980s this agent was the most frequently reported cause of photoallergic contact dermatitis (22). Many of the men sensitized to musk ambrette developed persistent light reactivity/chronic actinic dermatitis. In 1985, the International Fragrance Association recommended that musk ambrette not be

**Table 3** Salicylanilide-Related Antibacterials

| |
|---|
| Hexachlorophene (pHisoHex) |
| Dibromsalan (dibromosalicylanilide, DBS) |
| Multifungin (bromochlorosalicylanilide, BCSA) |
| Triclocarban (trichlorocarbanilide, TCC) |
| Bithionol (thiobisdichlorophenol) |
| Fenticlor (thiobischlorophenol) |
| Buclosamide (Jadit, butylchlorosalicylamide) |
| Chloro-2-phenylphenol (Dowicide 32) |
| Tetrachlorosalicylanilide (Irgasan BS200) |

utilized in products in contact with skin. In other products a concentration of 4% or less is recommended (23).

6-Methylcoumarin is a synthetic fragrance that caused an epidemic of photoallergic contact dermatitis when it was used in "sun-tanning" lotions in the late 1970s. The reactions were particularly severe, requiring hospitalization in many cases. The morphology of many of reactions suggested phototoxicity, but photoallergy was probably the underlying mechanism (Fig. 2). The agent was removed from the above lotions, and it is no longer recommended for use as a fragrance component (23,24).

Sandalwood oil is a "woodsy" smelling fragrance ingredient. It is a rarely reported photosensitizer (25).

## C. Sunscreen Agents

In the 1970s and 1980s people in the United States, Europe, and other areas of the world populated by Caucasians were made aware of the deleterious effects of solar radiation. They were encouraged to use sunscreens, and agents designed to protect against sun damage proliferated at a rapid rate. Use of such agents is continuing to increase. While sunscreens are effective and extremely safe products, side effects do occur; one of these is photoallergic contact dermatitis. While the incidence of such reactions is unknown, photoallergic contact der-

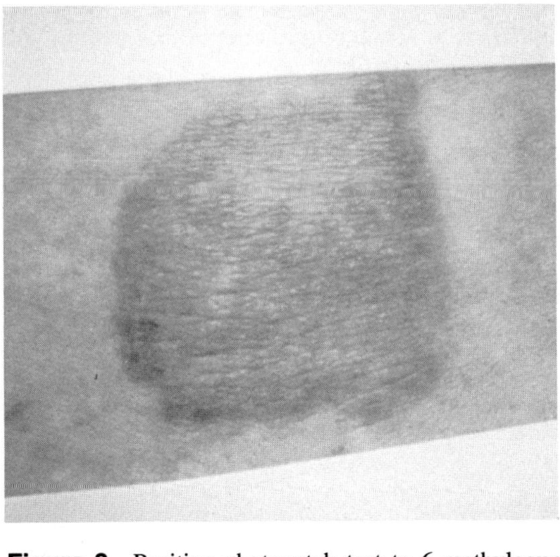

**Figure 2** Positive photopatch test to 6-methylcoumarin.

matitis to a sunscreen agent is certainly rare. The first agent to produce this effect was PABA. Recognition of this allergy and the perception, probably unwarranted, that it was common led to development of PABA-free sunscreens. Photoallergy to these newer agents has also been reported. In all likelihood sunscreen agents will be the most frequent cause of photoallergic contact dermatitis in the 1990s (17,18).

PABA was the first chemical used as a sunscreen. It was the most widely used active agent in sunscreens in the 1960s and early 1970s. PABA is presently an infrequent component of marketed sunscreens and has been replaced by its esters, particularly octyl-dimethyl PABA (padimate O). Amyl dimethyl PABA (padimate A) was utilized for a short period of time in some products, but its usage was associated with an immediate photoirritancy. Glycerol PABA, once thought to be a high-level sensitizer, was found to be contaminated with benzocaine, which was the true allergen in such reactions. Cross-reactivity between PABA and its esters has been reported. It is also possible that patients sensitive to these agents may react to related chemicals listed in Table 4.

The benzophenones are ultraviolet-absorbing agents with absorbance into the UVA range. Sulisobenzone (benzophenone 4) was the first agent in this class to be used in PABA-free sunscreens. It is a low-level photosensitizer. Oxybenzone (benzophenone-3) is the most commonly used benzophenone in sunscreens today and probably is the most commonly used active agent of all classes. Reports of photoallergic contact dermatitis to this agent are increasing as its use increases.

Butyl methoxydibenzoylmethane (avobenzone, PARSOL 1789) is a new addition to sunscreens marketed in the United States. It is an excellent UVA absorber and is being marketed for this activity in "broad-spectrum" sunscreens. It has been reported to cause photoallergic contact dermatitis in Europe and will likely do the same in the United States.

**Table 4** Chemicals That May Cause Dermatitis in Patients Who Are Allergic or Photoallergic to Sunscreens Containing PABA and PABA Derivatives

| | |
|---|---|
| Sunscreens | PABA |
| | Amyl dimethyl PABA (padimate A) |
| | Octyl dimethyl PABA (padimate O) |
| | Glycerol PABA |
| Diuretic | Hydrochlorothiazide |
| Antidiabetic agent | Sulfonylureas |
| Hair dye | Paraphenylenediamine |
| Anesthetic | Benzocaine, procaine |
| Anti-infective | Para-aminosalicylic acid |
| Cardiac drug | Procainamide |

Cinoxate (2-methoxy-ethyl-p-methoxy cinnamate) is an infrequently used agent in sunscreens sold in the United States. It has been reported to cause photoallergic contact dermatitis in Europe.

## D. Other Agents

Few systemic drugs that produce photosensitivity have been reported to cause photoallergic contact dermatitis when applied topically. In theory, this might occur with many such agents. The two most frequently reported are the phenothiazines, chlorpromazine hydrochloride (Thorazine) and promethazine (Phenergan). The photoallergic contact dermatitis reported for the former has been found in health care workers who have frequent skin contact while dispensing the medication. The reactions reported to promethazine were usually in the setting of use of a topical antipruritic (Phenergan Cream) in France. Care must be taken in reading the photopatch tests with these agents since they frequently induce false-positive phototoxic responses (26).

Thiourea (thiocarbamide) has been reported to cause an unusual photoallergic contact dermatitis reaction of the hands including the palms from contact with the agent in copy paper.

# III. PHOTOPATCH TESTING TECHNIQUES

Photopatch testing is patch testing with the addition of radiation to induce formation of the photoantigen. Both photoantigens and appropriate light sources are needed (Tables 5, 6). A suggested protocol for photobiological testing is shown in Table 7.

With very few exceptions, the most notable being with the use of diphenylhydramine hydrochloride, the radiation responsible for formation of the photoantigen and clinical photoallergic contact dermatitis falls within the UVA spectrum (320–400 nm). The ideal light source for photopatch testing should produce UVA radiation in a continuous spectrum of sufficient irradiance and field size to allow radiation of 20–25 antigen sites with a dosage of 5–10 J/cm$^2$

**Table 5** Light Sources for Photopatch Testing

| |
|---|
| Fluorescent black lights |
| Fluorescent PUVA lights |
| Mercury halide lamp with filter |
| Hot quartz lamp with filter[a] |
| Sunlight with filter |

[a]Discontinuous spectrum: less desirable.

**Table 6** Photopatch Test Antigens Used at Columbia–Presbyterian Medical Center

| | |
|---|---|
| Amyl dimethyl PABA | 5% in pet. or alcohol |
| Sulisobenzone (BZP-4) | 10% in pet. |
| Cinoxate | 1% |
| Thiourea (thiocarbamide) | 0.1% in pet |
| Dichlorophene | 1% in pet. |
| Triclosan | 2% in pet. |
| Hexachlorophene | 1% in pet. |
| Chlorhexidine | 1% aqueous |
| Sandalwood oil | As is |
| Chlorpromizine hydrochloride | 0.1% |
| Musk ambrette | 1% in pet. |
| PABA | 5% in pet. or alcohol |
| Tribromosalicylanilide | 1% in pet |
| Octyl dimethyl PABA | 5% in pet. or alcohol |
| Oxybenzone (BZP-3) | 3% in pet. |
| Promethazine | 1% in pet. |
| Bithionol | 1% in pet. |
| Musk ambrette | 1% in alcohol |
| Fenticlor | 1% in pet. |
| Butyl methoxydibenzoylmethane | 5% in pet. |
| 6-Methylcoumarin | 1% alcohol[a] |

pet., Petrolatum

[a]6-Methylcourmarin should only be applied 30–60 min before irradiation.

**Table 7** Suggested Protocol For Photobiological Testing

| Day | Procedure |
|---|---|
| 1 | Radiation for MEDs (with immediate readings)<br>Apply two sets of antigens<br>Cover sites |
| 2 | Read MEDs |
| 2 or 3 | Remove antigen patches<br>Irradiate one set of antigens<br>Cover sites |
| 4 or 5 | Read patch and photopatch tests (48 hr after patch irradiation) |
| > 5 | Second patch and photopatch reading |

within a reasonable time (30 min). The source should also produce little UVB or should be equipped with a filter to remove most of such radiation. A photometer/radiometer matched to the source is also required for proper dosimetry. Such source/radiometer matching is most easily accomplished by purchasing the two units from the same manufacturer. By far the most readily available source in the dermatologist office is the unit used for photochemotherapy (Table 5).

A radiation dosage between 5 and 10 J/cm$^2$ of UVA is used in photopatch tests. The largest dosage that does not by itself induce erythema in skin would be most likely to result in a positive test result. Since the minimal erythema dosage (MED) in the UVA range is between 20 and 60 J/cm$^2$, any dosage that can be conveniently delivered below this level can be used.

In addition to photopatch testing, it is recommended that the patient being tested also be evaluated for photosensitivity by determination of the MED.

One antigen, 6-methylcoumarin, is not applied on day 1 with the other antigens. This agent's ability to form a photoantigen disappears rapidly after application to skin. For this reason the antigen is applied on the day of patch irradiation (usually day 2). It is applied in duplicate in Finn chambers for a 30 min period before irradiation. Longer application times result in false-negative photopatch tests.

Reading the photopatch test is done as for routine patch tests. An area of significant controversy surrounds the differentiation of photocontact allergy from contact allergy; the system used by the North American Contact Dermatitis Group is shown in Table 8 (15). It is agreed that a positive response in the irradiated site and negative in the covered site is diagnostic of photocontact allergy, while equal positives in both irradiated and covered sites are diagnostic of contact allergy. The North American system allows for the diagnosis of both contact allergy and photocontact allergy when both sites are positive, and response in the irradiated site is more intense than in the covered site. In the system used by the Scandinavian group and the Mayo Clinic group, any reaction in the covered site results in diagnosis of contact allergy (21).

**Table 8** Reading the Photopatch Test

| | Reading | |
|---|---|---|
| Diagnosis | Irradiated Site | Unirradiated Site |
| No sensitivity | − | − |
| Photocontact allergy | + | − |
| Contact allergy | + | + |
| Photocontact and contact allergy | ++ | + |

As with patch testing, false-positive and false-negative reactions can occur in photopatch testing. One particularly common false-positive or photoirritant response is to the phenothiazine agents in the tray (chlorpromazine and promethazine). Some antigens will produce an immediate photoirritant response, manifested as erythema at the completion of the irradiation period. This is not clinically relevant and may be disregarded.

In addition to the photoallergens in the tray, patients can be tested to their own personal care products, particularly sunscreens and fragrance-containing cosmetics. Industrial cleansers as well as personal care cleansers that may be the source for antibacterial agents must be diluted appropriately for use in such testing.

## REFERENCES


1. Epstein JH. Phototoxicity and photoallergy. In: Pathak MA, et al, eds. Sunlight and Man. Tokyo: University of Tokyo Press, 1974:459–477.
2. Drugs that cause photosensitivity. Med Lett Drugs 1986; 28:51–52.
3. Halevy S, Grunwald MH, Feuerman EJ, Levni E. Lichenoid eruption due to hydrochlorothiazide. Diagnostic aid of macrophage migration inhibition factor (MIF) test. Ann Allergy 1986; 56:402–405.
4. Robinson HN, Morison WL, Hood AF. Thiazide diuretic therapy and chronic photosensitivity. Arch Dermatol 1985; 121:522–524.
5. Schreiber MM, Naylor LZ. Antihistamine photosensitivity. Arch Dermatol 1962; 86:58–62.
6. Emmett EA: Drug photoallergy. Int J Dermatol 1978; 17:370–379.
7. Chang TW. Cold urticaria and photosensitivity due to griseofulvin. JAMA 1965; 193:192–194.
8. Stern RS, Bigby M. An expanded profile of cutaneous reactions to nonsteroidal anti-inflammatory drugs. JAMA 1984; 252:1433–1437.
9. Serrano G, Bonillo J, Aliaga A, et al. Piroxicam-induced photosensitivity. In vivo and in vitro studies of its photosensitizing potential. J Am Acad Dermatol 1984; 11:113–120.
10. McKerrow KJ, Greig DE. Piroxicam-induces photosensitive dermatitis. J Am Acad Dermatol 1986; 15:1237–1241.
11. Halasz CLG. Photosensitivity to the nonsteroidal anti-inflammatory drug piroxicam. Cutis 1987; 39:37–39.
12. Lim HW, Buchness MR, Ashinoff R, Soter NA. Chronic actinic dermatitis. Arch Dermatol 126:317–322; 1990.
13. DeLeo VA, Harber LC. Contact photodermatitis. In: Fisher AA, ed. Contact Dermatitis, 3rd ed. Philadelphia: Lea & Febiger, 1986:454–469.
14. Cronin E. Contact Dermatitis. New York: Churchill Livingstone, 1980.
15. Emmett EA. Phototoxicity and photosensitivity reactions. In: Adams RM, ed. Occupational Skin Disease, 2nd ed. Philadelphia: WB. Saunders 1990.
16. Nater JP, DeGroot AC. Unwanted Effects of Cosmetics and Drugs Used in Dermatology, 2nd ed. New York: Elsevier, 1985.

17. Droomgoole SH, Maibach HI. Contact sensitization and photocontact sensitization of sunscreening agents. J Am Acad Dermatol 1990; 22:1068–1078.
18. Thune P. Contact and photocontact allergy to sunscreens. Photodermatology 1984; 1:5–9.
19. Suarez S, Maso M. DeLeo VA. Photocontact Derm (in press).
20. Menz MB, Sigfrid AM, Connolly SM. Photopatch testing: a six year experience. J Am Acad Dermatol, 1988; 18:1044–1047.
21. Thune P, Jansén C, Wennersten G, et al. The Scandinavian multicenter photopatch study 1980–1985: final report. Photodermatology 1988; 5:261–269.
22. Raugi GJ, Storrs FJ. Photosensitivity from men's colognes. Arch Dermatol 1979; 115:106.
23. International Fragrance Association. Code of Practice, Oct. 1985.
24. Jackson RT, Nesbitt LT, DeLeo VA. 6-methylcoumarin photocontact dermatitis. J Am Acad Dermatol 1980; 2:124–127.
25. Starke JC. Photoallergy to sandalwood oil. Arch Dermatol 1967; 96:62–63.
26. Schauder S. How to avoid phototoxic reactions in photopatch testing with chlorpromazine. Photodermatology 1985; 2:95–100.

15

# The Porphyrias

**Henry W. Lim**
*New York University School of Medicine, New York, New York*

**Shigeru Sassa**
*Rockefeller University, New York, New York*

## I. PORPHYRINS AND PORPHYRINOGENS

Porphyrins and porphyrinogens are cyclic tetrapyrroles (Fig. 1). Structures and trivial names of the major porphyrins occurring in nature are given in Table 1. Enzymes in the heme biosynthetic pathway utilize porphyrinogens as substrates (i.e., the reduced form of porphyrins), except for ferrochelatase, which utilizes protoporphyrin as substrate. Porphyrinogens that have not been utilized for enzymatic reactions undergo rapid oxidation under air to become porphyrins. Thus porphyrins are the usual forms of the cyclic tetrapyrroles found in biological specimens.

Free porphyrins do not have any useful physiological functions. They emit intense red fluorescence upon exposure to Soret band radiation (400–410 nm), and, in the presence of molecular oxygen, they cause photodynamic effects on tissues, cells, subcellular elements, and biomacromolecules via generation of reactive oxygen species and their effects on inflammatory mediators and cells. As discussed later, animals and humans treated with porphyrins show signs and symptoms of photosensitization, such as pruritus, erythema, and edema, which can lead to ulcers, pigmentary alterations, and scars of the skin.

In contrast to free porphyrins, metal-chelated porphyrins carry out critically important biological functions. For example, heme (i.e, ferrous protoporphyrin

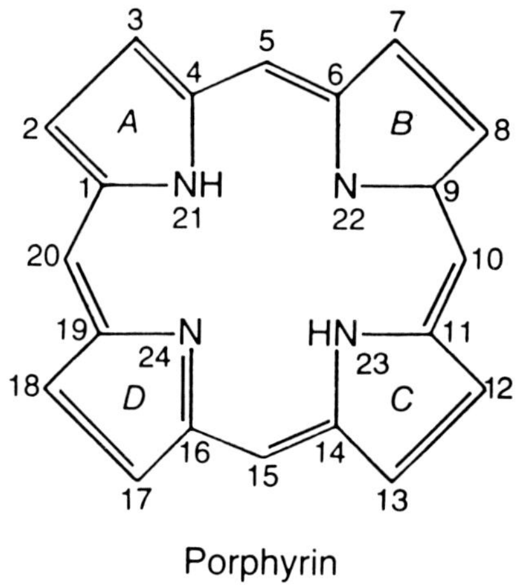


**Figure 1** Structure and nomenclature of the porphyrin macromolecule. Porphyrin is the basic structure of ring tetrapyrroles. See Table 1 for substituent groups for the major porphyrins occurring in nature.

[PROTO]) is the prosthetic group of a number of hemeproteins. Magnesium–PROTO and corrin are the precursors for chlorophyll and vitamin B12 synthesis, respectively.

## II. ENZYMES AND CHEMICAL INTERMEDIATES OF THE HEME BIOSYNTHETIC PATHWAY

The biochemical pathway for the formation of heme is illustrated in Figure 2. Eight enzymes are involved in the synthesis of heme. The first and the last three

**Table 1** Structures and Trivial Names of the Major Porphyrins Occurring in Nature

| Name | Substituent | | | | | | | |
|---|---|---|---|---|---|---|---|---|
| *IUPAC numeration* | *2* | *3* | *7* | *8* | *12* | *13* | *17* | *18* |
| Uroporphyrin I | A | P | A | P | A | P | A | P |
| Uroporphyrin III | A | P | A | P | A | P | P | A |
| Coproporphyrin I | M | P | M | P | M | P | M | P |
| Coproporphyrin III | M | P | M | P | M | P | P | M |
| Protoporphyrin IX | M | V | M | V | M | P | P | M |

M, $-CH_3$; A, $-CH_2COOH$; P, $-CH_2-CH_2COOH$; V, $-CH{=}CH_2$; IUPAC, International Union of Pure and Applied Chemistry.

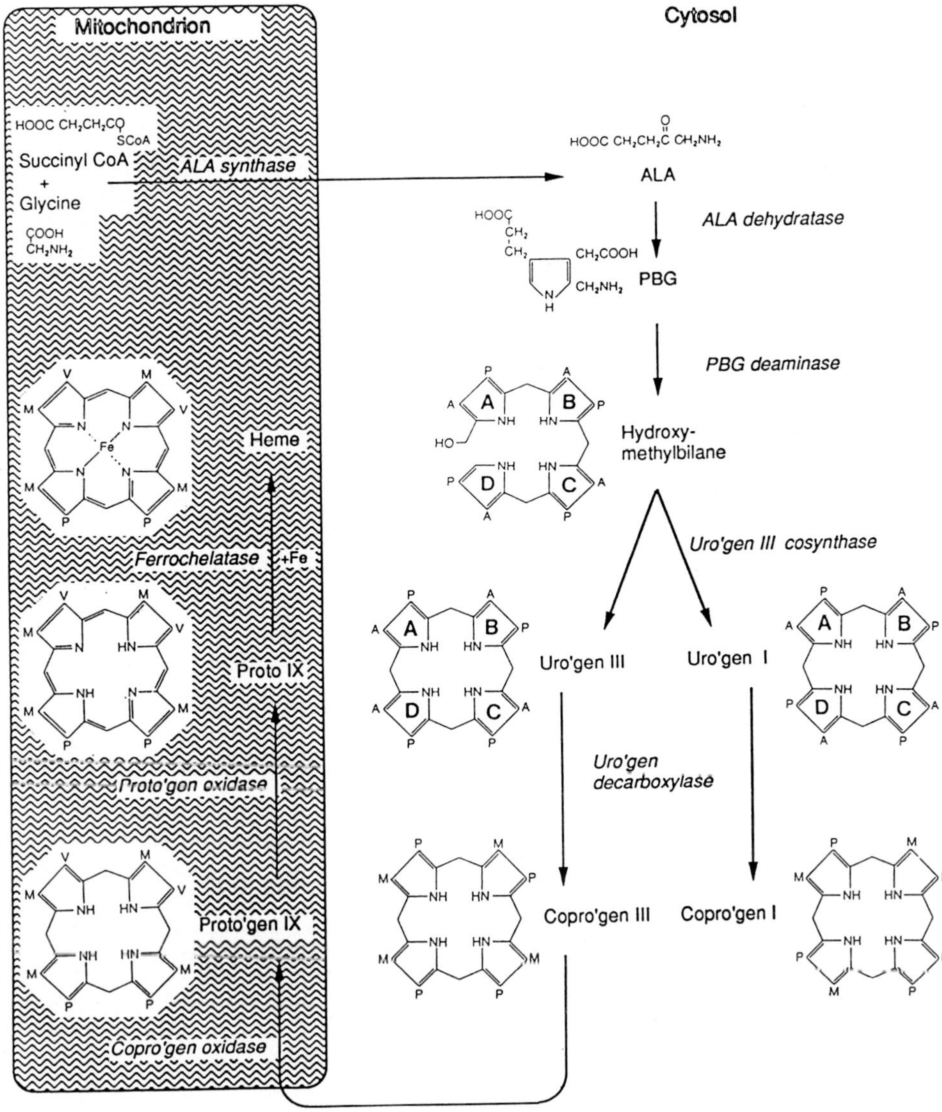


**Figure 2** The heme biosynthetic pathway. Subcellular distribution of enzymes and intermediates for heme synthesis is depicted. M, $-CH_3$; A, $-CH_2COOH$; P, $-CH_2-CH_2COOH$; V, $-CH=CH_2$.

of these enzymes are localized in the mitochondria; the intermediate enzymes are localized in the cytosol.

## A. Delta-Aminolevulinate Synthase

Delta-aminolevulinate (ALA) synthase is the first enzyme in the heme biosynthetic pathway and condenses glycine and succinyl CoA to form ALA (Fig. 2). The reaction requires pyridoxal 5'-phosphate. In mammalian cells, the enzyme is localized in the inner membrane of mitochondria. The enzyme is encoded by nuclear genes, synthesized on cytoplasmic ribosomes, and subsequently transferred into mitochondria.

There are erythroid and nonerythroid ALA synthase isoforms, which are encoded by two separate genes (1,2). The gene locus for the human erythroid ALA synthase has been assigned to Xp11.2, while that for the nonerythroid enzyme has been assigned to 3p.21 (2,3). It has been reported that ALA synthase activity is decreased in family members with an X-linked form of sideroblastic anemia (4). Recently, a point mutation in the erythroid-specific ALA synthase cDNA affecting the binding of pyridoxal 5'-phosphase was demonstrated in a patient with X-linked sideroblastic anemia (94), indicating that a mutation in the erythroid ALA synthase gene may occur in these individuals.

## B. Delta-Aminolevulinate Dehydratase

ALA dehydratase is a cytosolic enzyme that catalyzes the condensation of two molecules of ALA to form a monopyrrole, porphobilinogen (PBG), with the removal of two molecules of water (Fig. 2). The enzyme requires sulfhydryl compounds such as $\beta$-mercaptoethanol, cysteine, reduced glutathione, or dithiothreitol (5), as well as $Zn^{2+}$ for maximal activity (5,6). ALA dehydratase activity can be inhibited by lead, levulinic acid, hemin, or succinylacetone (4,6-dioxoheptanoic acid).

Human ALA dehydratase cDNA has an open reading frame of 990 bp, which codes for a protein of molecular mass (Mr) of 36,274 daltons (7). In contrast to ALA synthase, there are no tissue-specific isozymes of ALA dehydratase (8). On the other hand, there is a species specificity of ALA dehydratase consistent with phylogenetic development (8).

## C. Porphobilinogen Deaminase

Porphobilinogen (PBG) deaminase catalyzes the condensation of four PBG molecules to form hydroxymethylbilane (Fig. 2). The deaminase furnishes a straight chain tetrapyrrole hydroxymethylbilane, but it is not an enzyme for ring-closure (9). Hydroxymethylbilane is converted by uroporphyrinogen (UROGEN) III cosynthase to UROGEN III; as an alternative it can undergo a spontaneous, nonenzymatic conversion to UROGEN I. The only difference between type I and type III isomers is that substituents on positions 17 and 18 of the "D" ring of the tetrapyrrole are switched (Fig. 1, Table 1).

There are two isoforms of PBG deaminase in erythroid and in nonerythroid cells. An additional peptide of 17 amino acid residues is present at the amino terminus of the nonerythroid form of PBG deaminase, but otherwise the two isoforms are identical (10). These two isoforms are encoded by two distinct mRNAs. The two mRNAs arise from two overlapping transcription units in a single gene coding for human PBG deaminase, which spreads over 10 kb of DNA and is split into 15 exons. The first transcription unit (upstream) is active in all tissues and its promoter has certain structural features of a housekeeping promoter, while the second (located 3 kb downstream) is active only in erythroid cells. The gene locus encoding human PBG deaminase has been assigned to chromosome 11q23 → 11qter (11).

## D. Uroporphyrinogen III Cosynthase

UROGEN III cosynthase catalyzes the formation of UROGEN III from hydroxymethylbilane, or from PBG if PBG deaminase is also present (Fig. 2). Human UROGEN III cosynthase predicted from its cDNA has an open reading frame of 798 bp consisting of 263 amino acid residues, with an Mr of 28,607 daltons. There are no tissue-specific isozymes for this enzyme.

## E. Uroporphyrinogen Decarboxylase

A cytosolic enzyme, UROGEN decarboxylase, catalyzes the sequential removal of the four carboxylic groups of the acetic acid side chains in UROGEN to yield coproporphyrinogen (COPROGEN) (Fig. 2). This single enzyme catalyzes the four successive decarboxylation steps from the eight-carboxylate UROGEN to the four-carboxylate COPROGEN (12).

The cloned human UROGEN decarboxylase gene comprises 10 exons spread over 3 kb and contains two initiation sites (13). The two initiation sites are used in identical proportion in all tissues, giving rise to a unique mRNA (13). A gene locus encoding human UROGEN decarboxylase was assigned to pter → p21 region of human chromosome 1 (14).

## F. Coproporphyrinogen Oxidase

COPROGEN oxidase in mammalian cells is a mitochondrial enzyme that catalyzes the removal of the two carboxyl groups and the two hydrogens from the propionic groups of pyrrole rings A and B (Fig. 1) of COPROGEN to form vinyl groups at these positions (Fig. 2). The reaction thus yields a divinyl compound, protoporphyrinogen (PROTOGEN). This mitochondrial enzyme is situated in the intermembrane space of mitochondria. No cDNA cloning has been reported for this enzyme in mammalian cells.

## G. Protoporphyrinogen Oxidase

The oxidation of PROTOGEN to protoporphyrin (PROTO) is mediated by mitochondrial PROTOGEN oxidase, which catalyzes the removal of six hydrogen atoms from the porphyrinogen nucleus. This is the penultimate step in the heme biosynthetic pathway. No cDNA cloning has been reported for this enzyme in mammalian cells.

## H. Ferrochelatase

The final step of heme biosynthesis is the insertion of iron into PROTO. This reaction is catalyzed by the enzyme ferrochelatase (also termed heme synthase, heme synthetase, or protoheme–ferrolyase). Ferrochelatase activity in mammalian cells is localized in the inner membrane of mitochondria. Unlike other enzymatic steps in the heme biosynthetic pathway that utilize porphyrinogens, ferrochelatase utilizes PROTO as a substrate. The reduced form of iron ($Fe^{2+}$), but not the oxidized form ($Fe^{3+}$), is incorporated into PROTO by the enzyme (15).

A cDNA encoding human ferrochelatase was recently cloned and sequenced (16). The cDNA had an open reading frame of 1269 bp encoding a protein of 423 amino acid residues, with an Mr of 47,833 daltons. Amino acid sequencing showed that the mature protein consists of 369 amino acid residues corresponding to an Mr of 42,158 daltons, with a putative leader sequence of 54 amino acid residues.

# III. REGULATION OF NONERYTHROID ALA SYNTHASE IN THE LIVER

Biosynthesis of heme in the liver is controlled largely by the level of enzymatic activity of nonerythroid ALA synthase (17,18). The enzyme activity in normal liver cells is very low; however, the enzyme levels increase dramatically when the liver synthesizes more heme in response to exposure to various agents that increases the synthesis of microsomal cytochrome P450 (19). Heme suppresses the synthesis of nonerythroid ALA synthase (20) and induces microsomal heme oxygenase (21), resulting in the increased turnover of heme. Therefore, hepatic heme concentration is maintained by the balance between the synthesis of nonerythroid ALA synthase and heme oxygenase.

# IV. REGULATION OF ERYTHROID ALA SYNTHASE IN ERYTHROID CELLS

The amount of heme synthesized by erythropoietic bone marrow for new hemoglobin formation is equivalent to 85% of the total heme synthesized in the body, the rest being synthesized largely by the liver (22). There is a significant difference in the regulation of heme synthesis in erythroid cells, particularly at the level of ALA synthase (23,24). For example, heme inhibits the drug induction of nonerythroid ALA synthase in the liver, while it has little effect on erythroid ALA synthase in the bone marrow. Chemicals that induce nonerythroid ALA synthase do not induce the erythroid enzyme. As discussed earlier, erythroid and nonerythroid ALA synthases have been shown to be two separate gene products. In addition, a functional iron-responsive element has been demonstrated in the 5'-untranslated region of the human erythroid ALA synthase mRNA, whereas it is absent in the nonerythroid ALA synthase mRNA (25,26). These findings appear to account for the fact that the regulation of ALA synthase, and hence the regulation of heme synthesis between these two tissues is different.

## V. PATHOPHYSIOLOGY OF CUTANEOUS LESIONS IN THE PORPHYRIAS

The effect of porphyrins on reactive oxygen species, cells (erythrocytes, mast cells, neutrophils, and fibroblasts), and soluble mediators (the complement system, factor XII-dependent pathways, and the eicosanoids) are discussed here (Table 2). It is likely that interactions among these factors are responsible for the pathogenesis of cutaneous lesions in the porphyrias. While much has been learned about the pathophysiology of photosensitivity and sclerodermoid skin changes in porphyrias, the causes of periorbital mottled hyper- and hypopigmentation and hypertrichosis, seen most frequently in patients with porphyria cutanea tarda, remains to be elucidated.

**Table 2** Mediators in the Pathogenesis of Cutaneous Phototoxic Lesions in Porphyrias

| |
|---|
| Reactive oxygen species |
| Singlet oxygen |
| Superoxide anion |
| Superoxide radical |
| Hydroxyl radical |
| Cells |
| Mast cells |
| Neutrophils |
| Fibroblasts |
| Soluble mediators |
| The complement system |
| The eicosanoids |
| Factor XII-dependent pathways |

## A. Reactive Oxygen Species

Porphyrins such as uroporphyrin (URO), coproporphyrin (COPRO), and PROTO absorb light intensely in the 400–410 nm range, the co-called Soret band. This results in the generation of "excited state" porphyrins, which can transfer the absorbed energy to oxygen molecules, thereby generating reactive oxygen species, including singlet oxygen, superoxide anion, superoxide radical, and hydroxyl radical (27). Reactive oxygen species generated in the presence of porphyrins and light have been shown to induce photohemolysis of erythrocytes, release of mediators from mast cells, damage to hepatic and epidermal microsomal cytochrome P450, damage to lysosomal and mitochondrial membranes, and selective degradation of the guanine moiety in DNA (28). In vivo, beta-carotene, a scavenger of singlet oxygen, has been shown to be an effective treatment of erythropoietic protoporphyria (29). In a mouse model, it was demonstrated that superoxide anions played a significant role in the development of hematoporphyrin derivative-induced phototoxicity (30).

## B. Erythrocytes

Erythrocytes have been used extensively in the study of porphyrin–membrane interactions (28, 31). Radiation- and porphyrin-induced cross-linkage of proteins within membranes of erythrocytes has been demonstrated. Photohemolysis of erythrocytes in the presence of PROTO has been shown to be associated with peroxidation of cholesterol groups in cell membranes. Photohemolysis was observed only with PROTO, but not with URO or COPRO. This differential effect, also observed in studies using mast cells, polymorphonuclear cells, and fibroblasts, most likely reflects the different physicochemical properties of porphyrins. The lipophilic PROTO, a 2-carboxylate porphyrin, is more damaging to membranes than are COPRO and URO, which are a four-carboxylate and an eight-carboxylate porphyrin, respectively, and thereby more water soluble.

## C. Mast Cells

As mentioned previously, mast cell degranulation was noted in the exposed skin of patients with erythropoietic protoporphyria (EPP) (32). In vitro, release of mast cell-derived mediators was observed following exposure of mast cells to PROTO and radiation (33,34); in contrast, no mediator release was observed in the presence of URO and radiation (33). These findings may partly explain the differences in the clinical presentation of patients with EPP and porphyria cutanea tarda (PCT). The elevated PROTO in patients with EPP may induce the release of mast cell mediators in vivo following sun exposure, resulting in erythema, edema, urticarial lesions, and pruritus. The absence of these changes in patients with PCT may be explained by the lack of direct effect of URO on cutaneous mast cells.

Studies performed in animal models further confirmed the participation of mast cells in porphyrin-induced phototoxicity. In animals, porphyrin-induced phototoxicity was associated with elevated serum histamine levels and dermal mast cell degranulation, and the phototoxic response was suppressed by pretreatment with antihistamines. Furthermore, mice congenitally deficient in mast cells had a markedly diminished phototoxic response compared to mast cell-sufficient animals (28).

## D. Neutrophils

In vitro, exposure of human neutrophils to PROTO and radiation resulted in membrane damage (35). In animal models, the porphyrin-induced phototoxic response was associated with a dermal neutrophilic infiltrate, and the phototoxicity was markedly suppressed in leukopenic animals (36)

## E. Fibroblasts

The differential effect of various porphyrins on cells is clearly shown in studies using fibroblasts. An increase in collagen biosynthesis was observed following incubation of fibroblasts with URO. This effect was independent of irradiation (37) and may partly explain the sclerodermoid changes observed in patients with PCT, which can occur both in sun-exposed and sun-protected skin. In contrast, PROTO induced a photohemolysis of fibroblasts (38).

## F. The Complement System

Complement components have been detected at the vessel walls and dermoepidermal junction of skin of patients with porphyrias (39). Mast cell degranulation and infiltration of neutrophils were observed following irradiation of skin of patients with EPP, histologic changes consistent with those mediated by complement activation products. In vitro, irradiation of sera obtained from patients with EPP and PCT resulted in complement activation; similar results were seen following exposure of skin of these patients to Soret band radiation (40). In animal models, porphyrin-induced phototoxicity was suppressed in complement-depleted animals and in animals congenitally deficient in the fifth component of complement (41).

## G. The Eicosanoids

In vitro, incubation of mouse peritoneal macrophages or radiation-induced fibrosarcoma tumor cells with Photofrin II, followed by 630 nm radiation, resulted in a dose-dependent generation of $PGE_2$ (42). In a protoporphyric mouse model, the ability of epidermal eicosanoid-metabolizing enzymes to generate eicosanoids was markedly suppressed following irradiation (43).

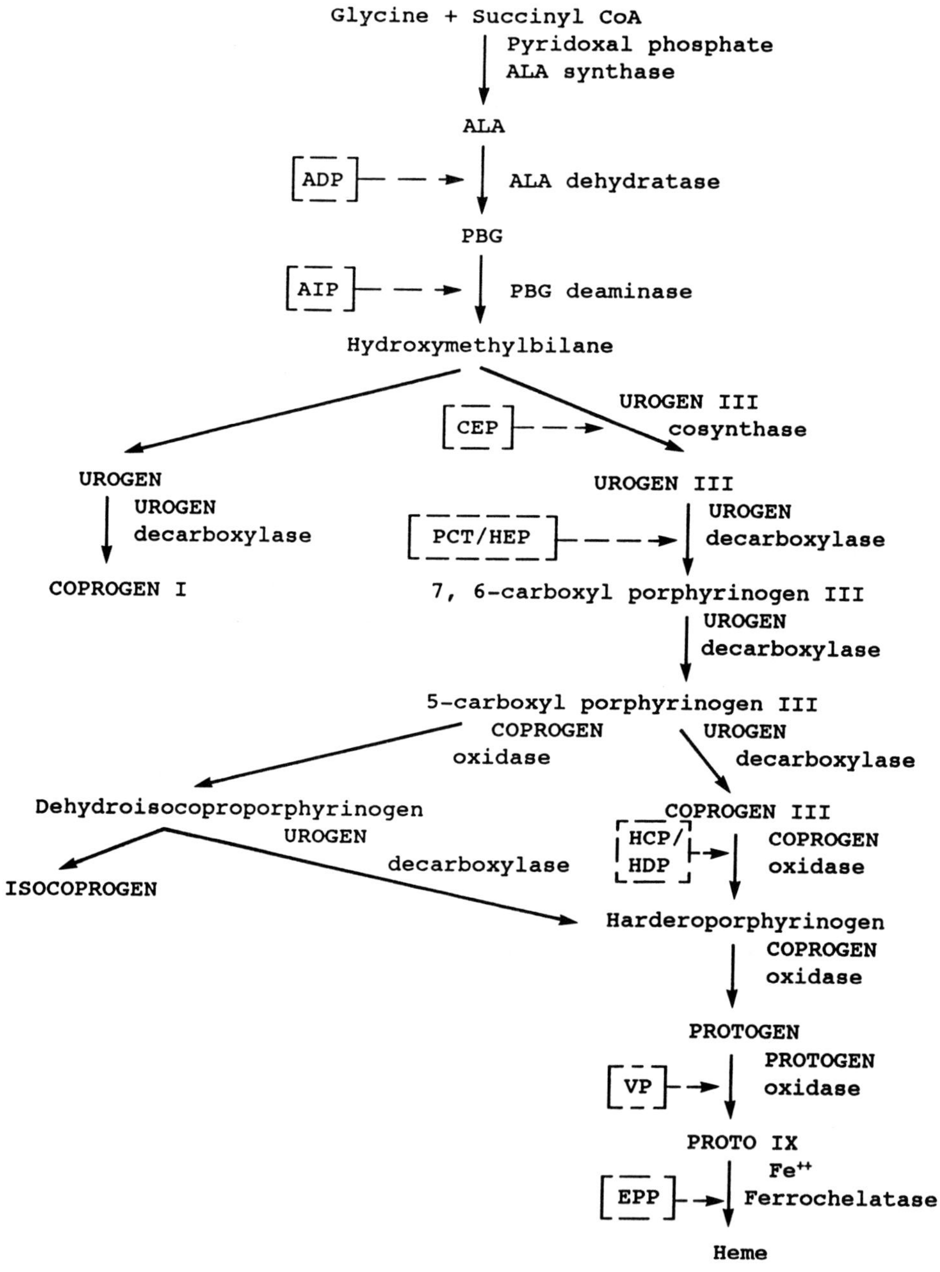
Glycine + Succinyl CoA
Pyridoxal phosphate
ALA synthase
ALA
ADP
ALA dehydratase
PBG
AIP
PBG deaminase
Hydroxymethylbilane
CEP
UROGEN III
cosynthase
UROGEN
UROGEN
decarboxylase
COPROGEN I
UROGEN III
PCT/HEP
UROGEN
decarboxylase
7, 6-carboxyl porphyrinogen III
UROGEN
decarboxylase
5-carboxyl porphyrinogen III
COPROGEN
oxidase
UROGEN
decarboxylase
Dehydroisocoproporphyrinogen
UROGEN
decarboxylase
ISOCOPROGEN
COPROGEN III
HCP/
HDP
COPROGEN
oxidase
Harderoporphyrinogen
COPROGEN
oxidase
PROTOGEN
VP
PROTOGEN
oxidase
PROTO IX
Fe++
EPP
Ferrochelatase
Heme

### H. Factor XII-Dependent Pathways of Coagulation

In vitro, activation of Hageman factor-dependent pathways in the presence of PROTO was demonstrated (44). In contrast, neither URO nor COPRO induced such activation. Whether this activation, which was independent of irradiation, contributes to the pathogenesis of porphyrin-induced phototoxity remains to be investigated further.

## VI. CLINICAL PORPHYRIAS

The porphryias will be discussed here in the order of the defective enzymes as they appear in the heme biosynthetic pathway (Fig. 3). It should be emphasized that in all patients suspected of having porphyrias, a complete porphyrin profile analysis, including quantitative porphyrin determinations in 24 hr urine, feces, plasma, and erythrocytes should be performed; all of these can be ordered through commercial laboratories. An office-based procedure of urinary Wood's light examination is only adequate as an initial screening test for those porphyrias associated with elevated urinary porphyrins (Table 3). Its modification includes acidification of the urine with glacial acetic acid, followed by the addition of talc or amyl alcohol (45).

All patients with photosensitive porphyrias should be advised to avoid sunlight; physical sunblocks offer the only complete protection: chemical sunscreens do not completely protect against the action spectrum of porphyrins (i.e., the Soret band).

### A. ALA Dehydratase Deficiency Porphyria

Deficiency of ALA dehydratase (less than 5% normal activity) is associated with symptoms similar to acute intermittent porphyria (AIP), also without any cutaneous manifestation. This condition is inherited as an autosomal recessive

---

**Figure 3** Biosynthesis of heme with sites of enzyme deficiencies in porphyrias. All porphyrinogens excreted are converted to their respective porphyrins. For simplicity, porphyrias associated with defective enzymes that catalyze multiple steps (PCT/HEP and UROGEN decarboxylase, HCP and COPROGEN oxidase) are only indicated once in this diagram.

ADP, delta-aminolevulinic acid (ALA) dehydratase deficiency porphyria; AIP, acute intermittent porphyria; CEP, congenital erythropoietic porphyria; COPROGEN, coproporphyrinogen; EPP, erythropoietic protoporphyria; HCP, hereditary coproporphyria; HDP, harderoporphyria; HEP, hepatoerythropoietic porphyria; ISOCOPROGEN, isocoproporphyrinogen; PBG, porphobilinogen; PCT, porphyria cutanea tarda; PROTO, protoporphyrin; PROTOGEN, protoporphyrinogen; UROGEN, uroporphyrinogen; VP, variegate porphyria.

trait (46). The four reported patients have markedly elevated ALA, with lesser elevated levels of COPRO, and URO in urine; PROTO is elevated in erythrocytes. Some patients also have elevated ALA, COPRO, and PROTO in plasma, and COPRO and PROTO in feces (Table 3) (47,48).

Molecular analysis of the ALA dehydratase defect was carried out in two patients. The first study was made in a German patient with an onset of symptoms in his teens (49,50). This patient was found to have two separate point mutations, one in each ALA dehydratase allele (50,51), which were compatible with compound heterozygosity. One point mutation was a substitution of A for G at nucleotide 820, which resulted in amino acid substitution from alanine to threonine at position 274. The other was a C to T transition resulting in an arginine to tryptophan substitution at position 240. The functional studies of these mutations demonstrated that the former yields an unstable enzyme, while the latter encodes a catalytically inactive enzyme (51). Compound heterozygosity was subsequently also demonstrated in another patient who developed the disease early in childhood (52). Since all heterozygotes of ALA dehydratase deficiency had no clinical consequences, these defects may be widely found as germ-line mutations. Heterozygous subjects for ALA dehydratase deficiency, however, may be at a high risk for the toxic effects of chemicals or metals known to affect ALA dehydratase activity adversely (53).

## B. Acute Intermittent Porphyria

### *1. Clinical Manifestations*

Acute intermittent porphyria (AIP) has a slight female predominance and usually has its onset after puberty. It is an autosomal dominant disorder due to a partial deficiency of PBG deaminase (Fig. 3). Patients with AIP present with episodic acute abdominal pain, vomiting, peripheral neuropathy, and paresis. Acute attacks can be precipitated by infection, starvation, and use of various medications, including barbiturates, estrogen, diphenylhydantoin, sulfonamides, and griseofulvin. Some patients may have hyponatremia due to inappropriate antidiuretic hormone secretion (54). AIP is not associated with cutaneous photosensitivity, since there is an overproduction of ALA and PBG but not of photosensitizing porphyrins.

### *2. Biochemical and Genetic Abnormalities*

ALA and PBG are elevated during acute attacks, and these porphyrin precursors usually remain elevated in between attacks. In most patients and latent gene carriers of the disease, erythrocyte PBG deaminase activity and protein are approximately 50% of the normal level (55), suggesting that both the patients and gene carriers are hetereozygous for the enzyme deficiency. These findings suggest that among the gene carriers, other factors trigger the clinical disease (22). The PBG deaminase defect in AIP patients is highly heterogenous; there are

**Table 3** Porphyrias: Enzyme Deficiencies and Porphyrin Profiles

| | | Predominant Porphyrins | | | |
|---|---|---|---|---|---|
| Enzyme Deficiency | Porphyria | RBC | Plasma | Urine | Feces |
| ALA dehydratase | ALA dehydratase deficiency porphyria (ADP) | ZnPROTO | ALA | ALA | COPRO, PROTO |
| Porphobilinogen deaminase | Acute intermittent porphyria (AIP) | — | — | ALA, PBG | — |
| Uroporphyrinogen III cosynthase | Congenital erythropoietic porphyria (CEP) | URO I > COPRO I | URO I > COPRO I | URO I > COPRO I | COPRO I > URO I |
| Uroporphyrinogen decarboxylase | Porphyria cutanea tarda (PCT) | — | URO, COPRO | URO I> III; 7-COOH-P III>I | ISOCOPRO |
| Uroporphyrinogen decarboxylase | Hepatoerythropoietic porphyria (HEP) | PROTO | URO, COPRO | URO, COPRO | COPRO ISO-CORPRO |
| Coproporphyrinogen oxidase | Hereditary coproporphyria (HCP) | — | COPRO | COPRO; in acute attacks also ALA, PBG | COPRO |
| Coproporphyrinogen oxidase | Harderoporphyria (HDP) | — | — | COPRO | HARDERO; COPRO |
| Protoporphyrinogen oxidase | Variegate porphyria (VP) | — | COPRO, PROTO | COPRO> URO: in acute attacks: ALA, PBG | PROTO > COPRO |
| Ferrochelatase | Erythropoietic protoporphyria | PROTO | PROTO | — | PROTO |

ALA, delta-aminolevulinic acid; 7-COOH-P, 7-carboxylate porphyrin; COPRO, coproporphyrin; HARDERO, harderoporphyrin; ISOCOPRO, isocoproporphyrin; PBG, porphobilinogen; PROTO, protoporphyrin; URO, uroporphyrin.

various independent mutations of PBG deaminase in AIP (56). While most patients with AIP show a cross-reactive immunologic material (CRIM)-negative mutation, a few other patients show a CRIM-positive mutation, where point mutations leading to exon skipping were found, resulting in dramatic changes in the catalytic activities of the enzyme (57, 58). In addition, some patients with AIP do not display the enzyme deficiency in erythrocytes, but their nonerythroid tissues show an approximately 50% PBG deaminase deficiency (59). At least in one patient with AIP, compound heterozygosity for PBG deaminase deficiency was documented retrospectively by studying surviving parents of the patient (60). The mutations in the parents of the proband were different from each other, and both of them corresponded to previously identified G → A changes in the coding region of the PBG deaminase mRNA.

Several restriction fragment length polymorphisms (RFLPs) of human PBG deaminase have been described (61,62). In certain families in whom enzymatic studies on erythrocyte PBG deaminase yielded equivocal results, RFLP studies have been useful in providing more definitive information in the identification of gene carriers (61,62).

### 3. Treatment

The principle of management is to suppress the induction of hepatic ALA synthase. This includes avoidance of precipitating factors, administration of carbohydrates, and infusion of hematin or heme arginate (a more stable form of hematin solution) (54).

## C. Congenital Erythropoietic Porphyria (Günther's Disease)

### 1. Clinical and Histologic Manifestations

Congenital erythropoietic porphyria (CEP) is transmitted as an autosomal recessive trait with onset usually in early childhood and, in some cases, in utero (63). Patients present with severe photosensitivity resulting in skin fragility, erosions, vesicles, marked scarring, mutilation of fingers, ears, and nose, and alopecia of the scalp. Erythrodontia due to porphyrin deposits is common, and facial hypertrichosis is occasionally observed. Other associated findings are hemolytic anemia with splenomegaly, thought to be secondary to photohemolysis of porphyrin-laden erythrocytes. Subepidermal blisters are seen on histologic examination.

### 2. Biochemical and Genetic Abnormalities

The defective enzyme in CEP is UROGEN III cosynthase, resulting in a nonenzymatic conversion of hydroxymethylbilane to UROGEN I and, in the presence of UROGEN decarboxylase, to COPROGEN I (Fig. 3). Therefore, CEP can be differentiated from other photosensitizing dermatoses by its childhood onset, severe and mutilating photosensitivity, erythrodontia, and elevated URO I and COPRO I in erythrocytes, plasma, and urine, and COPRO I in feces (Table 3).

The molecular defects in CEP responsible for the UROGEN III cosynthase deficiency have been examined by cloning and sequencing mutant cDNAs from patients (64). In one patient, two distinct mutations coexisted: a T → C change in codon 73 (Arg → Cys) and a C → T change in codon 53 (Leu → Pro). A second patient was homozygous for the same mutation in codon 53.

### 3. Treatment

Meticulous protection from sunlight is essential. The following therapeutic modalities have been used with varying degree of success: beta carotene (as a quencher of oxygen radicals), hematin infusion and transfusion of packed erythrocytes (to suppress porphyrinogenesis), and oral charcoal (to decrease porphyrin enterohepatic circulation) (65, 66).

## D. Porphyria Cutanea Tarda

### 1. Clinical and Histologic Manifestations

Porphyria cutanea tarda (PCT) is the most common type of porphyria. It begins usually in adulthood and can be classified into the sporadic form (type I) and the autosomally dominant inherited form (type II). Another form, termed type III, has also been recently described, which is characterized by familial occurrence of PCT with normal erythrocyte UROGEN decarboxylase activity (67,68). Type I PCT is frequently associated with exposure to various porphyrinogenic agents such as ethanol, estrogen, iron, and polychlorinated aromatic hydrocarbons. While PCT has been reported primarily in men, incidence in women has increased, probably as a reflection of the use of estrogen-containing birth control pills (69).

Patients with PCT usually complain of skin fragility and present with vesicles, erosions, crusts, and milia on sun-exposed areas, such as the dorsa of the hands and forearms (Fig. 4, see color plate). Other commonly seen findings include periorbital hypo- and hyperpigmentation and hypertrichosis with lanugo hair. Less frequently, sclerodermoid skin changes with calcification and ulcers can be seen in both sun-exposed and sun-protected areas (Fig. 5, see color plate). Unlike in patients with EPP, subjective discomfort such as a burning and stinging sensation following sun exposure is distinctly uncommon in PCT.

Biopsy of a vesicle reveals a characteristic subepidermal bulla with the dermal papillae rising irregularly from the floor of bulla into its cavity. There is only a sparse mononuclear cell infiltrate in the dermis. There are thickened capillary walls in the superficial dermis, which are best observed in sections stained with periodic acid-Schiff stain. These changes represent reduplication of the basal lamina as observed by electron microscopic techniques. On direct immunofluorescence, deposition of immunoglobulin and C3 has been reported at the dermal–epidermal junction as well as at blood vessel walls (39). Biopsy of a vesicle, which should include the edge of the lesion, is indicated to differentiate PCT from other blistering diseases such as bullous pemphigold (subepidermal bulla with eosinophils in the dermis), epidermolysis bullosa acquisita (subepidermal bulla, absence of thickened capillary walls), and pemphigus vulgaris (intraepidermal blister).

The association between PCT and infection with human immunodeficiency virus has been recognized since 1987 (70). It is possible that the infection may lead to impairment of hepatic function, resulting in the clincial manifestations of PCT.

Clinical presentations identical to those in PCT have been reported in patients undergoing hemodialysis and in those exposed to certain medications (e.g., furosemide, tetracycline, nalidixic acid, naprosyn) (71, 72). The term pseudoporphyria has been used to describe these conditions, since most of the patients have normal porphyrin profiles. However, it should be noted that an elevated plasma porphyrin level has been reported in some patients undergoing hemodialysis (71).

### 2. *Biochemical and Genetic Abnormalities*

The defective enzyme in PCT is UROGEN decarboxylase, which catalyzes the sequential decarboxylation of 8-carboxylate porphyrinogen (UROGEN) to yield 4-carboxylate porphyrinogen (COPROGEN) (Fig. 3). In sporadic, type I, PCT, the enzyme is defective primarily in the liver, while in the familial, type II, PCT, the defect is noted both in the erythrocytes and the liver (67). The activity of UROGEN decarboxylase is usually found at 50% of normal in type II patients. In type III PCT, UROGEN decarboxylase deficiency is present in the liver but not in erythrocytes (68).

Inherited deficiencies of UROGEN decarboxylase in patients with type II PCT were mostly CRIM-negative (22). The first mutation in familial PCT described was the replacement of glycine by valine at position 281 of the protein (73). In vitro measurements utilizing radiolabeled UROGEN decarboxylase revealed a t/2 or 12 hr for the mutant enzyme, compared with a t/2 of 102 hr for the normal enzyme. Subsequently, a splice-site mutation, resulting in exon 6 deletion, was also discovered in a pedigree with familial PCT (74). This mutation resulted in an enzyme protein with an abnormally short half-life and has been found in 5 of 22 pedigrees examined. Since this mutation has not been found in hepatoerythopoietic porphyria (HEP), it can be considered that a homozygous condition for this mutation may not be compatible with life.

The porphyrin profile of PCT reflects the defective UROGEN decarboxylase activity (Table 3). In urine, there are elevated levels of URO (isomers I>III) and 7-carboxylate porphyrin (isomers III>I), and lesser elevations of 6-, 5-, and 4-carboxylate porphyrins; therefore, Wood's light examination of the urine could serve as an office-based screening test. In stool, isocoproporphyrin (ISOCOPRO) III is the predominant porphyrin, with lesser amounts of URO, COPRO, and 7-carboxylate porphyrin. ISOCOPRO III is elevated because UROGEN decarboxylase is required for the conversion of its precursor dehydroisocoproporphyrinogen to harderoporphyrinogen (Fig. 3). Plasma porphyrin is elevated, while porphyrin in erythrocytes is normal. This profile reflects the hydrophilic nature of eight- and seven-carboxylate porphyrins.

Porphyrin profiles are essential in differentiating PCT from three other less common types of porphyria (i.e., variegate porphyria [VP], hereditary coproporphyria [HCP], and HEP), all of which can be clinically indistinguishable from PCT (Table 4). In VP, COPRO, rather than URO, is the predominant urinary porphyrin, urinary ALA and PBG are elevated during acute attacks, and fecal PROTO and COPRO levels are increased. In HCP, COPRO is elevated in the feces and, to a lesser degree, in the urine; urinary ALA and PBG are increased during attacks. In HEP, erythrocyte PROTO level is markedly elevated. It should be noted that ISOCOPRO is the hallmark of the UROGEN decarboxylase deficiency, and it is found in PCT and HEP, but not in other conditions.

Other important laboratory tests include a complete blood count (to assess hemoglobin and hematocrit levels prior to therapy), serum iron and total iron-binding capacity (to assess iron stores, which are increased in more than 30% of the patients), liver enzymes (due to the association with exposure to hepatotoxic agents), and fasting blood sugar (because of the increased incidence of diabetes).

### 3. Treatment

Patient education is an essential part of the management. Patients should be advised to avoid all known porphyrinogenic agents (e.g., ethanol, estrogen, iron, and chlorinated hydrocarbons). Avoidance of sunlight and the use of a physical sunblock (e.g., clothing, zinc oxide) are helpful. Patients should be reminded that regular sunscreens do not effectively absorb the Soret band.

The two most commonly used therapeutic modalities are phlebotomy and antimalarials. Phlebotomy is the treatment of choice, primarily due to removal of excessive iron (75). It can be done as an outpatient procedure. A unit of blood is removed weekly until the hemoglobin level reaches 10–11 g/dl or serum iron 50–60 $\mu$g/dl; in some cases, the treatment may have to be interrupted if patients become excessively symptomatic due to the anemia. Biochemical and clinical

**Table 4** Salient Differentiating Features of Porphyrin Profiles of PCT, VP, HCP, and HEP

| Porphyria | RBC | Urine | Feces |
|---|---|---|---|
| PCT | Normal | URO > COPRO | ISOCOPRO |
| VP | Normal | COPRO > URO | PROTO > COPRO |
| HCP | Normal | COPRO | COPRO |
| HEP | PROTO | URO > COPRO | ISOCOPRO |

improvements may be noted within 3–6 months. The latter is reflected in decreased skin fragility and blister formation. Remission periods range from 6 months to 10 years; recurrences can be treated with repeated phlebotomy.

Low-dosage chloroquine or hydroxychloroquine has been used successfully to treat patients with PCT in whom phlebotomy may not be appropriate (76). It is thought that these agents enhance porphyrin excretion by forming water-soluble complexes. The use of chloroquine phosphate base, 125 mg twice a week, is commonly reported in the literature; however, this preparation is not readily available in the United States. Hydroxychloroquine, 200 mg once or twice a week, is commonly used by the authors. As with phlebotomy, it takes a few months for biochemical and clinical improvements to be observed.

Other less frequently used therapeutic modalities include a combination of phlebotomy and low-dosage antimalarials, cholestyramine, iron chelators, high-dosage hydroxychloroquine, and erythropoietin (77–79).

## E. Hepatoerythropoietic Porphyria

Patients with HEP present with early-onset photosensitivity, which may progress to scarring and sclerodermoid skin changes. Erythrodontia, splenomegaly, and ectropion have been observed. At least 17 cases have been reported worldwide. HEP is inherited as an autosomal recessive pattern and is associated with marked decrease of UROGEN decarboxylase activity to $<10\%$ of normal level. The prophyrin profile is identical to that of PCT, with the exception that PROTO is significantly elevated in erythrocytes (Tables 3 and 4) (80,81).

Some HEP patients had a CRIM-positive UROGEN decarboxylase defect. In three patients with HEP, the half-lives of the enzyme were abnormally short. This was due to a point mutation (G → A) at position 860 of mRNA, resulting in a Gly → Glu replacement at position 281 (82). More recently, another point mutation, a G → A at position 517, resulting in a Glu → Lys replacement was described in a separate patient with HEP (83). Since these mutations are extremely rare in patients with type II (familial) PCT, HEP may not be a homozygous state of the UROGEN decarboxylase defect that is responsible for type II PCT, as had been originally speculated.

## F. Hereditary Coproporphyria

### *1. Clinical and Histologic Manifestations*

HCP is inherited in an autosomal dominant fashion. The predominant clinical presentation is similar to that of AIP and VP: acute abdominal and neurologic symptoms. Photosensitivity occurs in 20% of patients (84). While no skin histologic changes have been described, study of direct immunofluorescence in one patient showed deposition of IgG at the dermoepidermal junction (39).

### *2. Biochemical and Genetic Abnormalities*

Patients with HCP have 50% deficiency of COPROGEN oxidase activity, resulting in elevated urinary and fecal levels of COPRO III, the substrate for COPROGEN oxidase (Table 3, Fig. 3) (85). As in AIP and VP, elevated urinary ALA and PBG levels are observed during acute attacks.

### *3. Treatment*

Management includes avoidance of sunlight and measures similar to those in AIP (i.e., avoidance of known precipitating factors, carbohydrate loading, and hematin infusion).

## G. Harderoporphyria

Harderoporphyria has been identified only in three siblings who presented with jaundice and hemolytic anemia at birth. It is considered to be a homozygous form of HCP, based on the observation that these patients have an identical enzyme defect to that in HCP (i.e., COPROGEN oxidase), but have only 10% of activity; parents of the probands had 50% enzymatic activity (86).

## H. Variegate Porphyria

### *1. Clinical and Histologic Manifestations*

This porphyria with an autosomal dominant inheritance is most prevalent in the South African white population, and most of them are descendents of a pair of Dutch settlers. The cutaneous manifestations are similar to those of PCT: skin fragility, bullae, periorbital hypertrichosis, and mottled hyperpigmentation. As with AIP, visceral and neurologic symptoms are common, which can also be precipitated by infections and medications. However, the presence of cutaneous lesions differentiates VP from AIP. Histologic changes are similar to those in PCT.

### *2. Biochemical and Genetic Abnormalities*

The defective enzyme in VP is PROTOGEN oxidase, the penultimate enzyme in heme biosynthetic pathway (Fig. 3) (87). This results in elevated urinary URO and COPRO, with a ratio of $<1$, which serves to differentiate VP from PCT, in which the ratio is $>3$ (Tables 3,4). In feces, the predominant porphyrin in VP is PROTO and, to a much lesser extent, COPRO; in PCT, ISOCORPO is the predominant porphyrin. Urinary ALA and PBG are usually elevated only during acute attacks, while in AIP they are also often elevated between attacks.

### *3. Treatment*

Avoidance of precipitating factors, hematin infusion, and carbohydrate loading have all been used.

## I. Erythropoietic Protoporphyria

### 1. Clinical and Histologic Manifestations

Patients with EPP usually complain of a burning and stinging sensation upon exposure to sunlight; this is followed by the development of erythema, edema, and urticarial lesions (Fig. 6, see color plate). On occasion, petechiae and purpura may also occur, although, unlike in PCT, vesicles and erosions are uncommon in EPP. Waxy thickening of chronically sun-exposed skin (e.g., the dorsa of hands) and pitted scars can be seen. These symptoms commonly have their onset in childhood.

Approximately 10% of patients have been reported to have mild, nonhemolytic anemia or cholelithiasis (88). A few cases of hepatic failure, probably secondary to depositions of PROTO in the liver, have been reported (89).

On histologic examination the deposition of periodic acid–Schiff positive, diastase-resistant materials around the blood vessel wall in the papillary dermis is noted. On ultrastructural examination, this is associated with reduplication of the basal lamina. Depositions of IgG, IgM, and C3 have been reported at the dermal–epidermal junction and around blood vessels (39).

The clinical differential diagnosis includes polymorphous light eruption, solar urticaria, and other types of porphyria. Elevated porphyrin levels and thickening of capillary walls are not observed in polymorphous light eruption and solar urticaria. Porphyrin profiles are essential in differentiating EPP from other types of porphyria (see below).

### 2. Biochemical and Genetic Abnormalities

Patients with EPP have decreased activity of ferrochelatase, which catalyzes the insertion of ferrous iron into PROTO, resulting in elevated levels of PROTO (Fig. 3). The activity has been reported to be 50% of normal in various cell types (90). The fact that PROTO is a lipophilic, two-carboxylate porphyrin accounts for its characteristic distribution seen in EPP: elevated levels in erythrocytes, plasma, and stool but normal levels in the urine (Table 3).

The mode of inheritance is autosomal dominant with a variable degree of clinical expression. An inheritance mode consistent with autosomal recessive, and interdeterminant traits were also observed, suggesting that genetic heterogeneity probably exists in EPP (52).

The first study describing the molcular defect of ferrochelatase in an EPP patient has been reported (92). In a proband with EPP, ferrochelatase activity, an immunochemically quantifiable protein, and mRNA content of ferrochelatase were about 50% of the normal levels. In contrast, the rate of transcription of ferrochelatase mRNA in the proband's cells was normal, suggesting that decreased ferrochelatase mRNA was due to an unstable transcript. cDNA clones encoding ferrochelatase in the proband, isolated by the amplification of its mRNA using the polymerase chain reaction, were found to be classified either into those encoding the normal protein or into those encoding an abnormal protein that lacked exon 2 of the ferrochelatase gene. This finding indicated that the proband was heterozygous for the ferrochelatase defect. Genomic DNA analysis revealed that the abnormal allele had a point mutation, C → T, near the acceptor site of intron 1. This point mutation appeared to be responsible for the posttranscriptional splicing abnormality, resulting in an aberrant transcript of ferrochelatase in this patient.

### 3. *Treatment*

Sun protection is essential. A quencher of singlet oxygen, beta-carotene, at oral dosages ranging from 30–90 mg/day in children and 60–180 mg/day in adults is commonly used (29). A minimal plasma level of 600 $\mu$g/dl should maintained. Effects may not be noticeable for 1–3 months; therefore, in a temperate climate, the medication should be started in early spring and may be discontinued during the winter months. Beta-carotene is safe and well-tolerated; the most common side effect is yellow discoloration of the skin especially noticeable on the palms and soles of fair-skinned patients.

EPP-associated cholelithiasis can be managed surgically. Management of hepatic failure is exceedingly difficult; therapeutic modalities that have been tried include cholestyramine (to interfer with enterohepatic circulation of PROTO), iron (to enhance the conversion of PROTO to heme), and liver transplantation (89, 93).

## ACKNOWLEDGMENTS

This study was supported in part by a Merit Review from the Department of Veterans Affairs, USPHS grants AR-39749 (Skin Disease Research Center), DK-32890, DK-39264, and Suntory Fund for Biomedical Research.

## REFERENCES

1. Riddle RD, Yamamoto M, Engel JD. Expression of delta-aminolevulinate synthase in avian cells: Separate genes encode erythroid-specific and nonspecific isozymes. Proc Natl Acad Sci USA 1989; 86: 792–796.
2. Bishop DF, Henderson AS, Astrin KH. Human delta-aminolevulinate synthase: assignment of the housekeeping gene to 3p21 and the erythroid-specific gene to the X chromosome. Genomics 1990; 7: 207–214.
3. Sutherland GR, Baker E, Callen DF, Hyland VJ, May BK, Bawden MJ, Healy HM, Borthwick IA. 5-Aminolevulinate synthase is at 3p21 and thus not the primary defect in X-linked sideroblastic anemia. Am J Hum Genet 1988; 43: 331–335.
4. Aoki Y, Muranaka S, Nakabayashi K, Ueda Y. Delta-aminolevulinic acid synthetase in erythroblasts of patients with pyridoxine-responsive anemia. Hypercatabolism caused by the increased susceptibility to the controlling protease. J Clin Invest 1979; 64: 1196–1203.

5. Anderson PM, Desnick RJ. Purification and properties of delta-aminolevulinate dehydratase from human erythrocytes. J Biol Chem 1979; 254: 6924–6930.
6. Tsukamoto I, Yoshinaga T, Sano S. Zinc and cysteine residues in the active site of bovine liver delta-aminolevulinic acid dehydratase. Int J Biochem 1980; 12: 751–756.
7. Wetmur JB, Bishop DF, Cantelmo C, Desnick RJ. Human delta-aminolevulinate dehydratase: nucleotide sequence of a full length cDNA clone. Proc Natl Acad Sci USA 1986; 83:7703–7707.
8. Chang CS, Sassa S, Doyle D. An immunological study of delta-aminolevulinic acid dehydratase: specificity consistent with the phylogeny of species. Biochim Biophys Acta 1984; 797: 297–301.
9. Battersby AR, Fookes CJR, Matcham CWJ, McDonald E. Biosynthesis of the pigments of life: formation of the macrocycle. Nature 1980; 285: 17–21.
10. Grandchamp B, Verneuil H de, Beaumont C, Chretien S, Walte RO, Nordmann Y. Tissue-specific expression of porphobilinogen deaminase. Two isoenzymes from a single gene. Eur J Biochem 1987; 162: 105–110.
11. Wang AL, Arredondo Vega FX, Giampietro PF, Smith M, Anderson WF, Desnick RJ. Regional gene assignment of human porphobilinogen deaminase and esterase A4 to chromosome 11q23 → 11qter. Proc Natl Acad Sci USA 1981; 78: 5734–5738.
12. De Verneuil H, Sassa S, Kappas A. Purification and properties of uroporphyrinogen decarboxylase from human erythrocytes. A single enzyme catalyzing the four sequential decarboxylations of uroporphyrinogens I and III. J Biol Chem 1983; 258: 2454–2460.
13. Romeo P-H, Raich N, Dubart A, Beaupain D, Pryor M, Kushner J, Cohen-Solal M, Goossens M. Molecular cloning and nucleotide sequence of a complete human uroporphyrinogen decarboxylase cDNA. J Biol Chem 1986; 261:9825–9831.
14. McLellan T, Pryor MA, Kushner JP, Eddy RL, Shows TB. Assignment of uroporhyrinogen decarboxylase (UROD) to the pter → p21 region of human chromosome 1. Cytogenet Cell Genet 1985; 39: 224–227.
15. Porra RJ, Jones OTG. Studies on ferrochelatase. 1. Assay and properties of ferrochelatase from a pig liver mitochondrial extract. Biochem J 1963; 87: 181–185.
16. Nakahashi Y, Taketani S, Okuda M, Inoue K, Tokunaga R. Molecular cloning and sequence analysis of cDNA encoding human ferrochelatase. Biochem Biophys Res Commun 1990; 173: 748–755.
17. Granick S, Sassa S. Delta-aminolevulinic acid synthetase and the control of heme and chlorophyll synthesis. In: Vogel HJ, ed. Metabolic Pathway, vol 5. Metabolic Regulation. New York: Academic Press, 1971: 77.
18. Granick S. The induction in vitro of the synthesis of delta-aminolevulinic acid synthetase in chemical porphyria: a response to certain drugs, sex hormones, and foreign chemicals. J Biol Chem 1966; 241: 1359–1375.
19. Sassa S, Kappas A. Genetic, metabolic, and biochemical aspects of the prophyrias. In: Harris H, Hirschhorn K, eds. Advances in Human Genetics. New York: Plenum, 1981: 121.
20. Sassa S, Granick S. Induction of delta-aminolevulinic acid synthetase in chick embryo liver cells in culture. Proc Natl Acad Sci USA 1970; 67: 517–522.

21. Tenhunen R, Marver HS, Schmid R. The enzymatic catabolism of hemoglobin: stimulation of microsomal heme oxygenase by hemin. J Lab Clin Med 1970; 75: 410–425.
22. Kappas A, Sassa S, Galbraith RA, Nordmann Y. The porphyrias. In: Scriver CR, Beaudet AL, Sly WS, Valle D, eds. The Metabolic Basis of Inherited Disease. New York: McGraw-Hill, 1989: 1305.
23. Sassa S, Granick JL, Eisen H, Ostertag W. Regulation of heme synthesis in mouse Friend virus-transformed cells in culture. In: Murphy MJ Jr, ed. In Vitro Aspects of Erythropoiesis. New York: Springer-Verlag, 1978: 135.
24. Fujita H, Yamamoto M, Yamagami T, Hayashi N, Sassa S. Erythroleukemia differentiation. Distinctive responses of the erythroid-specific and the nonspecific delta-aminolevulinate synthase mRNA. J Biol Chem 1991; 266: 17494–17502.
25. Dandekar T, Stripecke R, Gray N, Goossens B, Constable A, Johansson HE, Hentze MW. Identification of a novel iron-responsive element in murine and human erythroid delta-aminolevulinic acid synthase mRNA. EMBO J 1991; 10: 1903–1909.
26. Cox TC, Bawden MJ, Abraham NG, Bottomley SS, May BK, Baker E, Chen LZ, Sutherland GR. Erythroid 5-aminolevulinate synthase is located on the X chromosome. Am J Hum Genet 1990; 46: 107–111.
27. Spikes JD. Photodynamic reactions in photomedicine. In: Regan JD, The Science of Photomedicine. New York: Plenum Press, 1982: 113.
28. Lim HW. Pathophysiology of cutaneous lesions in porphyrias. Semin Hematol 1989; 26: 114–119.
29. Mathews-Roth MM, Pathak MA, Fitzpatrick TB, et al. Beta-carotene as an oral photoprotective agent in erythropoietic protoporphyria. JAMA 1974; 228: 1004–1008.
30. Athar M, Elmets CA, Bickers DR, Mukhtar H. A novel mechanism for the generation of superoxide anions in hematoporphyrin derivative-mediated cutaneous photosensitization. Activation of the xanthine oxidase pathway. J Clin Invest 1989; 83: 1137–1143.
31. Goldstein BD, Harber LC. Erythropoietic protoporphyria: lipid peroxidation and red cell membrane damage associated with photohemolysis. J Clin Invest 1972; 51: 892–899.
32. Schnait FG, Wolff K, Konrad K. Erythropietic protoporphyria—submicroscopic events during the acute photosensitivity flare. Br J Dermatol 1975; 92: 545–557.
33. Lim HW, Wasserman SI, Gigli I. Differential effects of protoporphyrin and uroporphyrin on murine mast cells. J Invest Dermatol 1987; 88: 281–286.
34. Yen A, Gigli I, Barrett KE. Dual effects of protoporphyrin and long wave ultraviolet light on histamine release from rat peritoneal and cutaneous mast cells. J Immunol 1990; 144: 4327–4332.
35. Sandberg S, Glette J, Hopen J, Solberg CO, Romslo I. Porphyrin-induced photodamage to isolated human neutrophils. Photochem Photobiol 1981; 34: 471–475.
36. He D, Soter NA, Lim HW. The late phase of hematoporphyrin derivative-induced phototoxicity in mice: release of histamine and histologic changes. Photochem Photobiol 1989; 50: 91–95.
37. Varigos G, Schiltz JR, Bickers DR. Uroporphyrin I stimulation of collagen

biosynthesis in human skin fibroblasts. A unique dark effect of porphyrin. J Clin Invest 1982; 69: 129–135.
38. Wakulchik SD, Schiltz JR, Bickers DR. Photolysis of portoporphyrin-treated human fibroblasts in vitro. J Lab Clin Med 1980; 96: 158–167.
39. Epstein JH, Tuffanelli DH, Epstein WL. Cutaneous changes in the porphyrias. Arch Dermatol 1973; 107: 689–698.
40. Lim HW, Poh-Fitzpatrick MB, Gigli I. Activation of the complement system and generation of chemotactic activity in vivo in patients with porphyrias. J Clin Invest 1984; 74: 1961–1965.
41. Lim HW, Hagan M, Gigli I. Phototoxicity induced by hematoporphyrin derivative in C5-deficient, mast cell-deficient and leukopenic mice. Photochem Photobiol 1986; 44: 175–180.
42. Henderson BW, Donovan JM. Release of prostaglandin $E_2$ from cells by photodynamic treatment in vitro. Cancer Res 1989; 49: 6896–6900.
43. He D, Lim HW. Irradiation of protoporphyric mice induces down-regulation of epidermal eicosanoid metabolism. J Invest Dermatol 97: 488–494, 1991.
44. Becker CG, Wagner M, Kaplan A, Silverberg M, Grady RW, Liem H, Müller-Eberhard U. Activation of factor XII-dependent pathways in human plasma by hematin and protoporphyrin. J Clin Invest 1985; 76: 413–419.
45. Cripps DJ, Peters HA. Fluorescing erythrocytes and porphyrin screening tests on urine, blood and stool. Arch Dermatol 1967; 96: 712–720.
46. Doss M, Benkmann HG, Goedde HW. Delta aminolevulinic acid dehydrase (porphyrinogen synthase) in two families with inherited enzyme deficiency. Clin Genet 1986; 30: 191–198.
47. Fujita H, Sassa S, Lundgren J, Holmberg L, Thunell S, Kappas A. Enzymatic defect in a child with hereditary hepatic porphyria due to homozygous delta-aminolevulinic acid dehydratase deficiency: immunochemical studies. Pediatrics 1987; 80: 880–885.
48. Hassoun A, Verstraeten L. Biochemical diagnosis of an hereditary aminolaevulinate dehydratase deficiency in a 63-year-old man. J Clin Chem Clin Biochem 1989; 27: 781–786.
49. Doss M, von Tiepermann R, Schneider J, Schmid H. New type of hepatic porphyria with porphobilinogen synthase defect and intermittent acute clinical manifestation. Klin Wochenschri 1979; 57: 1123–1127.
50. Ishida N, Fujita H, Noguchi T, Doss M, Kappas A, Sassa S. Message amplification phenotyping of an inherited delta-aminolevulinate dehydratase deficiency in a family with acute hepatic porphyria. Biochem Biophys Res Commun 1990; 172: 237–242.
51. Ishida N, Fujita H, Fukuda Y, Noguchi T, Doss M, Kappas A, Sassa S. Cloning and expression of the defective genes from a patient with delta-aminolevulinate dehydratase porphyria. J Clin Invest 1992; 89: 1431–1437.
52. Plewinska M, Thunell S, Holmberg L, Wetmur JG, Desnick RJ. Delta-aminolevulinate dehydratase deficient porphyria: identification of the molecular lesions in a severely affected homozygotes. Am J Hum Genet 1991; 49: 167–174.
53. Sassa S, Fugita H, Kappas A. Genetic and chemical influences on heme biosynthe-

sis. In: Kotyk A, Skoda J, Paces V, Kostka V, eds. Highlights of Modern Biochemistry. Zeist: VSP Int. Science Publishers, 1989: 329.
54. Sack GH Jr. Acute intermittent porphyria. JAMA 1990; 264: 1290–1293.
55. Nordmann Y, De Verneuil H, Deybach JC, Delfau MH, Grandchamp B. Molecular genetics of porphyrias. Ann Med 1990; 22: 387–391.
56. Delfau MH, Picat C, De Rooij F, Voortman G, Deybach JC, Nordmann Y, Grandchamp B. Molecular heterogeneity of acute intermittent porphyria: identification of four additional mutations resulting in the CRIM-negative subtype of the disease. Am J Hum Genet 1991; 49: 421–428.
57. Desnick RJ, Ostasiewicz LT, Tishler PA, Mustajoki P. Acute intermittent porphyria: characterization of a novel mutation in the structural gene for porphobilingen deaminase. Demonstration of noncatalytic enzyme intermediates stabilized by bound substrate. J Clin Invest 1985; 76: 865–874.
58. Delfau MH, Picat C, De Rooij FW, Hamer K, Bogard M, Wilson JH, Deybach JC, Nordmann Y, Grandchamp B. Two different point G to A mutations in exon 10 of the porphobilinogen deaminase gene are responsible for acute intermittent porphyria. J Clin Invest 1990; 86: 1511–1516.
59. Grandchamp B, Picat C, Mignotte V, Wilson JH, Te Velde K, Sandkuyl L, Romeo PH, Goossens M, Nordmann Y. Tissue-specific splicing mutation in acute intermittent porphyria. Proc Natl Acad Sci USA 1989; 86: 661–664.
60. Beukeveld GJJ, Wolthers BG, Nordmann Y, Deyback JC, Grandchamp B, Wadman SK. A retrospective study of a patient with homozygous form of acute intermittent porphyria. J Inherited Metab Dis 1990; 13: 673–683.
61. Llewellyn DH, Elder GH, Kalsheker NA, Marsh OW, Harrison PR, Grandchamp B, Picat C, Nordmann Y, Romeo PH, Goossens M. DNA polymorphism of human porphobilinogen deaminase gene in acute intermittent porphyria. Lancet 1987; 2: 706–708.
62. Lee, JS, Anvret M, Lindsten J, Lannfelt L, Gellerfors P, Wetterberg L, Floderus Y, Thunell S. DNA polymorphisms within the porphobilinogen deaminase gene in two Swedish families with acute intermittent porphyria. Hum Genet 1988; 79: 379–381.
63. Ippen H, Fuchs T. Congenital porphyria. Clin Haematol 1980; 9: 323–344.
64. Deybach J-C, De Verneuil H, Boulechfar S, Grandchamp B, Nordmann Y. Point mutations in the uroporphyrinogen III synthase gene in congenital erythropoietic porphyria (Günther's disease). Blood 1990; 75: 1763–1765.
65. Piomelli S, Poh-Fitzpatrick MB, Seaman C, Skolnick LM, Berdon WE. Complete suppression of the symptoms of congenital erythropoietic porphyria by long term treatment with high level transfusions. N Engl J Med 1986; 314: 1029–1031.
66. Pimstone NR, Gandhi SN, Mukerji SK. Therapeutic efficacy of oral charcoal in congenital erythropoietic porphyria. N Engl J. Med 1987; 316: 390–393.
67. Elder GH, Roberts AG, de Salamanca RE. Genetics and pathogenesis of human uroporphyrinogen decarboxylase defects. Clin Biochem. 1989; 22: 163–168.
68. Held JL, Sassa S, Kappas A, Harber LC. Erythrocyte uroporphyrinogen decarboxylase activity in porphyria cutanea tarda: a study of 40 consecutive patients. J Invest Dermatol 1989; 93: 332–334.
69. Grossman ME, Bickers DR, Poh-Fitzpatrick MB, DeLeo VA, Harber LC. Porphyr-

ia cutanea tarda. Clinical features and laboratory findings in 40 patients. Am J Med 1979; 67: 277–286.
70. Cohen P. Porphyria cutanea tarda in human immunodeficiency virus-seropositive men: case report and literature review. J AIDS 1991; 4: 1112–1117.
71. Poh-Fitzpatrick MB, Bellet N, DeLeo VA, Grossman ME, Bickers DR. Porphyria cutanea tarda in two patients treated with hemodialysis for chronic renal failure. N Engl J Med 1978; 299: 292–294.
72. Suarez SM, Cohen PR, DeLeo VA. Bullous photosensitivity to naproxen: "pseudoporphyria." Arthritis Rheum 1990; 33: 903–908.
73. Garey JR, Hansen JL, Harrison LM, Kennedy JB, Kushner JP. A point mutation in the coding region of uroporphyrinogen decarboxylase associated with familial porphyria cutanea tarda. Blood 1989; 73: 892–895.
74. Garey JR, Harrison LM, Franklin KF, Metcalf KM, Radisky ES, Kushner JP. Uroporphyrinogen decarboxylase: a splice site mutation causes the deletion of exon 6 in multiple families with porphyria cutanea tarda. J Clin Invest 1990; 86:1416–1422.
75. Ippen H. Treatment of porphyria cutanea tarda by phlebotomy. Semin Hematol 1977; 14: 253–260.
76. Kordac V, Papezova R, Semradova M. Chloroquine in the treatment of porphyria cutanea tarda. N Engl J Med 1977; 296:949.
77. Grossman ME, Poh-Fitzpatrick MB. Porphyria cutanea tarda. Diagnosis, management, and differentiation form other hepatic porphyrias. Dermatol Clin 1986; 4: 297–309.
78. Petersen CS, Thomsen K. High-dose hydroxychloroquine treatment of porphyria cutaneous tarda. J Am Acad Dermatol 1992; 26: 614–619.
79. Anderson KE, Goeber DE, Carson RW, Lee SM, Stead RB. Erythropoietin for the treatment of porphyria cutanea tarda in a patient on long-term hemodialysis. N Engl J Med 1990; 322: 315–317.
80. Lim HW, Poh-Fitzpatrick MB. Hepatoerythropoietic porphyria: a variant of childhood-onset porphyria cutanea tarda. J Am Acad Dermatol 1984; 11: 1103–1111.
81. Toback AC, Sassa S, Poh-Fitzpatrick MB, Schechter J, Zaider E, Harber LC, Kappas A. Hepatoerythropoietic porphyria: clinical, biochemical, and enzymatic studies in a three generation family lineage. N Engl J Med 1987; 316: 645–650.
82. Verneuil H de, Grandchamp B, Beaumont C, Picat C, Nordmann Y. Uroporphyrinogen decarboxylase structural mutant (Gly281 → Glu) in a case of porphyria. Science 1986; 234: 732–734.
83. Romana M, Grandchamp B, Dubart A, Amselem S, Chabret C, Nordmann Y, Goossens M, Romeo PH. Identification of a new mutation responsible for hepatoerythopoietic porphyria. Eur J Clin Invest 1991; 21: 225–229.
84. Brodie NJ. Hereditary coproporphyria. Q J Med 1977; 46: 229–242.
85. Grandchamp B, Nordmann Y. Decreased lymphocyte and coproporphyrinogen-free oxidase activity in hereditary coproporphyria. Biochem Biophys Res Commun 1977; 74: 1089–1095.
86. Nordmann Y, Grandchamp B, Verneuil H de, Phung L, Cartigny B, Fontaine G. Harderoporphyria: a variant of hereditary coproporphyria. J Clin Invest 1983; 72: 1139–1149.

87. Brenner DA, Bloomer JR. The enzymatic defect in variegate porphyria. Studies with human cultured skin fibroblasts. N Engl J Med 1980; 302: 765–769.
88. Schmidt H, Snitker G, Thomsen K, Lintrup J. Erythropoietic protoporphyria: a clinical study based on 29 cases in 14 families. Arch Dermatol 1974; 110: 58–64.
89. Bloomer JR. Pathogenesis and therapy of liver disease in protoporphyria. Yale J Biol Med 1979; 52: 39–48.
90. Sassa S, Zalar GL, Poh-Fitzpatrick MB, Anderson KE, Kappas A. Studies in porphyria. Functional evidence for a partial deficiency of ferrochelatase activity in mitogen-stimulated lymphocytes form patients with erythropoietic protoporphyria. J Clin Invest 1982; 69: 809–815.
91. Norris PG, Nunn AV, Hawk JLM, Cox TM. Genetic heterogeneity in erythropoietic protoporphyria: a study of the enzymatic defect in nine affected families. J Invest Dermatol 1990; 95: 260–263.
92. Nakahashi Y, Fujita H, Taketani S, Ishida N, Kappas A, Sassa S. The molecular defect of ferrochelatase in a patient with erythropoietic protoporphyria. Proc Natl Acad Sci USA 1992; 89: 281–285.
93. Shehade SA, Chalmers RJ, Prescott RJ. Predictable and unpredictable hazards of erythropoietic protoporphyria. Clin Exp Dermatol 1991; 16: 185–187.
94. Cotter PD, Baumann M, Bishop DF. Enzymatic defect in "X-linked" sideroblastic anemia: Molecular evidence for erythroid δ-aminolevulinate synthase deficiency. Proc Natl Acad Sci USA 1992; 89: 4028–4032.

# 16

# Photoexacerbated Dermatoses

**Mary Ruth Buchness**
*Department of Veterans Affairs Medical Center and New York University School of Medicine, New York, New York*

Numerous dermatoses can be exacerbated by sunlight but cannot be classified into a simple category. They can be classified into autoimmune diseases, genodermatoses, infectious diseases, nutritional deficiencies, neoplasms, and primary dermatologic diseases. In some cases, the action spectrum of photosensitivity has been clearly determined; in others, conflicting results were reported; and in still others, a summer exacerbation originally thought to be due to ultraviolet radiation now appears to be due to other factors, such as heat, humidity, or the use of sunscreens. Ultraviolet radiation will lead to an exacerbation of some diseases by the Koebner reaction, either in areas of ultraviolet-induced erythema or in areas simultaneously affected by other photodermatoses. Examples include the induction of psoriatic lesions within a sunburned area or within lesions of polymorphous light eruption.

This chapter includes a discussion of the diseases of the skin that are induced or exacerbated by ultraviolet light exposure and that have not been reviewed in other chapters.

## I. AUTOIMMUNE DISEASES

### A. Lupus Erythematosus

Photosensitivity, defined as a skin rash as a result of unusual reaction to sunlight, documented by history or observed by physician, is included as a criterion distinct from malar rash and discoid lupus in the 1982 American Rheumatism Association (ARA) criteria for systemic lupus erythematosus (SLE) (1). Between 15 and 19% of black patients with SLE are photosensitive compared to 32–71% of white patients with SLE (2–4) (Fig. 1, see color plate). The presence of photosensitivity correlates with the incidence of antibodies to the cytoplasmic antigen Ro (SS-A) (4,5), with photosensitivity being present in up to 90% of Ro-positive patients (5). In one series, Ro-positivity was determined to be a greater predictor of photosensitivity in white patients. In white patients, 87% of Ro-positive and 54% of Ro-negative patients were photosensitive, while in black patients, 4% of Ro-positive and 55% of Ro-negative individuals were photosensitive. These findings led the authors to suggest that Ro-positive black patients may have a blocking factor that protects them against photosensitivity (4).

Three specific subsets of patients with LE merit further discussion. Subacute cutaneous lupus erythematosis (SCLE), which is more common in women and in white individuals, is characterized by annular erythematous plaques with scale on the chest, back, and extremities. There are no follicular plugs, and the lesions heal without scars or atrophy. There are two morphologic variants: a papulosquamous pattern (Fig. 2, see color plate) and an annular pattern, in which the lesions coalesce to produce polycyclic, gyrate lesions with central clearing (6). The incidence of photosensitivity ranges from 52 to 87% in this group of patients (6–8). Fifty percent of all patients with SCLE fulfill the ARA criteria for SLE (6). A lower incidence and less severe manifestations of renal and neurologic disease are present in SCLE patients than in those with SLE (6). SCLE patients tend to be anti-Ro positive (7), and antinuclear antibody (ANA) negative (9).

The second subset of LE is neonatal LE, which occurs in infants born to women with anti-Ro antibodies. The pathogenesis involves transplacental transfer of maternal IgG antibodies to Ro (10). The clinical manifestations include lesions of cutaneous LE, which may first develop in response to sun exposure and generally resolve by 6 months of age (8), with or without congenital heart block (11). The congenital heart block, unlike the skin lesions, is not reversible.

Discoid lupus erythematosus (chronic cutaneous LE), the third subset of LE, is characterized by plaques with erythema, telangiectases, follicular plugs, and scale. The lesions resolve leaving an atrophic, scarred, depigmented plaque that may have a peripheral rim of hyperpigmentation (Fig. 3). It is more common in women and in black patients. Discoid lesions occur primarily on the sun-exposed areas of the face, scalp, and ears and are found on non-sun-exposed areas in only 7% of patients (12). Photosensitivity, defined as worsening of preexisting lesions, onset of new lesions after sun exposure, or an exacerbation in the spring or summer, is present in 50–70% of patients (2, 12, 13). Two studies have shown that approximately 6.5% of patients with DLE will progress to SLE (12, 14).

Until recently, the action spectrum for the induction of LE skin lesions was believed to be in the ultraviolet B (UVB) range, with anecdotal reports of disease flares after exposure to ultraviolet-A (UVA) (15). Fifteen patients (13 with DLE, 2 with SCLE) were irradiated with eight times their minimal erythema doses (MED) over a range of wavelengths from 300 to 500 nm (15). Nine developed

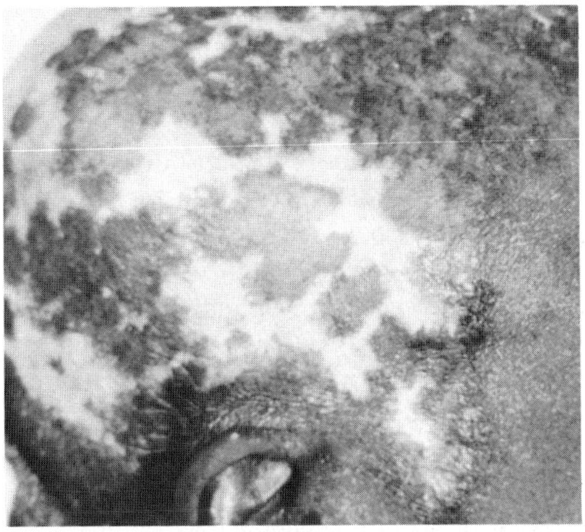

**Figure 3** Discoid lupus erythematosus. Erythema, depigmentation, hyperpigmentation, and alopecia involving most of the scalp.

either persistent erythema or lupus-like lesions only in the area irradiated with 300 nm. In another series, provocative testing of 21 patients with SLE and 4 with DLE with UVB resulted in photosensitivity in 9 patients, 5 of whom developed papules and plaques histologically consistent with LE within the irradiated areas (2). A third study demonstrated that irradiation of skin of clinically photosensitive patients with disseminated DLE with wavelengths between 250 and 330 nm showed no differences in MED tests between patients and controls. Irradiation with 8–10 MEDs or with 5–8 MEDs and the addition of 5 MEDs to the same site later, with wavelengths between 250 to 305 mm, resulted in the appearance of lupus-like lesions at the test sites (17). No differences in the erythema threshold between 23 patients with DLE, 5 with SLE, and normal controls were detected in another study; however, 1 patient with "subacute" LE had a decreased erythema threshold and prolonged erythema (18).

In contrast to the above findings, the action spectrum of LE was found to extend into the UVA range in a study of 128 patients, 53% of whom developed lesions on exposure to UVA and UVB, 14% on exposure to UVA alone, and 33% on exposure to UVB. Lesions were inducible in 64% of patients with SCLE, 42% with DLE, and 25% with SLE (8).

On histologic examination, lesional skin in LE is characterized by hyperkeratosis with follicular plugs, a thin and flat stratum malphigii, focal hydropic degeneration of basal cells, a patchy infiltrate composed predominantly of lymphocytes with a tendency towards a periappendageal distribution, edema, venular vasodilation, extravasation of red blood cells in the upper dermis, and thickening of the basement membrane zone (19). Not all of the criteria need be present to make a diagnosis of cutaneous LE. It is not possible to differentiate reliably the subsets of cutaneous LE on a histologic basis (19, 20).

Direct immunofluorescence tests show granular deposits of IgG and/or IgM at the basement membrane zone in lesional skin of 90% of patients with SLE and DLE (22) and 60% of patients with SCLE (6). Therefore, the presence of these deposits can be used to make a diagnosis when the histopathologic appearance is equivocal. Immunoglobulins in the same distribution are found in nonlesional skin of 50% of patients with SLE and in less than 5% of patients with DLE (22).

The first step in the treatment of cutaneous LE is avoidance of ultraviolet light. With the recent evidence that the action spectrum of disease involves both UVA and UVB, a broad-spectrum sunscreen is recommended. Discoid lesions, as well as limited SCLE, will respond to topical or intralesional administration of corticosteroids in most cases (23).

Patients who are refractory to treatment with sunscreens and topical or intralesional corticosteroids may require treatment with antimalarials, at least on a seasonal basis. Before the initiation of therapy, and at 6 month intervals, an ophthalmologic examination is essential, since one of the rare side effects of the antimalarials is irreversible macular degeneration that can result in blindness. This appears to be less common with the use of hydroxychloroquine sulfate than with chloroquine phosphate, and may be related more to daily dose than to cumulative total dose.

Therapy should begin with 200–400 mg/day of hydroxychloroquine sulfate. If a response is not seen in 6–8 weeks, quinacrine hydrochloride 100 mg/day may be added. An attempt should be made to taper or to discontinue the antimalarials as early as possible. Cosmesis may be of benefit in some patients: makeup to cover depigmented lesions and wigs to cover scalp lesions. Excision of lesions is contraindicated because the disease will generally recur within the scar.

## B. Pemphigus

### *1. Pemphigus Foliaceous*

Pemphigus comprises a group of diseases characterized histologically by acantholysis of the epidermis. In pemphigus foliaceous, the level of acantholysis is through the stratum granulosum (24). Patients present with flaccid bullae on the scalp, face, and upper trunk. Occasionally the lesions generalize. Rupture of the bullae leads to eczematous-appearing lesions. Evidence that pemphigus foliaceous may be exacerbated by ultraviolet light is the report of the induction of a Nikolsky's sign in six of seven patients with pemphigus foliaceous after UVB irradiation (25). This response could not be duplicated after 5 days of treatment with an antimalarial drug. The Nikolsky's sign was not produced by friction, infrared radiation, occlusive dressings, or a topically applied irritant. Two of the seven patients had a clinical history of exacerbation by sunlight.

#### 2. *Pemphigus Erythematosus (Senear-Usher Disease)*

Pemphigus erythematosus is clinically a less severe form of pemphigus foliaceous (24), with lesions localized to the butterfly area of the face and scattered elsewhere (26). Direct immunofluorescence tests of skin from patients show deposits of immunoglobulins both intercellularly and at the dermoepidermal junction; the ANA is positive (27). Lesions have been induced by irradiation of skin with UVB in two patients (28, 29).

#### 3. *Pemphigus Vulgaris*

New lesions were induced by UVB in each of four patients with pemphigus vulgaris tested in one study. Biopsy specimens of the UVB-induced lesions showed the characteristic histopathologic appearance of pemphigus vulgaris and either the appearance of or an increase in epidermal intercellular staining for immunoglobulins and complement proteins (30).

### C. Bullous Pemphigoid

Two patients with bullous pemphigoid developed lesions histologically diagnostic of bullous pemphigoid in UVB-irradiated sites (30). Other authors have reported patients who developed lesions of bullous pemphigoid while receiving phototherapy for unrelated diseases (31, 32).

## II. GENODERMATOSES

### A. Familial Benign Chronic Pemphigus (Hailey-Hailey Disease)

Familial benign chronic pemphigus is inherited in an autosomal dominant pattern and is characterized by flaccid vesicles and bullae that rupture easily to leave eroded, crusted plaques. Lesions generally occur on the sides and the nape of the neck and in intertriginous areas. Lesions are clinically exacerbated by ultraviolet light (33). In addition, lesions with the characteristic histologic appearance of familial benign chronic pemphigus have been induced in seven patients by UVB radiation (34) and in two heterozygous carriers by UVA radiation (35). Lesions have, however, also been induced by heat, cold, topical applications of allergens, and bacteria (36), so the induction of lesions by ultraviolet radiation is most likely a Koebner phenomenon.

### B. Keratosis Follicularis (Darier's Disease)

Darier's disease is an autosomal dominantly inherited disorder. Patients present clinically with crusted, greasy, brownish papules prominent on the seborrheic areas of the face, scalp, and trunk. The disease begins in childhood or at adolescence and can flare after exposure to sunlight (37). Both UVA and UVB radiation as well as other mechanical and physical trauma can induce lesions (38).

## C. Bloom's Syndrome (Congenital Telangiectatic Erythema)

Bloom's syndrome (Fig. 4, see color plate) is inherited in an autosomal recessive manner. The patients present in infancy with a telangiectatic erythema of the butterfly area of the face and occasionally of the dorsal aspects of the hands and forearms. The initial lesions are induced by sun exposure (39). Other features include stunted growth, characteristic facies (40), and a tendency towards spontaneous and ultraviolet C (UVC)-induced chromosomal breakage and sister chromatid exchanges, leading to an increased risk of neoplasia (41, 42). Negative phototests to unspecified wavelengths of UV radiation were reported in three individuals with Bloom's syndrome (40).

## D. Rothmund-Thomson Syndrome (Poikiloderma Congenitale)

Rothmund-Thomson syndrome is a syndrome of autosomal recessive inheritance that affects girls more often than boys (43). In his original description, Rothmund, an ophthalmologist, described a reticulate erythematous eruption progressing to hyperpigmentation with onset at 3–6 months of age, associated with juvenile cataracts (43). Similar cases, but without cataracts, were later described by Thomson under the name poikiloderma congenitale (44). Finally, Taylor concluded that the entities described by Rothmund and Thomson were the same (45). A review of 107 cases of the syndrome (46) showed that 33% of patients were photosensitive and 100% developed poikiloderma. Other features, in descending order of frequency included short stature (62%), absent or sparse scalp, eyelash, and eyebrow hair (60%), juvenile cataracts (47%), bony defects (42%), small hands (36%), hypogonadism (29%), defective dentition (27%), nail dystrophy (22%), and keratotic lesions (19%). The degree of photosensitivity ranged from immediate erythema to bulla formation. One patient was phototested and showed an increased sensitivity to UVA and a normal response to UVB.

## E. Kindler Syndrome

Kindler syndrome (47) is characterized by photosensitivity, the formation of acral bullae in infancy and early childhood, generalized poikiloderma, diffuse cutaneous atrophy, photosensitivity, acral hyperkeratosis, webbing of the fingers and toes, nail dystrophy, and oral mucosal lesions. The photosensitivity is manifested by facial erythema after minimal sun exposure, blister formation in the summer, and susceptibility to severe sunburn. The action spectrum of the photosensitivity is unknown. An autosomal recessive mode of inheritance is believed to be most likely.

## F. Cockayne's Syndrome

Cockayne's syndrome (48, 49) is a premature aging syndrome inherited in an autosomal recessive fashion. The onset is before age 2. The clinical manifestations include photosensitivity that occurs as malar erythema and sunburn followed by pigmentation and scarring. Early and severe growth retardation, characteristic bird-like facies, disproportionately large hands and feet and long limbs, multiple neurologic abnormalities, ocular lesions, and premature death from atherosclerosis are other features. The action spectrum of the photosensitivity has not been determined.

The biochemical defect in Cockayne's syndrome has been identified as an increase in cell killing and sister chromatid exchange induced by irradiation of fibroblasts with UVC (50). Excision and postreplication repair rates are normal in these patients (50). In spite of normal DNA repair rates, a specific defect in active gene repair results in the failure of RNA synthesis to recover, and hypersensitivity to ultraviolet radiation (51).

## G. Xeroderma Pigmentosum

Xeroderma pigmentosum (XP) is a genetically heterogeneous group of autosomal recessive disorders characterized by defective excision repair of DNA damaged by UVC radiation in fibroblasts in vitro (52) and by sunlight in epidermal cells in vivo (53). Cell fusion studies have been used to demonstrate the genetic heterogeneity of the defect in XP. When a fibroblast from one patient with XP is fused with the fibroblast from another patient, the defects of both cells are corrected (complemented), and the cells are said to be in different complementation groups (54). Nine complementation groups have been described, labeled A–I, although recent evidence suggests that the single members of groups H and I belong in group D and C, respectively (51). In addition, an XP variant is characterized by a defect in postreplication repair or in a later stage of excision repair. The clinical onset of the cutaneous manifestations is delayed in the XP variant.

Xeroderma pigmentosum (Fig. 5, see color plate) affects 1 in 250,000 people (54). Acute photosensitivity with erythema, blisters, and edema that develop after minimal sun exposure usually is present in early childhood. All patients develop freckles on sun-exposed areas, usually before the age of 2 (54, 55), and these areas are interspersed with areas of hypopigmentation and telangiectases. The skin becomes atrophic, dry, and scaly, with multiple actinic keratoses and verrucous papules (55). Patients ultimately develop skin tumors, most commonly basal cell carcinomas, followed, in descending order of frequency, by squamous cell carcinomas, keratoacanthomas, angiomas, fibromas, and sarcomas. In rare cases, internal malignant conditions also develop. In one series, the median age for the development of nonmelanoma skin cancers was 8 years (55); 5% of patients developed melanoma. At least 40% of patients had eye findings such as neoplasia, inflammation, or scars of the lids, conjunctivae, corneas, and irides. The same DNA repair defect has been found in conjunctival tissue as in skin (55). There is a 20,000 times increased risk of the development of squamous cell carcinoma of the tip of the tongue in patients with XP.

XP may be associated with neurologic abnormalities. In the deSanctis-Cacchione syndrome (56), clinical manifestations include the cutaneous changes of XP, microcephaly and progressive mental retardation, retarded growth and sexual development, deafness, choreoathetosis, ataxia, and quadriparesis. Not all of the neurologic manifestations need be present, however. Of the neurologic manifestations, mental retardation is the most common, being present in 80% of patients (55).

Phototesting of patients with XP showed abnormal responses in most patients to wavelengths from 290 to 340 nm, with delayed erythema and papular reactions at 72 hr. The most effective wavelengths for producing this response are in the UVB range, with a peak at 293 nm (57, 58).

## H. Trichothiodystrophy

Trichothiodystrophy is an autosomal recessive disease (X-linked in a single case) in which there is a reduced cystine/cysteine and sulfur content of the hair resulting in sparse, brittle, dry scalp, eyelash, and eyebrow hair (59). There is a variable association with neuroectodermal defects, leading to a question of whether trichothiodystrophy represents a subclass of the ectodermal dysplasias (59). Photosensitivity occurs as an exaggerated sunburn reaction in one-half of affected families (60). The action spectrum of the photosensitivity has not been determined, but fibroblasts and lymphocytes from photosensitive patients with trichothiodystrophy show decreased rates of unscheduled DNA synthesis in response to irradiation with UVC (60). Photosensitive trichothiodystrophy patients have been shown to share the XP group D mutation (60).

## I. Hartnup Disease

Hartnup disease is a syndrome of autosomal recessive inheritance with a "pellagra-like" photosensitivity eruption and reversible cerebellar ataxia that develops in association with fever or mental retardation and aminoaciduria (61). The basic metabolic defect is thought to be a failure to absorb tryptophan from the gastrointestinal tract and from the renal tubule (62). Because tryptophan is metabolized to niacin, the clinical expression may be the result of niacin deficiency, as is seen in pellagra. The full expression of Hartnup disease depends on multifactorial events, since it has been shown that not all persons who carry the gene mutation have the disease (63). The action spectrum of photosensitivity has not been determined.

## III. INFECTIOUS DISEASES

### A. Herpes Simplex Labialis

Lesions of herpes simplex labialis tend to appear 2–3 days after the patient sustains a sunburn. Lesions have been produced experimentally in 6 of 10 individuals with 4–6 MEDs of ultraviolet light from a sunlamp, 23–94 hr after irradiation. Herpes simplex cultures were positive in three of four cases tested (64). Because the lesions were sharply localized to the irradiated area, the author postulated that virus is shed constantly from the nerve endings and that clinical lesions are produced only in an area of skin rendered susceptible.

### B. Other Infectious Agents

There have been two reports of varicella (65, 66), three of vaccinia (67, 68), and isolated reports of other viral exanthemata occurring initially in a photodistribution (66, 69) after a sunburn. Erythema multiforme in a photodistribution was reported in a 25-year-old man who developed herpes simplex 2 days after a sunburn and erythema multiforme 14 days later (70). A possible mechanism for this photolocalization is increased capillary permeability in the solar-damaged areas, leading to localization of virus or other antigens in the sunburned areas (65, 66).

## IV. NUTRITIONAL DEFICIENCIES

### A. Pellagra

Pellagra (Fig. 6) consists of a triad of dermatitis, diarrhea, and dementia. It is due to a dietary deficiency of niacin or its precursor, the essential amino acid, tryptophan. Pellagra is seen most often in persons on a diet of maize; in those with inadequate animal protein intake such as vegetarians, alcoholics, and anorectics; after bowel bypass surgery; and in those with Hartnup disease (71, 72). An erythematous to hyperpigmented dermatitis occurs on the dorsal aspects of the hands and on the V area of the neck, where it is called Casal's necklace. The involved skin becomes indurated and hyperpigmented. Mucous membrane changes include dry, fissured lips; red, smooth, dry buccal mucosa; oral aphthous ulcers; and changes in the tongue, which is usually swollen and magenta, but occasionally is black and atrophic (71).

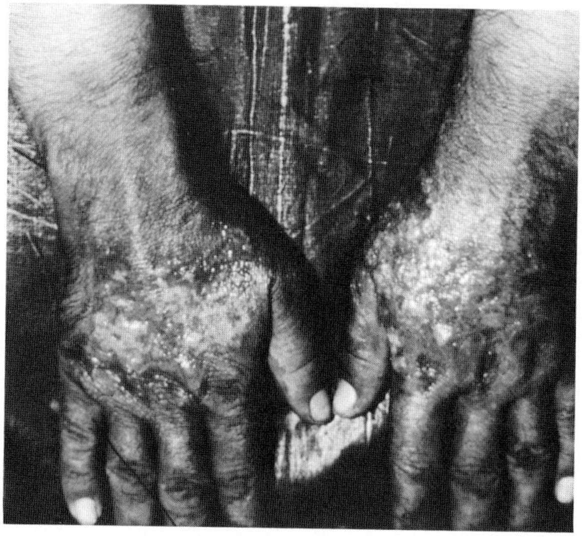

**Figure 6** Pellagra. Note the sharp localization of this oozing, hyperpigmented dermatitis to the sun-exposed dorsal aspects of the hands and arms.

The action spectrum of photosensitivity is unknown. The mechanism may involve a secondary deficiency of coenzyme I (oxidized nicotinamide adenine dinucleotide, NAD) and coenzyme II (reduced form of nicotinamide adenine dinucleotide, NADH) which are produced from niacin. NAD and NADH are cofactors important to the repair of ultraviolet-induced damage in the epidermis. In carcinoid tumor metastatic to the liver, tryptophan is diverted from the production of niacin to the production of serotonin, which has led to a deficiency of niacin and dermatitis indistinguishable from pellagra in one patient (73).

## B. Pyridoxine (Vitamin B6) Deficiency

Pyridoxine is necessary for the endogenous synthesis of nicotinic acid from tryptophan. Therefore, pyridoxidine deficiency leads to pellagra-like manifestations including dermatitis on sun-exposed areas (71).

## V. NEOPLASMS: MYCOSIS FUNGOIDES

Twelve patients with previously diagnosed mycosis fungoides were phototested. All patients had papules, plaques, or tumors. Although only 4 had a history of photodermatitis, 7 of the 12 had an abnormal MED in the UVA range and 5 in the UVB range (74). The previously reported cases of photosensitive mycosis fungoides may have represented actinic reticuloid (75).

## VI. PRIMARY DERMATOLOGIC DISEASES

### A. Hydroa Aestivale/Hydroa Vacciniforme

Hydroa aestivale and hydroa vacciniforme (Fig. 7, see color plate) are thought to be the same disease, with the former being a less severe, nonscarring form of the latter (76). Since most of the findings related to this entity have been reported under the name "hydroa vacciniforme," this term will be used. Hydroa vacciniforme is a rare idiopathic disease of children and is more common in girls than in boys (77). It flares in the spring and summer with the development of tense vesicles or bullae on an erythematous base on sun-exposed areas, followed by the appearance of necrosis and varioliform scars (77). There may be spontaneous remission when patients are in their late teens (77).

The interpretation of earlier studies on the action spectrum in hydroa vacciniforme is impossible because earlier series did not make a distinction between children with erythropoietic protoporphyria, which presents with clinical manifestations similar to those of hydroa vacciniforme (78). Phototests in patients with a definite diagnosis of hydroa vacciniforme have yielded inconsistent results. In one series, 3 of 10 patients had a decreased MED test to UVA, but a normal response to UVB (79). A patient was reported with an abnormal MED to UVA in whom lesions could be produced by repetitive doses of UVA. Tolerance to this effect was induced by multiple UVB exposures (80). Other investigators have reported induction of lesions with repetitive UVA exposures (81). Tolerance was induced in one case by pigmentation occurring after treatment with psoralens plus UVA. In contrast to the above results, one group of investigators found a bullous response to intense radiation within the wavelengths 280–420 nm and no response to UVA alone (77).

### B. Atopic Dermatitis

Five atopic girls were described who developed distinctive photodistributed erythematous papules in the summer. However, phototests with UVA, UVB, and visible light showed normal responses (82).

## C. Transient Acantholytic Dermatosis (Grover's Disease)

Transient acantholytic dermatosis is a disease of middle-aged and elderly men in whom discrete, pruritic, edematous papules or papulovesicles appear, most frequently on the chest and back. Histopathologic features consist of acantholysis with or without dyskeratosis. The original description included six patients, three of whom first developed lesions in sun-exposed areas after intense sun exposure. However, since a fourth patient developed lesions at the site of a contact dermatitis, Grover suggested that the disease was subject to the Koebner phenomenon and not due to photosensitivity per se (83).

## D. Disseminated Superficial Actinic Porokeratosis

Disseminated superficial actinic porkeratosis presents in adult white patients with an average age of 36. There is a female predominance. The lesions are annular with a thin keratotic rim and a slightly atrophic center. They are found on sun-exposed areas only, most often on the extensor aspects of the legs. Forty-eight percent of patients experience summer exacerbations. The lesions are not reproduced by mechanical injury or by extremes of temperature, so the Koebner reaction is thought not to occur (84). The disease is thought to be inherited as an autosomal dominant trait with reduced penetrance (85). The action spectrum of the photosensitivity has not been determined.

## E. Lichen Planus Actinicus

Lichen planus actinicus is seen in Middle Eastern countries on the sun-exposed areas of dark-skinned people ages 21–30. A lesion begins as a bluish–brown annular or oval macule that develops a slightly raised, pale margin, giving the illusion of central atrophy. The disease has winter remissions. The histopathologic appearance is identical to lichen planus and the lesions respond to Grenz ray therapy. Lesions were induced in a single patient with exposure to UVB but not to UVA (86).

## F. Photosensitive Psoriasis

A certain percentage of patients with psoriasis have exacerbations after sun exposure or in the summer. In one study, a questionnaire was mailed to 2,000 patients with psoriasis, and 5.5% of the patients reported photosensitivity (87). A series of 14 patients was reported with lesions localized to the face and the dorsal aspects of the hand and a seasonal deterioration in spring and summer (88). These patients improved after treatment with antimalarials. Photosensitive psoriasis has been treated with psoralens plus UVA photochemotherapy (89).

## G. Reticular Erythematous Mucinosis

In reticular erythematous mucinosis, patients develop reticulate, erythematous, indurated plaques on the chest, upper back and abdomen. Two of the four original patients had the onset of lesions with sun exposure, and all of them responded to the administration of antimalarials. The histologic picture consists of a normal epidermis, a perivascular infiltrate of round cells, and deposits of Alcian blue-positive, mucicarmine-negative material in the dermis (90).

In a review of the world's literature, 50% of 53 patients had exacerbation or onset of disease after sun exposure (91). Phototests show inconsistent results (92, 93). Lesions were reproduced in one patient by irradiation with 2.0–2.5 MEDs of UVB or UVC (93).

## REFERENCES

1. Tan EM, Cohen AS, Fries JF, Masi AT, McShane DJ, Rothfield NF, Schaller JG, Talal N, Winchester RJ. The 1982 revised criteria for the classification of systemic lupus erythematosus. Arthritis Rheum 1982; 25: 1271–1277.
2. Epstein JH, Tuffanelli DL, Dubois EL. Light sensitivity and lupus erythematosus. Arch Dermatol 1965; 91: 483–486.
3. Ward MM, Studenski S. Clinical manifestations of systemic lupus erythematosus. Identification of racial and socioeconomic influences. Arch Intern Med 1990; 150: 849–853.
4. Sutej PG, Gear AJ, Morrison RCA, Tikly M, deBeer M, dos Santos L, Scher R. Photosensitivity and anti-Ro (SS-A) antibodies in black patients with systemic lupus erythematosus (SLE). Br J Rheum 1989; 28: 321–324.
5. Mond CB, Peterson MGE, Rothfield NF. Correlation of anti-Ro antibody with photosensitivity rash in systemic lupus erythematosus patients. Arthritis Rheum 1989; 32: 202–204.
6. Sontheimer RD, Thomas JR, Gilliam JN. Subacute cutaneous lupus erythematosus. A cutaneous marker for a distinct lupus erythematosus subset. Arch Dermatol 1979; 115: 1409–1415.
7. Sontheimer RD, Maddison PJ, Reichlin M, Jordon RE, Stasny P, Gilliam JN. Serologic and HLA associations in subacute cutaneous lupus erythematosus, a clinical subset of lupus erythematosus. Ann Intern Med 1982; 97: 664–671.
8. Lehman P, Hölzle E, Kind P, Goerz G, Plewig G. Experimental reproduction of skin lesions in lupus erythematosus by UVA and UVB radiation. J Am Acad Dermatol 1990; 22: 181–187.
9. Provost TT. Commentary: neonatal lupus erythematous. Arch Dermatol 1983; 119: 619–622.
10. Scott JS, Maddison PJ, Taylor PV, Esscher E, Scott O, Skinner RP. Connective-tissue disease, antibodies to ribonucleoprotein, and congenital heart block. N Engl J Med 1983; 309: 209–212.
11. McCune AB, Weston WL, Lee LA. Maternal and fetal outcome in neonatal lupus erythematosus. Ann Intern Med 1987; 106: 518–523.

12. Callen JP. Chronic cutaneous lupus erythematosus. Clinical, laboratory, therapeutic, and prognostic examination of 62 patients. Arch Dermatol 1982; 118: 412–416.
13. Prystowsky SD, Herndon JH, Gilliam JN. Chronic cutaneous lupus erythematosus (DLE): a clinical and laboratory investigation of 80 patients. Medicine 1975; 55: 183–191.
14. Millard LG, Rowell NR. Abnormal laboratory test results and their relationship to prognosis in discoid lupus erythematosus. Long term follow up study of 92 patients. Arch Dermatol 1979; 115: 1055–1058.
15. Stern RS, Docken W. An exacerbation of SLE after visiting a tanning salon. JAMA 1986; 255:3120.
16. Freeman RG, Knox JM, Owens DW. Cutaneous lesions of lupus erythematosus induced by monochromatic light. Arch Dermatol 1969; 100: 677–682.
17. Cripps DJ, Rankin J. Action spectra of lupus erythematosus and experimental immunofluorescence. Arch Dermatol 1973; 107: 563–567.
18. Baer RL, Harber LC. Photobiology of lupus erythematosus. Arch Dermatol 1965; 92: 124–128.
19. Lever WF, Schaumberg-Lever G. Histopathology of the Skin. Philadelphia: JB Lippincott 1983; 90: 494–506.
20. Jerden MS, Hood AF, Moore GW, Callen JP. Histopathologic comparison of the subsets of lupus erythematosus. Arch Dermatol 1990; 126: 52–55.
21. Burnham TK, Teblett TR, Fine G. The application of fluorescent antibody technique to the investigation of lupus erythematosus and various dermatoses. J Invest Dermatol 1963; 41: 451–456.
22. Harrist TJ, Mihm MC Jr. The specificity and clinical usefulness of the lupus band test. Arthritis Rheum 1980; 23: 479–490.
23. Callen JP. Therapy of cutaneous lupus erythematosus. Med Clin North Am 1982; 66: 795–805.
24. Perry HO, Brunsting LA. Pemphigus foliaceous. Arch Dermatol 1965; 91: 10–23.
25. Cram DL, Winklemann RK. Ultraviolet-induced acantholysis in pemphigus. Arch Dermatol 1965; 92: 7–13.
26. Usher B, Senear FE. An unusual type of pemphigus combining features of lupus erythematosus. Arch Dermatol Syphilol 1926; 13: 761–781.
27. Chorzelski T, Jablonska S, Blaszcyzyk M. Immunopathological investigations in the Senear-Usher syndrome (co existence of pemphigus and lupus erythematosus). Br J Dermatol 1968; 80: 211–217.
28. Deschamps P, Pedailles S, Michel M, Leroy D. Photoinduction of lesions in a patient with pemphigus erythematosus. Photodermatology 1984; 1: 38–41.
29. Jacobs SE. Pemphigus erythematosus and ultraviolet light. Arch Dermatol 1965; 91: 139–141.
30. Cram DL, Fukuyama K. Immunohistochemistry of ultraviolet-induced pemphigus and pemphigoid lesions. Arch Dermatol 1972; 106: 819–824.
31. Preesman AH, Toonstra J, Van der Putte SC, De Geer DB, Van Weelden H, Van Vloten WA. UVB induced bullous pemphigoid restricted to mycosis fungoides plaques. Clin Exp Dermatol 1990; 15: 363–366.
32. Abel EA, Bennett A. Bullous pemphigoid. Occurrence in psoriasis treated with psoralens plus long wave ultraviolet radiation. Arch Dermatol 1972; 115: 988–989.

33. Hailey H. Familial benign chronic pemphigus. South Med J 1953; 46: 763–765.
34. Suhonen R, Niemi KM. Ultraviolet light in familial benign chronic pemphigus. J Cutan Pathol 1979; 6: 414–417.
35. Cram DL, Muller SA, Winklemann RK. Ultraviolet-induced acantholysis in familial benign pemphigus. Arch Dermatol 1967; 96: 636–641.
36. Chorzelski T. Experimentally induced acantholysis in Hailey's benign pemphigus. Dermatologica 1962; 124: 21–30.
37. Shelley WB, Arthur RP, Pillsbury DM. A view of keratosis follicularis (Darier's disease) as a neoplastic process. Arch Dermatol 1959; 80: 122–128.
38. Penrod JN, Everett MA, McCreight WG. Observations on keratosis follicularis. Arch Dermatol 1960; 82: 367–370.
39. Bloom D. Congenital telangiectatic erythema resembling lupus erythematosus in dwarfs. Am J Dis Child 1954; 88: 754.
40. German J. Bloom's syndrome. I. Genetical and clinical observations in the first twenty-seven patients. Am J Hum Genet 1969; 21: 196–227.
41. Chaganti RSK, Schonberg S, German J. A many-fold increase in sister chromatid exchanges in Bloom's syndrome lymphocytes. Proc Natl Acad Sci USA 1974; 71: 4508–4512.
42. Kurihara T, Tatsumi K, Takahashi H, Inove M. Sister chromatid exchanges induced by ultraviolet light in Bloom's syndrome fibroblasts. Mutat Res 1987; 183: 197–202.
43. Rook A, Davis R, Stevanovic D. Poikiloderma congenitale—Rothmund-Thomson syndrome. Acta Derm Venereol (Stockh) 1959; 39: 392–420.
44. Thomson MS. An hitherto undescribed familial disease. Br J Dermatol 1923; 35: 455.
45. Taylor WB. Rothmund's syndrome—Thomson's syndrome. Arch Dermatol 1957; 75: 236–244.
46. Berg E, Chuang T, Cripps D. Rothmund-Thomson syndrome. A case report, phototesting, and literature review. J Am Acad Dermatol 1978; 17: 332–338.
47. Forman AB, Prendiville JS, Esterly NB, Hebert AA, Duvic M, Horiguchi Y, Fine J. Kindler syndrome: report of two cases and review of the literature. Pediatr Dermatol 1989; 6: 91–101.
48. Cockayne EA. Dwarfism with retinal atrophy and deafness. Arch Dis Child 1936; 11: 1–8.
49. MacDonald WB, Fitch KD, Lewis K. Cockayne's syndrome. A hereditofamilial disorder of growth and development. Pediatrics 1960; 25: 997–1007.
50. Marshall RR, Arlett CF, Harcourt SA, Broughton BA. Increased sensitivity of cell strains from Cockayne's syndrome to sister-chromatid exchange induction and cell killing by UV light. Mutat Res 1980; 69: 107–112.
51. Lehmann AR, Norris, PG. DNA repair deficient photodermatoses. Semin Dermatol 1990; 9: 55–62.
52. Cleaver JE. Defective repair replication of DNA in xeroderma pigmentosum. Nature 1968; 218: 652–656.
53. Epstein JH, Fukuyama K, Reed WB. Defect in DNA synthesis in skin of patients with xeroderma pigmentosum demonstrated in vivo. Science 1970; 168: 1477–1478.

54. Robbins JH, Kraemer K, Lutzner MA, Festoff BW, Coon HG. Xeroderma pigmentosum: an inherited disease with sun sensitivity, multiple cutaneous neoplasms, and abnormal DNA repair. Ann Intern Med 1974; 80: 221–248.
55. Kraemer KH, Lee MM, Scotto J. Xeroderma pigmentosum: cutaneous, ocular, and neurologic abnormalities in 830 published cases. Arch Dermatol 1987; 123: 241–250.
56. DeSanctis C, Cacchione A. L'idioza xerodermica. Riv Sper Freniatri 1932; 56: 269–292.
57. Cripps DJ, Ramsay CA, Ruch DM. Xeroderma pigmentosum. Abnormal monochromatic action spectrum and autoradiographic studies. J Invest Dermatol 1971; 56: 281–286.
58. Ramsay CA, Giannelli F. The erythemal action spectrum and deoxyribonucleic acid repair synthesis in xeroderma pigmentosum. Br J Dermatol 1975; 92: 49–56.
59. Itin PH, Pittelkow MR. Trichothiodystrophy: review of sulfur deficient brittle hair syndromes and association with the ectodermal dysplasias. J Am Acad Dermatol 1990; 22: 705–717.
60. Stefanini M, Lagomarisini P, Giorgi R, Nuzzo F. Complementation studies in cells from patients affected by trichothiodystrophy with normal or enhanced UV photosensitivity. Mutat Res 1987; 191: 117–119.
61. Baron DN, Dent CE, Harris H, Hart EW, Jepson JB. Hereditary pellagra-like skin rash with temporary cerebellar ataxia, constant renal amino-aciduria and other bizarre biochemical features. Lancet 1956; 2: 241–248.
62. Milne D. The metabolic disorder in Hartnup disease. Q J Med 1960; 29: 407–421.
63. Scriver CR, Mahon B, Levy HL, Clow CL, Reade TM, Kronick J, Lemieux B, Laberge C. The Hartnup phenotype: Mendelian transport disorder, multifactorial disease. Am J Hum Genet 1987; 40:401–412.
64. Spruance SL. Pathogenesis of herpes simplex labialis: experimental induction of lesions with UV light. J Clin Microbiol 1985; 22: 366–368.
65. Castrow FF, Wolf JE. Photolocalized varicella (letter). Arch Dermatol 1973; 107: 628.
66. Gilchrest B, Baden HP. Photodistribution of viral exanthems. Pediatrics 1974; 54: 136–138.
67. Pace BF, Owens DW. Photosensitivity eruption following smallpox vaccination. Cutis 1969; 5: 850–853.
68. Coskey RJ, Bryan HG. Photosensitivity secondary to smallpox vaccination. Cutis 1970; 6: 761.
69. Cherry JD. Photodistribution of viral exanthems (letter). Pediatrics 1974; 54: 574.
70. Weston WL, Huff C. The photodistribution of erythema multiforme. Arch Dermatol 1980; 116: 477.
71. Barthelemy H, Chouvel B, Cambazard F. Skin and mucosal manifestations in vitamin deficiency. J Am Acad Dermatol 1986; 15: 1263–1274.
72. Rapaport MJ. Pellagra in a patient with anorexia nervosa. Arch Dermatol 1985; 121: 255–257.
73. Bridges JM, Gibson JB, Loughridge LW, Montgomery DAD. Carcinoid syndrome with pellagrous dermatitis. Br J Surg 1957; 45: 117–122.
74. Volden G, Thune PO. Light sensitivity in mycosis fungoides. Br J Dermatol 1977; 97: 279–284.

75. Ive FA. Mycosis fungoides presenting as light sensitivity. Br J Dermatol 1964; 76: 145.
76. Senear FE, Fink HW. Hydroa vacciniforme seu aestivale. Arch Dermatol Syphilol 1923; 7: 145–162.
77. Bickers DE, Demar LK, DeLeo V, Poh-Fitzpatrick MB, Aronberg JM, Harber LC. Hydroa vacciniforme. Arch Dermatol 1978; 114: 1193–1196.
78. McGrae JD, Perry HO. Hydroa vacciniforme. Arch Dermatol 1963; 87: 124–131.
79. Sonnex TS, Hawk JLM. Hydroa vacciniforme: a review of ten cases. Br J Dermatol 1988; 118: 101–108.
80. Halasz CLG, Leach EE, Walther RR, Poh-Fitzpatrick MB. Hydroa vacciniforme: induction of lesions with ultraviolet A. J Am Acad Dermatol 1983; 8: 171–176.
81. Hann SK, Im S, Park Y, Lee S. Hydroa vacciniforme with usually severe scar formation: diagnosis by repetitive UVA phototesting. J Am Acad Dermatol 1991; 25: 401–403.
82. Frain-Bell W, Scatchard M. The association of photosensitivity and atopy in the child. Br J Dermatol 1971; 85: 105–110.
83. Grover RW. Transient acantholytic dermatosis. Arch Dermatol 1970; 101: 426–434.
84. Chernosky ME, Freeman RG. Disseminated superficial actinic porokeratosis (DSAP). Arch Dermatol 1967; 96: 611–624.
85. Chernosky ME, Anderson DE. Disseminated superficial actinic porokeratosis: clinical studies and experimental production of lesions. Arch Dermatol 1969; 99: 401–407.
86. Van der Schroeff JG, Schothorst AA, Kanaar P. Induction of actinic lichen planus with artificial UV sources. Arch Dermatol 1983; 119: 498–500.
87. Ros AM, Eklund G. Photosensitive psoriasis. An epidemiologic study. J Am Acad Dermatol 1987; 17: 752–758.
88. Bielicky T, Kvicalova E. Photosensitive psoriasis. Dermatologica 1964; 129: 339–348.
89. Ros AM, Wennersten G. PUVA therapy for photosensitive psoriasis. Acta Derm Venereol (Stockh) 1987; 67: 501–505.
90. Steigleder GK, Gartmann H, Linker U. REM syndrome: reticular erythematosus mucinosis (round-cell erythematosis), a new entity? Br J Dermatol 1974; 91: 191–199.
91. Braddock SW, Davis CS, Davis RB. Reticular erythematous mucinosis and thrombocytopenic purpura. J Am Acad Dermatol 1988; 19: 859–868.
92. Bleehen SS, Slater DN, Mahood J, Church RE. Reticular erythematous mucinosis: light and electron microscopy, immunofluorescence and histochemical findings. Br J Dermatol 1982; 106: 9–18.
93. Morison WL, Shea CR, Parrish JA. Reticular erythematous mucinous syndrome. Report of two cases. Arch Dermatol 1979; 115: 1340–1342.

# V

# Photoprotection

17

# Topical and Systemic Photoprotection of Human Skin Against Solar Radiation

**Madhu A. Pathak**
*Harvard Medical School, Boston, Massachusetts*

Although people strongly believe that sunlight is healthy, we cannot deny or ignore the acute and chronic (cumulative) effects of sunlight. A normal healthy person should have two concerns about the deleterious effects of sun exposure: the acute effects (manifested in the form of sunburn, hyperpigmentation or tanning, and drug-induced phototoxicity), and the potential long-term risk of repeated sun exposure leading to actinic damage manifested in the form of dermatoheliosis (e.g., wrinkling, irregular thickening and thinning of the epidermis, telangiectasia, hyperpigmentation and hypopigmentation, keratoses), and the induction of both nonmelanoma and melanoma skin cancer (Table 1) (1–3).

In addition, in certain individuals abnormal reactions (e.g., vesicular, eczematous, urticarial, or papular eruptions) can selectively result from exposure to ultraviolet B (UVB) (290–320 nm), ultraviolet A (UVA) (320–400 nm) or, to a limited extent, visible radiation (400–760 nm). These photodermatoses have been discussed in Chapters 10–16.

These harmful effects of ultraviolet radiation depend upon the length (duration) and the frequency of exposures, the intensity of sunlight based on geographic locations (latitude), air quality, seasonal variations, and reactivity of skin based on genetically determined constitutive skin color (intrinsic) and the facultative (induced) pigmentation response of white, light-brown, or dark-skinned individuals. We summarize here the essential guidelines for protecting human skin against harmful effects of solar radiation (3–5).

**Table 1** UVR or Sunlight-Induced Human Skin Disorders

| |
|---|
| Cataracts (unprotected eyes) |
| Acute Effects |
| Sunburn reaction |
| Tanning reaction (neomelanogenesis) |
| Drug-induced phototoxic and photoallergic reactions (e.g., psoralens, phenothiazines, nalidixic acid) |
| Light-induced photosensitivity diseases (polymorphous light eruption, lupus erythematosus, porphyrias, etc.) |
| Chronic Effects |
| Aging and wrinkling of skin (photoaging or dermatoheliosis) |
| Cancers (basal and squamous cell carcinomas and melanomas) |
| Alterations of immune responses |

## I. SOLAR RADIATION AND HUMAN SKIN

The solar radiation at the earth's surface is subdivided into three spectral bands: 1. ultraviolet radiation (290–400 nm),* 2. visible radiation (400–760 nm) and 3. near-infrared radiation of wave-lengths longer than 760 nm. Depending on the geographic location and the season, it has been estimated that sunlight provides about 20 to 60 $W/m^2$ of ultraviolet radiation between 290 and 400 nm (6).

Tables 2 and 3 provide solar irradiance (in terms of percentage of the total irradiance in $W/m^2$) before and after attenuation by the earth's atmosphere, which is rich in ozone (6). The ultraviolet radiation from both sunlight and artificial sources is subdivided for convenience into three bands from the longer to shorter wavelengths. These are referred to as UVA, UVB, and ultraviolet C (UVC) regions (Tables 2, 3). The UVA region extends from 320 to 400 nm. It is also referred to as near-ultraviolet radiation or long-wave ultraviolet radiation. UVA causes the tanning reaction, both the immediate tanning or immediate pigment-darkening reaction (IPD) seen during exposure, and the delayed tanning reaction seen approximately 2 or 3 days after exposure. This is referred to as neomelanogenesis (7). UVA can also evoke the sunburn reaction, but the erythemogenic or sunburn-producing capacity of UVA is weaker than UVB radiation; UVA is about 600–1000 times less effective than UVB radiation in inducing sunburn reaction. About 20–60 $J/cm^2$ of UVA dose is required to produce a minimally perceptible erythema; this dose is referred to as minimal erythema dose (MED) for UVA radiation (Table 4). UVA-induced erythema is

*Nanometer (nm) is a wavelength unit (equals $10^{-9}$ meter); radiant energy (joules) = watt × sec; radiant flux = rate of delivery of radiant energy expressed as watt (W) or milliwatt (mW); irradiance = $W/m^2$; radiation exposure ($J/cm^2$ or $mJ/cm^2$) represents exposure dose and equals irradiance × time (sec).

**Table 2** Wavelength Distribution of Solar Irradiance Prior to Attenuation by the Earth's Atmosphere

| Wavelength region (nm) | Irradiance ($W/m^2$) | Percentage of total |
|---|---|---|
| UVC (< 280) | 6.4 | 0.5 |
| UVB (280-320) | 21.1 | 1.5 |
| UVA (320-400) | 85.7 | 6.3 |
| Visible (400-760) | 532 | 38.9 |
| Infrared (> 760) | 722 | 52.8 |

Source: Ref. 6.

seen soon after exposure, and the intensity of erythema is optimal at approximately 10–12 hr after exposure; the UVA-exposed sites may remain erythematous up to 24–36 hr. UVA radiation (but not the UVB radiation) causes the immediate tanning or IPD reaction that represents photo-oxidation of preexisting melanin. By 48–96 hr after exposure, the IPD reaction is followed by a delayed tanning reaction that involves new and increased synthesis of melanin manifested by proliferation of melanocytes, increased synthesis of tyrosinase enzyme, and increased formation and transfer of melanosomes by melanocytes to keratinocytes. Human skin exposed to UVA radiation also shows changes in DNA (formation of thymine dimers) and proteins (cross-linking reaction). UVA-induced changes are caused by an indirect mechanism involving the formation of reactive oxygen species (singlet oxygen, superoxide anion, hydroxyl radicals) mediated by endogenous photosensitizers such as riboflavin, quinones, and NADH. [These reactive oxygen species selectively induce cell membrane damage, lipid peroxidation, and damage to proteins of the epidermis and dermis in the form of cross-links.] Upon repeated exposure, these biochemical events are manifested as wrinkling or photoaging changes.

**Table 3** Percentage Solar Irradiance at Sea Level for Two Zenith Angles

| Wavelength region (nm) | Zenith angle 0 degrees | Zenith angle 20 degrees |
|---|---|---|
| UVC (< 280) | 0.0 | 0.0 |
| UVB (280-320) | 0.5 | 0.3 |
| UVA (320-400) | 6.3 | 5.0 |
| Visible (400-760) | 38.9 | 41.0 |
| Infrared (> 760) | 54.3 | 54.0 |

**Table 4** Minimal Erythema Dose (MED) and Minimal Melanogenic Dose (MMD) for Individuals of Skin Types I–VI[a] for UVA (320–400 nm) and UVB (290–320 nm) Radiation

| | UVA | | UVB | | |
|---|---|---|---|---|---|
| Skin Type | MED (J/cm$^2$) | MMD (J/cm$^2$) | MED (mJ/cm$^2$) | MMD (mJ/cm$^2$) | Recommended SPF |
| I | 20–35 | — | 20–30 | — | |
| II | 30–45 | 15–20 | 25–35 | 15–25 | 15–30 |
| III | 40–55 | 20–30 | 30–50 | 17–25 | 15–30 |
| IV | 50–70 | 20–40 | 45–60 | 20–30 | 15–30 |
| V | 65–80 | 30–50 | 60–100 | 30–35 | 10–15 |
| VI | 80–100 | 30–50 | 80–150 | 40–80 | 6–10 |

[a]Based on nearly 300 test subjects (7).
*Source:* Ref. 7.

The UVB band extends from 290 to 320 nm. It is often referred to as sunburn-producing radiation, mid-ultraviolet radiation, or erythemal band. Exposure to UVB radiation is the principal cause of the sunburn reaction. It is also the most effective band in stimulating pigmentation (melanogenesis or tanning) in the skin. In individuals with skin types I–IV, the MED of UVB is approximately 20–60 mJ/cm$^2$ (Table 4). The intensity of UVB radiation decreases with increasing latitude from 0 to 90 degrees north and south. Fair-skinned individuals living on the equatorial belt (0–10 degrees N) may obtain one MED dose in about 10 min (11 am to 1 pm); those living in subtropical regions (10–25 degrees N,S) may obtain one MED dose in 10–15 min, and those living between 25 and 35 degrees N,S may obtain one MED dose in about 15–20 min during the summer months. Exposure of human skin to multiple MEDs (8–15 times MED) for a prolonged period can cause a marked sunburn reaction that can be painful and lead to severe edema and blistering response. The UVB-induced erythema reaction is maximal in intensity between 20 and 24 hr after exposure and appears to be caused by the generation of eicosanoids. Exposure to UVB radiation results in immediate damage to DNA, RNA, and proteins of epidermis and dermis. The damage to DNA in the form of thymine dimers and strand breaks can contribute to mutation and skin carcinogenesis.

Thus, both UVA and UVB spectral regions are erythemogenic, melanogenic, and cause changes in DNA, RNA, and proteins. However, the amount of UVB energy required to produce these effects is 600–1000 times less than that required for the UVA region (Table 4). It should be realized, however, that the amount of solar UVA energy reaching the surface of the earth is about 20 times greater than that of UVB (5.0–6.0 mW/cm$^2$ UVA vs. 0.3–0.5 mW/cm$^2$ UVB). Therefore, the long-term cumulative effects of UVA radiation may be as important as the effects of UVB radiation. The combined effects of UVA and UVB are usually additive. In addition, UVA radiation is usually involved in evoking drug-induced, phototoxic (erythema, edema, blisters, etc), and photoallergic (eczematous, papular) reactions. Individuals must, therefore, be protected against immediate and delayed harmful effects of both UVB and UVA radiation.

UVC radiation (germicidal, short-ultraviolet radiation, or far-ultraviolet radiation) comprises wavelengths shorter than 290 nm (200–290 nm). Efficient absorption by molecular oxygen and ozone at the top surface of the earth's atmosphere (i.e., stratosphere) removes essentially all the UVC radiation. Although UVC radiation from sunlight does not reach the surface of the earth, individuals can encounter this spectrum from artificial UVR sources (e.g., germicidal lamps, high-pressure and low-pressure mercury arc lamps used in industry). Radiation in the UVC band from artificial sources (e.g., welding) is very efficient in causing erythema of normal skin (MED about 15–25 $mJ/cm^2$) and photokeratitis of the eye. It can stimulate melanogenesis but far less effectively than that stimulated by UVB or UVA radiation. The erythema reaction caused by UVC is maximum in intensity within 6–8 hr after exposure, and only rarely does it produce a blistering reaction after the skin is exposed to multiple (10–100) MEDs. This is because UVC is essentially absorbed by stratum corneum and the granular-cell layer of the epidermis, and it does not penetrate appreciably to the dermis.

Visible radiation (400–760 nm) is generally innocuous to a normal individual; however, in the presence of certain chemicals (e.g., porphyrins, certain dyes, etc.), this radiation can be damaging to the skin. Infrared radiation can be innocuous in low doses, but it can be damaging to skin proteins and accelerate or potentiate the harmful effects of UVR.

## II. HUMAN SKIN TYPES AND THEIR REACTIVITY TO ULTRAVIOLET RADIATION

There is considerable variability in the reaction of human skin to the effects of solar radiation (2–4,7,8). Different people are variably susceptible to sun-induced skin damage. This susceptibility is influenced by genetic factors that control the gene function for melanin pigmentation and DNA repair of the photodamaged DNA. For this reason, individuals of different color ranging from fair (white) to dark skin are classified in six sun-reactive skin types (I–VI). This inherent variability is based largely on the genetically controlled constitutive skin color of skin (epidermal melanin content of skin not habitually exposed to sunlight) and on the facultative capacity of the skin to darken or tan in response to ultraviolet radiation (facultative skin color or delayed tanning response). In general, the susceptibility of human skin to sun-induced acute and chronic damage is directly related to the ultraviolet radiation intensity of the sun (geographic location and proximity to the equator), duration and habits of sun exposure of the individual, and inversely related to his or her tanning ability and capacity to repair photodamaged DNA. Therefore, a personal history of sunburning and suntanning (ability to stimulate pigmentation) for the past few years following the first 45–60 min exposure to midday summer sun (equivalent to approximately 3 MED) is very helpful to classify people into the six sun-reactive skin types, I–VI, and these are briefly described in Table 5 (2–4,7). Individuals of skin types I, II, and III are more susceptible to acute and chronic skin damage (sunburn, photoaging, skin cancer, etc.) than individuals of skin types IV, V, and VI.

It is important to realize that both UVB and UVA radiation are erythemogenic, and, all individuals, irrespective of their skin types, will exhibit sunburn response provided they receive sun exposure exceeding their MED values (Table 4). Individuals who sunbathe for 60 min or more acquire sunburn due to the additive effects of both UVB and UVA radiation. Both UVB and UVA radiation also stimulate neomelanogenesis (skin tanning), induce skin cancer, and promote skin aging (3–5,7).

**Table 5** Human Skin Types and Their Reactivity to Sunlight

| Skin type | Constitutive color of unexposed skin[a] | Sensitivity to UV[a] | Sunburn and tanning history[a] |
|---|---|---|---|
| I | White | Very sensitive + + + + | Always burns easily; never tans, shows no IPD |
| II | White | Very sensitive + + + + | Always burns easily; tans minimally and with difficulty, IPD+ |
| III | White | Sensitive + + + | Burns moderately; tans gradually and uniformly (light brown), IPD+ |
| IV | Pale brown | Moderately sensitive + + | Burns minimally; always tans well (moderate brown), IPD+ |
| V | Brown | Minimally sensitive + to ± | Rarely burns, tans profusely (dark brown), IPD+ |
| VI | Dark brown or black | Insensitive or least sensitive | Never burns; deeply pigmented (black), tans profusely, IPD+ |

[a]Based on first 45–60 min (2–3 MED) of sun exposure of untanned skin (type I or II) after winter season or no previous sun exposure. Individuals of all skin types will exhibit a sunburn reaction if exposed to 75–100 mJ/cm$^2$ of UVB radiation. IPD = Immediate pigment darkening reaction
*Source:* Ref. 7.

## III. NATURAL DEFENSES OF SKIN AGAINST SUNLIGHT

To survive the insults of actinic damage, human skin has evolved at least five basic defensive mechanisms (3,5):

1. The process of epidermal hyperplasia leading to the formation of compact and cohesive stratum corneum of varying thickness (15–50$\mu$m or more) containing ultraviolet-absorbing keratin proteins; these cells not only absorb and reflect the impinging ultraviolet radiation but also attenuate it by scattering.
2. The formation and transfer of melanized melanosomes from melanocytes to basal and suprabasal keratinocytes. The colloidal melanin and melanosomes present in the epidermis not only absorb UV radiation but also attenuate radiation by scattering. The melanin pigment also acts as a free-radical scavenger and an optical filter that shields nuclear DNA of keratinocytes and the dermal proteins, collagen and elastin, from ultraviolet-induced harmful alterations.
3. The preferential accumulation of carotenoid pigments in subcutaneous tissue, which allows this beta-carotene pigment to diffuse to both the epidermis and dermis to act as a membrane stabilizer and a quencher against the damaging forms of reactive $O_2$ species (singlet oxygen or $^1O_2^{*-}$, $OH^{\cdot}$, etc.) that are generated by ultraviolet radiation. Humans accumulate carotenoid pigments through dietary intake (green vegetables, carrots, peppers, etc.).
4. The presence of superoxide dismutase and glutathione peroxidase–reductase enzymes in the epidermis acting as selective deactivating systems for the reactive form of $O_2$ (superoxide anion or $O_2^{\cdot -}$) generated by ultraviolet radiation, and thus protecting the peroxidation reaction-induced damage to cell membranes, epidermal and dermal proteins (3,5).
5. The excision repair capacity of cutaneous cells to repair ultraviolet radiation-induced damage in DNA appropriately by an error-free DNA replicating mechanism. Xeroderma pigmentosum, an autosomal recessive disease, is the prototype disease in which the defective excision repair due to altered endonuclease activity leads to an early onset of neoplastic changes.

Although each of these five defensive mechanisms that exist in mammalian skin are important and complement each other in minimizing ultraviolet radiation-induced damage, the two major defensive components that play a significant role in protection against the harmful effects of ultraviolet radiation are the compact multicell layers of the stratum corneum and the presence of the melanin filter in the form of colloidal melanin and melanosome granules in basal, suprabasal cells and stratum corneum of the epidermis that shields the nuclear DNA from harmful alterations. Of the two defensive barriers to ultraviolet radiation, the melanin filter is the most important, inasmuch as humans with normal stratum corneum but without melanin (e.g., albinos) succumb to repeated ultraviolet exposures with early onset of chronic solar damage (actinic elastosis or dermatoheliosis) and skin cancer despite the capacity of the albino skin to respond normally by the process of hyperplasia and subsequent thickening of the stratum corneum.

## IV. SUNSCREENS

### A. Sun Protection Factor

The effectiveness of a sunscreen is based on its sun protection factor (SPF) value (2–5,9,10). The SPF represents protective value or potency of a sunscreen, and it is defined as a ratio of an exposure dose of the erythemogenic ultraviolet radiation required to produce the minimal erythema reaction through the applied sunscreen product to the UV dose required to produce the same minimal erythema reaction without topical applications of the sunscreen (3,4). For SPF evaluation of a sunscreen product in the form of a lotion, cream, or a gel, the U.S. Food and Drug Administration (FDA) (9) requires a minimum of 20 fair-skinned volunteers and application of the test product or products at 2 $\mu l/cm^2$ or 2 $mg/cm^2$ on the back of the test subject. The recommended ultraviolet source for eliciting erythema reaction is usually a high-pressure xenon-arc lamp (solar simulator equipped with 150 W xenon-arc lamp) that provides spectral output similar to sunlight in 290–400 nm ultraviolet region. A series of exposures (six to eight) are given to determine the dose of UVB required to produce a MED response of the untreated and sunscreen-treated skin sites. The SPF of a product can be calculated as follows: SPF = MED ($mJ/cm^2$) of sunscreen-protected skin divided by the MED ($mJ/cm^2$) of nonprotected skin.

### B. Ultraviolet-Absorbing Chemical Agents Used in Sunscreen Formulations

Until the mid-1970s, sunscreens were regarded as cosmetic products intended to promote a tan and to moisturize dry skin during and after sunbathing. The prevention of sunburn reaction was not obligatory. In 1978, the U.S. Food and Drug Administration (FDA) classified sunscreens as "drugs" intended to protect the structure and function of the human skin against actinic damage manifested in the form of sunburn, solar elastosis, and possibly skin cancer. The use of sun-protective topical preparations containing approved chemicals that were safe and effective was recommended for the general public. These chemical agents (9, 11–13) formulated in the form of solutions (clear or milky lotions), gels, creams, or ointments absorb or filter out 95% or more of UVB radiation and, to a limited extent, UVA radiation, and prevent or minimize the deleterious effects of ultraviolet radiation on human skin. Protection is provided by the chemical agents through processes involving absorption, reflection, and scattering of solar radiation impinging on the skin.

## C. Classification of Sunscreens

The available photoprotective agents recognized as sunscreens can be grouped into two broad categories: topical sunscreens and oral sunscreens. Topical sunscreens can be differentiated into two subcategories: chemical sunscreens and physical sunscreens. Chemical sunscreens are usually colorless preparations that are used topically and contain one or more ultraviolet-absorbing chemicals listed in Table 6. Topical sunscreens are formulated to screen out, either totally or partially, ultraviolet radiation impinging on the skin ($\lambda > 290$ nm); prevent ultraviolet radiation-induced photochemical and biochemical reactions discussed previously in this chapter; and provide water-resistant or waterproof qualities so that the applied sunscreen does not wash off after sweating or swimming. These ultraviolet-absorbing sunscreens are usually applied in the form of an invisible (10–20 $\mu$m thick or 2$\mu$l/cm$^2$) film and are cosmetically acceptable to most normal individuals and patients provided they are nonirritant, nonsensitizing, stable to ultraviolet radiation, nonvolatile, nonstaining, noncomedogenic, nonmutagenic, and noncarcinogenic. The most widely used chemical sunscreens contain p-aminobenzoic acid (PABA) or derivatives of PABA (amyldimethyl PABA or padimate A, and octyldimethyl PABA or padimate O), benzophenones (oxybenzone and dioxybenzone), cinnamates (octylmethyl cinnamate and cinnoxate), salicylates (homomenthyl salicylate), anthranilates, and Parsol 1789. To date, over 21 such chemicals classified under category I have been declared by the U.S. FDA as reasonably safe and effective agents in the protection of skin against acute and chronic effects of solar radiation; the chemicals now commonly used in sunscreen formulations are listed in Table 6.

### 1. *Topical Chemical Sunscreens*

Most chemical sunscreens (e.g., PABA, PABA esters, or cinnamates) are narrow-band, UVR-absorbing sunscreens that prevent the penetration of sunburn-producing UVB (290-320 nm) radiation into skin. For clinical use, they are usually classified as PABA sunscreens, non-PABA sunscreens containing one or two chemicals, and combination sunscreens containing PABA ester, benzophenone, and other UVA-absorbing chemicals (Tables 7–9). Newer sunscreens marketed in the 1980s have been designed to provide a higher degree of protection (SPF>15–30) over a broader spectrum of ultraviolet radiation extending up to 400 nm, and, hence, can be classified as "broad-spectrum sunscreens." Many of them are formulated in water-resistant bases. These sunscreens contain

**Table 6** Sunscreen Agents of Category I (Safe and Effective)

| Compound | % Concentration permitted | Absorbance range (nm) |
|---|---|---|
| PABA | | |
| p-Aminobenzoic acid | 5.0 –15.0 | 260–313 |
| Glyceryl aminobenzoate | 3.0 –5.0 | 260–315 |
| Amyl p-dimethylamino benzoate (Padimate-A) | 1.0 –5.0 | 290–315 |
| 2-Ethylhexyl-o-dimethylamino benzoate (Padimate-O) | 1.4 –8.0 | 290–315 |
| Cinnamates | | |
| 2-Ethoxy-ethylhexyl-p-methoxycinnamate (cinnoxate) | 1.0 –3.0 | 270–328 |
| Diethanolamine-p-methoxycinnamate | 8.0–10.0 | 280–310 |
| Ethylhexyl-p-methoxycinnamate | 2.0 –7.5 | 290–380 |
| Benzophenones | | |
| 2,2-Dihydroxy-4-methoxybenzophenone (dioxybenzone) | 3.0 | 260–380 |
| 2-Hydroxy-4-methoxybenzophenone (oxybenzone) | 2.0 –6.0 | 270–350 |
| 2-Hydroxy-4-methoxybenzophenone 5-sulphonic acid (sulisobenzone) | 5.0 –10.0 | 270–360 |
| Salicylates | | |
| 2-Ethylhexyl-salicylate | 3.0 –5.0 | 250–320 |
| Triethanolamine salicylate | 5.0 –12.0 | 260–320 |
| 3,3,5-Trimethylcyclohexyl salicylate (homosalate) | 4.0 –15.0 | 290–315 |
| Miscellaneous | | |
| 2-Ethylhexyl-2-cyano-3,3-diphenylacrylate | 7.0 –10.0 | 290–360 |
| Ethyl-4-bis-(hydroxypropyl)-aminobenzoate | 1.0 –5.0 | 280–330 |
| Digalloyl trioleate | 2.0 –5.0 | 270–320 |
| Lawsone + dihydroxyacetone | 0.25–3.0 | 290–340 |
| Menthylanthranilate | 3.5 –5.0 | 290–320 |
| 2-Phenyl-benzimidazole-5-sulphonic acid | 1.0 –4.0 | 290–320 |
| Physical Sunscreens | | |
| Red veterinary petrolatum | 30 –100 | 260–380 |
| Titanium dioxide | 2.0 –25.0 | UV scatter |

Chemicals of category II include 2-ethyl-hexyl-4-phenylbenzophenone-2'-carboxylic acid; 3-(4-methylbenzylidene)-camphor; and sodium-3,4-dimethylphenyl glyoxalate. These chemicals are not commonly used for over-the-counter sunscreen products unless approved by the FDA.

*Source*: Ref. 9.

**Table 7** Examples of Sunscreen Formulations Containing One or Two Sunscreen Ingredients Available in the United States

| Trade Name (Manufacturer) | SPF | Active Ingredients |
|---|---|---|
| Coppertone (Plough) | 2 | Octyl salicylate |
| Bain de Soleil (Bain de Soleil) | 2,4 | Padimate O |
| Eclipse (Eclipse Labs) | 5 | Padimate O |
| Eclipse (Eclipse Labs) | 10 | Glyceryl PABA |
| Supershade (Plough) | 8,15 | Parsol MCX[a], oxybenzone |
| Coppertone (Plough) | 4,6,8,15 | Padimate O, oxybenzone |
| Shade (Plough) | 4,6 | Padimate O, oxybenzone |
| PreSun (Westwood) | 8,15 | Padimate O, oxybenzone |
| Water Babies (Plough) | 15 | Parsol MCX, oxybenzone |
| Sundown (Johnson & Johnson) | 4,6,8 | Padimate O, oxybenzone |
| Block Out (Carter Products) | 15 | Padimate O, oxybenzone |
| Photoplex (Herbert Labs) | 15+ | Padimate O, avobenzone[b] (Parsol 1789) |

[a]Parsol MCX = ethylhexyl p-methoxycinnamate.

[b]Avobenzone = Parsol 1789 = [1-4(-tert-butylphenyl-3-(4-methoxyphenyl)-propane-1,3-dione] or t-butyl methoxydibenzoyl methane.

two or more ultraviolet-absorbing chemicals (e.g., padimate O plus oxybenzone or padimate O plus Parsol 1789) and some light-scattering agents such as micronized titanium dioxide or zinc oxide (Tables 7–9).

### 2. *Topical Physical Sunscreens*

Physical sunscreens or ultraviolet blockers are chemicals that do not necessarily absorb ultraviolet radiation but, because of their particle size (30 100 nm) and physical nature, reflect or scatter the ultraviolet radiation. Physical sunscreens

**Table 8** Examples of Sunscreen Formulations Containing Three Sunscreen Ingredients Available in the United States

| Trade Name (Manufacturer) | SPF | Active Ingredients |
|---|---|---|
| Solbar (Person and Covey) | 50 | Parsol MCX[a], octocrylene[b], oxybenzone |
| PreSun for Kids (Westwood) | 39[c] | Parsol MCX, octyl salicylate, oxybenzone |
| PreSun 29 | 29[c] | Parsol MCX, octyl salicylate, oxybenzone |
| Bain de Soleil (Bain de Soleil) | 30 | Padimate O, Parsol MCX, oxybenzone |
| Ultrashade (Plough) | 23[c] | Padimate O, Parsol MCX, oxybenzone |
| Total Eclipse (Eclipse Labs) | 15 | Padimate O, octyl salicylate, oxybenzone |
| Sundown (Johnson & Johnson) | 15[c] | Padimate O, Parsol MCX, oxybenzone |

[a]Parsol MCX = ethylhexyl-p-methoxycinnamate.

[b]Octocrylene = 2-ethylhexyl-2-cyano-3,3-diphenylacrylate.

[c]Sweat- and water-resistant (suitable for children).

*Source:* Refs. 3, 11, 12, 13.

**Table 9** Examples of Sunscreen Formulations Containing Four Sunscreen Ingredients Available in the United States

| Trade Name (Manufacturer) | SPF | Active Ingredients |
|---|---|---|
| Coppertone (Plough) | 30[a] | Padimate O, Parsol MCX[b], octylsalicylate, oxybenzone |
| Sundown (Johnson & Johnson) | 30[a] | Parsol MCX, octyl salicylate, oxybenzone, titanium dioxide |
| | 20 | Padimate O, Parsol MCX, octyl salicylate, oxybenzone |
| Cancer Garde (Eclipse Labs) | 30[a] | Padimate O, Parsol MCX, oxybenzone, titanium dioxide |
| T/I Screen (T/I Pharmaceuticals) | 30[a] | Parsol MCX, octocrylene[c], octyl salicylate, oxybenzone |
| Block Out (Carter Products) | 30[a] | Parsol MCX, Padimate O, octyl salicylate, oxybenzone |
| Supershade (Plough) | 44[a] | Parsol MCX, Padimate O, homosalate, oxybenzone |

[a]Waterproof sunscreens.
[b]Parsol MCX = Ethylhexyl p-methoxycinnamate.
[c]Octocrylene = 2-ethylhexyl-2-cyano-3,3-diphenylacrylate.
*Source:* Refs. 3, 11, 12, 13.

are usually opaque and may be colorless or colored formulations, which, when applied topically to the skin as a thin, even film will primarily attenuate the impinging radiation by processes known as reflection and scattering. When incorporated in an oily base at 2–20% concentration, they adhere well to the skin and are not easily washed off after sweating or swimming. These physical sunscreens may include titanium dioxide, talc, zinc oxide, ferrous oxide or ferric chloride, silicone-treated titanium dioxide, chromium oxide, cobalt oxides, kaolin, ichthyol, starch, and other ingredients. Physical sunscreens are cosmetically unacceptable to many people and can often be occlusive, comedogenic, and messy to use. These sunscreens are, however, essential for those patients who are unusually sensitive to UV and visible radiation. Fair-skinned individuals with normal skin who are constantly exposed to the sun (e.g., lifeguards, roofers, farmers, and sailors) and children with tender skin should be advised to apply these physical sunscreens regularly to limited areas over the bridge of the nose, shoulders, lips, and/or helix of the ears when direct or indirect sun exposure for a long duration is anticipated.

## D. Photoprotection Against UVA

Until recently, most sunscreens were formulated to filter out UVB radiation. UVA radiation (320–400 nm) was purposely not screened out because it was believed to be innocuous, as well as beneficial for its ability to stimulate the tanning reaction in the skin. Although UVA radiation is less erythemogenic, less damaging than UVB (290–320 nm) radiation, and even stimulates immediate and delayed tanning reactions, it is becoming increasingly clear that this spectrum of sunlight cannot be considered innocuous. UVA radiation penetrates epidermis and deep into the dermis and is a harmful spectrum (2,3,14–24). UVA radiation has been shown to produce many potentially deleterious effects in mammalian skin, including human skin. UVA radiation causes sunburn (epidermal and dermal cell damage); damage to melanocytes that results in proliferation of melanocytes and augmentation of melanin pigmentation; damage to DNA in the form of cyclobutyl pyrimidine dimers and inhibition of DNA synthesis; inhibition of RNA synthesis and protein synthesis; depletion of epidermal Langerhans cells; dermal damage to fibroblasts, mast cells, endothelial cells, collagen, and elastic fibers (eventually contributing to solar elastosis); and damage to cell mambranes involving lipid peroxidation. UVA is the action spectrum of many drug-induced phototoxic and photoallergic reactions, chronic actinic dermatitis, and solar urticaria. Long-term UVA exposures have also contributed to epidermal tumors (squamous cell carcinoma) and augmentation of UVB photocarcinogenesis in hairless mice (2,3,14–24).

To minimize the harmful cumulative effects of UVA radiation, certain UVA-absorbing chemicals are added to the UVB-absorbing sunscreens. These include certain benzophenones (oxybenzone, dioxybenzone, sulisobenzone); dibenzoyl methanes (e.g., Parsol 1789, Eusolex-8020, Parsol DAM); and certain physical sunscreen agents (e.g., zinc oxide or titanium dioxide). Parsol 1789 [1-(4-tert-butylphenyl-3-(4-methoxyphenyl) propane-1,3-dione] has received FDA approval for use as a UVA blocker in a specific formulation (Photoplex) at a concentration of 3%. Thus, UVA-induced increased pigmentation and psoralen-induced phototoxic reactions (8-methoxypsoralen and 4,5',8-trimethylpsoralen-induced phototoxic reactions) caused by topical application or oral ingestion of psorlen and subsequent exposure to 320–380 nm radiation can be minimized by sunscreens containing certain benzophenone derivatives as well as by Parsol 1789 (3,5,13,19,25,26). Drug-induced phototoxic reactions (e.g., chlorothiazides, demethylchlortetracycline, nalidixic acid, amiodarone, and various psoralens) can be prevented by the use of topical sunscreens containing UVA-absorbing chemicals.

## V. GUIDELINES FOR RECOMMENDING TOPICAL SUNSCREENS IN HEALTH AND LIGHT-INDUCED DISEASES

Figure 1 and Table 4 show the recommended SPF values for individuals of skin types I–VI. People with fair skin, blue eyes, red hair, with or without freckles, and who burn and peel easily but tan poorly (skin types I and II) should be

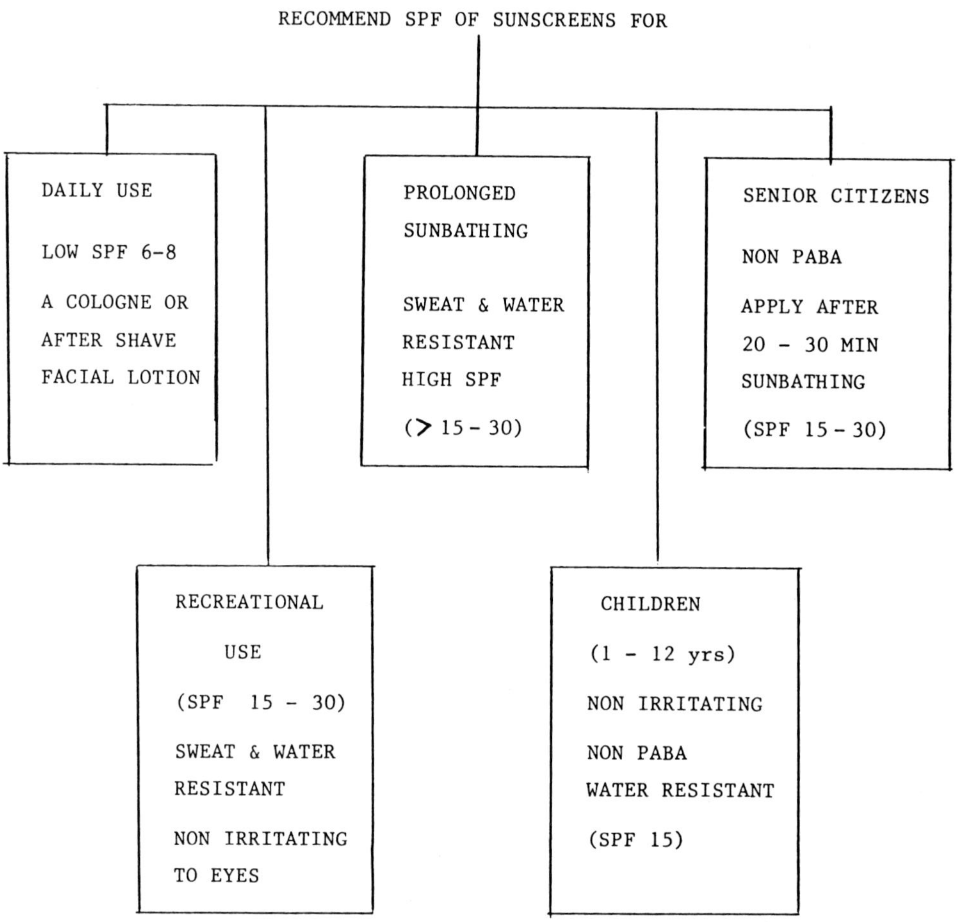


**Figure 1** Recommended SPF values of sunscreens for various uses.

prescribed sunscreens that have high protection values (Tables 7–9) (SPF $\geq$ 15). Individuals with skin types III and IV, who burn moderately or minimally but tan well, may be urged to use sunscreens with an SPF of 10–15. Sunscreens may be applied 15–30 min before sunbathing and should be reapplied after swimming or during prolonged sunbathing. Reapplication of the product is to ensure continuous protection against overexposure to the sun. A major achievement in the area of sunscreen technology in recent years has been the development of water-resistant or waterproof and highly substantive products that tend to remain adsorbed or bonded on the skin and resist wash-off after swimming or profuse sweating. To avoid constant reapplication, water-resistant polymer-base sunscreens (acrylate and methylacrylate base) have been introduced and should be strongly recommended to individuals with skin types I–III.

The deleterious effects of sun exposure are cumulative and often irreversible. Prevention against sunburn and the development of keratoses and carcinoma in skin with solar degeneration (photoaging) can be effectively achieved with the daily topical application of water-resistant sunscreens (Tables 7–9) on regular basis. Children do not like alcoholic lotions of PABA or PABA esters and should be prescribed water-resistant, sweat-resistant, polymer-based non-PABA sunscreen lotions or opaque sunscreens (Tables 8, 9) that are not irritating to skin and eyes.

Since ultraviolet radiation also impairs the skin's cellular immune response, which may, in the long run, facilitate the development of skin tumors, questions have been raised whether the regular use of topical sunscreens can ameliorate certain immunosuppressive effects of UVB radiation. Laboratory studies in mice (personal discussion with Dr. M. L. Kripke, Houston, Texas) do suggest that the use of topical sunscreens (SPF 15 or higher) has the capacity to protect mice from immunosuppression by UVR. In this regard, clinical experience with well controlled PUVA treatment using oral 8-methoxypsoralen and UVA radiation is worth mentioning. A short course of oral PUVA (two times per week for three to four weeks or longer) induces significant photoprotection through increased pigmentation and increased thickening of stratum corneum. Such treatments can ameliorate certain immunosuppressive effects of UVR. Photoprotective effect of oral PUVA therapy has been found to be beneficial in a number of photosensitivity diseases (polymorphous light eruption, solar urticaria, persistant light reaction, actinic reticuloid, etc.)

A variety of compounds such as aloe (an extract of aloe vera), jojoba oil (an extract from a desert shrub), and vitamin E are receiving increasing endorsement through seasonal articles and advertisements in magazines and newspapers as soothing agents and home remedies for alleviating the discomforts of sunburn reaction. Some of these products are very popular at beach resorts and vacation islands in tropical and semitropical places and have been in use as "folk remedies," "moisturizers," and lubricants that instantly alleviate burning, itching, and discomfort associated with overexposure to sunlight. The efficacy of such agents has not been evaluated under well-controlled experimental conditions involving randomized, double-blind, clinical trials. Our experience, confirmed by a controlled double-blind study, does not lend credibility to such anecdotal claims. Repeated topical application of antioxidants such as $\alpha$-tocopherol acetate and ascorbic acid at high concentrations (2–10%) may help to minimize the sunburn reaction. We have observed a definite reduction of the sunburn reaction and the formation of sunburn cells only when the skin has been protected with repeated (four to five) consecutive applications of $\alpha$-tocopherol acetate-(2%) or ascorbic acid- (5–10%) enriched formulations at intervals of 30 min. Any exposure greater than 3 MED (75 $mJ/cm^2$) will, however, result in sunburn reaction even after repeated applications.

## VI. ADVERSE REACTIONS TO TOPICAL SUNSCREENS

Contact sensitization, phototoxicity, and photoallergic reactions may develop from the use of sunscreens (3,5,11,27). A list of sunscreen agents involved in inducing contact dermatitis and photocontact allergic reactions is presented in Table 10. Glyceryl PABA seems to be one of the most common causative agents and generally its use has now been discontinued. Cross-reactivity may occur among PABA, benzocaine, procaine, and p-phenylenediamine hair dyes. Photocontact allergy to other PABA esters (padimate A or amyl-p-dimethyl PABA, and padimate O or octyldimethyl PABA) has been observed by few investigators. PABA may also cross-react with sulfanilamide and thiazides. This potentially harmful reaction may be of some medical significance in patients receiving diuretics. Phototoxicity to padimate O alone has not been well documented in a large number of users. Recent information suggests photocontact sensitivity to some sunscreen formulations may be due to 6-methylcoumarin or musk ambrette (an aromatic ingredient in aftershave lotions) and not to the actual sunscreening chemical agent(s). These fragrant substances are photoallergic agents and may cause light-induced papular or eczematous eruptions in skin exposed to these substances. Some formulations contain parabens and lanolin derivatives, which

**Table 10** Chemicals Used in Sunscreens Known to Cause Contact Dermatitis and Photocontact Allergic Reactions

| Chemical | Type of Reaction Reported |
|---|---|
| PABA | CA, PCA, CPTX |
| Glyceryl PABA | CA, PCA, CPTX |
| Padimate O | PCA |
| Padimate A | PCA, CPTX |
| Oxybenzone (benzophenone-3) | PCA |
| Sulisobenzone | CA |
| Dioxybenzone | CA |
| Homomenthyl salicylate | CA |
| Digalloyl trioleate | PCA |
| Menthylanthranilate | CA |
| 4-Isopropyldibenzoyl methane (Eusolex 8020) | PCA |
| 2-Ethoxy-ethyl-p-methoxy cinnamate | PCA, CPTX |
| Butyl methoxydibenzoyl methane (Parsol 1789) | CA, PCA |

CA, contact allergy; PCA, photocontact allergy; CPTX, contact phototoxicity.
*Source*: Refs. 11, 27.

can also act as contact sensitizers. Besides PABA and PABA esters, several other sunscreen chemicals have been implicated as causative agents for contact sensitization reaction. Oxybenzone, dioxybenzone, sulisobenzone, homomenthylsalicylate, menthylanthranilate, 4-isopropyl methane, 2-ethoxyethyl-p-methoxycinnamate, 3-(4-methyl-benzylidine) or Eusolex 6300, and Parsol 1789 have been reported to cause photosensitivity reactions (3,5,11,27). It should be emphasized that preservatives, fragrances, and emulsifiers are also responsible for contact allergies.

## VII. ORAL SUNSCREENS (SYSTEMIC PHOTOPROTECTIVE AGENTS)

For long-term and preventive use, many people do not like to apply topical sunscreens on a daily basis. For these individuals, the development of orally administered photoprotectants to reduce unavoidable sun exposure and subsequent reaction of skin to UVB and UVA radiation has long been hoped for but has not proved successful (2,3,5,28). Agents such as PABA; para-aminosalicylic acid; antihistaminic compounds such as triprolidine exhibiting UVR absorption in the 238–290 nm range; acetylsalicylic acid (aspirin); vitamins A, C (ascorbic acid), and E ($\alpha$-tocopherol); certain unsaturated fatty acids that are easily photo-oxidized; and certain steroids have been suggested as systemic photoprotective agents. The effectiveness of most of these agents in preventing UVB-induced sunburn reaction after single administration is mostly anecdotal. When they were tested by us in a well-controlled study protocol, all these systemic agents exhibited no convincing partial or total in vivo photo-protective properties. Orally administered antioxidants (e.g., $\alpha$-tocopherol, $\beta$-carotene) upon daily use for 4–6 weeks appear to provide limited protection of less than 2.5 SPF values due to their ability to act as antioxidants and free radical scavengers. Although acetylsalicylic acid (aspirin) has excellent analgesic properties and may influence early sunburn erythema reaction because of its inhibitory effects on the cyclo oxygenase pathway, it has failed to provide any significant anti-inflammatory and photoprotective effect against delayed erythema reaction, which shows a peak response 24 hr after exposure.

The topical application or intradermal injection of indomethacin, a pharmacologic agent known to inhibit prostaglandin synthesis ($PGE_2$ and $PGF_2\alpha$), has been reported to decrease the intensity of UVB-induced erythema in the skin of humans and guinea pigs (5,29). Indomethacin is efficacious in blanching the existing UVB-induced erythema, and its anti-inflammatory action is due to its ability to inhibit the cyclooxygenase pathway involved in the generation of prostaglandins. Morison and associates (29) determined the effects of indomethacin administered topically, intradermally, and orally on the delayed phase of erythema response. Although UVB-induced erythema could be blanched, indomethacin did not diminish the delayed erythema response to UVA or the delayed phototoxic reaction produced by PUVA. It should be stressed that indomethacin has a little or no inhibitory effect on UVA-induced erythema and has virtually no effect on UVA-induced damage to DNA or to epidermal cells (sunburn cells).

Carotenoid pigments can exert a protective action against photosensitivity in humans as well as in animals and plant systems. The oral administration of $\beta$-carotene is recommended for the amelioration of photosensitivity reactions caused by visible radiation (400–760 nm) in patients with erythropoietic protoporphyria, erythropoietic porphyria, and variegate porphyria (2,3,5,28). Although $\beta$-carotene absorbs light primarily in the visible spectrum (360–500 nm), with a maximum absorption at 450–475 nm, its effectiveness is due to its ability to quench singlet oxygen generated by photoexcited porphyrins and is not due to its photoabsorbant properties. Usual dosages are 30–60 mg day for children 1–8 years old and 120–180 mg day or more for adults. It requires 1–2 months for patients to notice an increased tolerance to light. At present, the recommendation for treatment is the oral ingestion of $\beta$-carotene regulated to maintain blood levels between 600 and 800 $\mu$g/dl. From the results of our studies and published data, it would seem that there is no evidence of toxicity from the ingestion of large amounts of pure $\beta$-carotene for prolonged periods of time.

## ACKNOWLEDGMENT

This investigation was supported by grant No. 5-RO1-CA-05003-32, awarded by the U.S. National Cancer Institute, Department of Health and Human Services, and Plough Inc., Memphis, TN.

## REFERENCES

1. Giese AC. Living With Our Sun's Ultraviolet Rays. New York: Plenum, 1976.
2. Pathak MA. Sunscreens: topical and systemic approaches for protection of human skin against harmful effects of solar radiation. J Am Acad Dermatol 1982; 1:285–312.
3. Pathak MA, Fitzpatrick TB, Greiter F, Kraus EW. Preventive treatment of sunburn, dermatoheliosis, and skin cancer with sun protective agents. In: Fitzpatrick TB, Eisen AA, Wolff K, Freedberg IM, Austen KF, eds. Dermatology in General Medicine, 3rd ed. New York: McGraw-Hill, 1987; 1507–1522.
4. Pathak MA. Intrinsic photoprotection in human skin. In: Lowe NJ, Shaath NA, eds. Sunscreens: Development, Evaluation, and Regulatory Aspects. New York: Marcel Dekker, 1990: 73–83.
5. Pathak MA. Topical and systemic approaches for the prevention of acute and chronic sun-induced skin reactions. Dermatol Clin 1986; 4:321–324.
6. Fredrick JE, Snell HE, Haywood EK. Solar ultraviolet radiation at the earth's surface. Photochem Photobiol 1989; 50:443–450.

7. Pathak MA. Activation of melanocyte system by ultraviolet radiation and cell transformation. Ann NY Acad Sci 1985; 435:328–339.
8. Fitzpatrick TB. Soleil et peau. J Med Esthet 1975; 2:33–38.
9. Sunscreen drug products for over-the-counter human drugs; proposed safety, effectiveness, and labeling conditions. Washington, DC: Department of Health, Education, and Welfare, Food and Drug Administration, August 25, 1978, p 38206.
10. Greiter F. Sonnenschutzfaktor-entstehung, methodik. Perfum Kosmet 1974; 55:70–76.
11. Pathak MA. Sunscreens: topical and systemic protection against solar radiation for human skin. In: Favre A, Tyrrell R, Cadet J, eds. From Photophysics to Photobiology: Photochemistry and Photophysics. Amsterdam: Elsevier, 1987: 447–461.
12. Shaath NA. The chemistry of sunscreens. In: Lowe NJ, Shaath NA, eds. Sunscreens: Development, Evaluation and Regulatory Aspects. New York: Marcel Dekker, 1990: 211–233.
13. Lowe NJ. Sunprotection factors: Comparative techniques and selection of ultraviolet sources. In: Lowe NJ, Shaath NA, eds. Sunscreens: Development, Evaluation and Regulatory Aspects. New York: Marcel Dekker, 1990: 379–393.
14. Pathak MA. Sunlight and melanin pigmentation. In: Smith KC, ed. Photochemical and Photobiological Reviews, vol. 1. New York: Plenum Press, 1976: 211–223.
15. Parrish JA, Anderson RR, Urbach F, Pitts DA. UVA: Biological Effects of Ultraviolet Radiation with Emphasis on Human Responses to Long-Wave Ultraviolet. New York: Plenum Press, 1978: 107–175.
16. Paul BS, Parrish JA. The interaction of UVA and UVB in the production of threshold erythema. J Invest Dermatol 1982; 78:371–374.
17. Gilchrest BA, Soter NA, Hawk JLM, et al. Histologic changes associated with ultraviolet A-induced erythema in normal human skin. J Am Acad Dermatol 1983; 9:213–219.
18. Urbach F, Gange RW. The Biological Effects of UVA Radiation New York: Prager Publishers, 1986.
19. Lowe NJ. UVA photoprotection. In: Lowe NJ, Shaath NA, eds. Sunscreens: Development, Evaluation and Regulatory Aspects. New York: Marcel Dekker, 1990: 459–468.
20. Freeman SE, Hacham H, Gange RW, Maytum DJ, Sutherland JC, Sutherland BM. Wavelength dependence of pyrimidine dimer formation in DNA of human skin irradiated in situ with ultraviolet light. Proc Natl Acad Sci USA 1989; 86:5605–5610.
21. Van der leun JC. Animal experiments in study of photocarcinogenesis. In: Favre A, Tyrrell R, Cadet J, eds. From Photophysics to Photobiology: Photochemistry and Photophysics. Suppl. Amsterdam: Elsevier, 1987: 353–360.
22. Freeman SE, Gange RW, Sutherland JC, Sutherland BM. Pyrimidine dimer formation in human skin. Photochem Photobiol 1987; 46:207–212.
23. Kligman LH. Photoaging: manifestations, prevention and treatment. In: Gilchrest BA, ed. Dermatologic Clinics: The Aging Skin. Philadelphia: WB Saunders, 1986: 4:517–528.
24. Pathak MA. Photoaging (dermatoheliosis) and biochemical approaches in the amelioration of photoaging changes. Photochem Photobiol 1989; 49:79–80.

25. Kaidbey K, Gange RW. Comparison of methods for assessing photoprotection against ultraviolet A in vivo. J Am Acad Dermatol 1987; 16:346–353.
26. Gange RW, Soparkar A, Matzinger E, Dromgoole SH, Sefton J, DeGryse R. Efficacy of a sunscreen containing butyl methoxydibenzoylmethane against ultraviolet A radiation in photosensitized subjects. J Am Acad Dermatol 1986; 15:494–499.
27. Dromgoole SH, Maibach HI. Contact sensitization and photocontact sensitization of sunscreening agents. In: Lowe NJ, Shaath NA, eds. Sunscreens: Development, Evaluation and Regulatory Aspects. New York: Marcel Dekker, 1990: pp. 313–340.
28. Harber LC, DeLeo VA, Prystowsky JH. Intrinsic and extrinsic photoprotection against UVB and UVA radiation. In: Lowe NJ, Shaath NA, eds. Sunscreens: Development, Evaluation and Regulatory Aspects. New York: Marcel Dekker, 1990: pp. 359–377.
29. Morison WL, Paul BS, Parrish JA. The effect of indomethacin on longwave-ultraviolet-induced delayed erythema. J Invest Dermatol 1977; 68:130–133.

# VI

# Light as Therapeutic Agent

# 18

# Phototherapy with Ultraviolet B

**Serge A. Coopman**
*University of Leuven, Leuven, Belgium*
**Robert S. Stern**
*Harvard Medical School, Boston, Massachusetts*

Since ancient times, the beneficial influence of sunlight radiation on a broad variety of diseases has been recognized, and a variety of therapies based on these observations have been developed. The history of modern phototherapy began in the late 19th century with the work of Niels Finsen. Finsen observed that sunlight, and, more specifically, the ultraviolet radiation, had a beneficial effect on lupus vulgaris. In 1903 he received the Nobel Prize in Physiology and Medicine for this work.

Early in the 20th century, so-called helio- or actinotherapy for psoriasis enjoyed great popularity, especially in Europe. The first treatments for psoriasis with an artificial ultraviolet source utilized a carbon arc lamp developed by Finsen. The more practical medium-pressure mercury arc lamps, which provide higher ultraviolet output, replaced the carbon arc lamp between the World Wars. In 1925, Goeckerman, at the Mayo Clinic, advocated a regimen that combined tar baths (or other applications of tar) with ultraviolet for the treatment of psoriasis. This approach has enjoyed wide acceptance in many dermatologic centers. After World War II, the development of fluorescent ultraviolet lamps provided the third innovation in sources of ultraviolet for dermatologic therapy. Their emission spectrum was modified in the following decade with the use of several envelope-coating materials.

Today, ultraviolet B (UVB) phototherapy (290–320 nm) is widely used in the treatment of a variety of skin diseases. The basic data on ultraviolet radiation and the mechanisms of its interaction with the skin are discussed more extensively in Chapters 1–2 of this book. This chapter presents a general overview of the specific treatment modalities and indications for UVB phototherapy in dermatologic disease.

# I. UVB PHOTOTHERAPY OF PSORIASIS

## A. Action Spectrum and Mechanisms

Psoriasis is a chronic skin disease of unknown origin characterized by an increased epidermal cell turnover. It has long been known that many patients with psoriasis experience improvement of their disease in the summer months. This observation lead to many empiric forms and regimens of therapeutic irradiation with artificial ultraviolet sources that usually emit a broad array of wavelengths. More recently, several studies attempted to define the relative therapeutic efficacy of the different wavelengths in clearing psoriasis.

Parrish et al. compared several wavelengths in the range of 254–313 nm for their capacity in resolving psoriasis relative to their erythemogenic effect (1). Wavelengths shorter than 295 nm displayed no therapeutic effect, even at erythemogenic doses. Other observations confirmed the lack of therapeutic efficacy of ultraviolet C (UVC) (200–290 nm) either alone (2) or in combination with shorter wavelengths of UVB (3). Several studies suggested that the action spectrum for erythema and therapeutic response are closely parallel within the UVB and ultraviolet A (UVA) (320–400 nm) portions of the electromagnetic spectrum (4,5). In these wavelengths, erythemogenic doses are generally needed to clear psoriasis (6). The absolute values of the doses needed (in $J/cm^2$) are higher by 1000 for UVA compared to UVB. Because of the long times required to deliver sufficient UVA, UVA without a photosensitizer is generally considered impractical as a therapy for psoriasis.

The effective action spectrum can be expected to be shaped by optical factors and more specifically by the extent to which the wavelengths of therapeutic interest penetrate to the basal layers of the epidermis or below. Wavelengths shorter than 295 nm, which are inefficient therapeutically, are mostly absorbed by the upper epidermis. On the other hand, longer wavelengths of UVB may yield a better therapeutic result because of their deeper penetration into psoriatic plaques. Studies by Fisher (7) suggest that wavelengths around 313 nm have optimal efficacy in the treatment of psoriasis. Several new types of lamps that have been developed preferentially emit these ultraviolet wavelengths.

The mechanistic pathways involved in phototherapy for psoriasis are not known. In animal as well as in human studies, a biphasic alteration of DNA, RNA, and protein synthesis was demonstrated: a transient decrease in macromolecule synthesis was followed by a rebound increase (8–10). These events are thought to be accompanied by a temporary return to more normal cell kinetics of the psoriatic keratinocyte. Other mechanisms may also contribute to the therapeutic action, such as immunologic modifications (11) and interference with the cascade of chemical mediators important in the maintenance of the inflammatory or hyperproliferative state.

## B. General Considerations: Patient Evaluation, Selection Criteria, and Contraindications

Evaluation of the patient with psoriasis who is a candidate for UVB phototherapy includes several essential steps in order to select individuals who are likely to respond and to minimize the risk of adverse events from the therapy.

From each patient, a complete medical history is needed. Special attention should be paid to previous responses to sun exposure and/or artificial ultraviolet light, possible photosensitivity reactions, signs or symptoms of lupus erythematosus, and a history of cutaneous neoplasms. If any suspicion about lupus exists, serologic tests should be obtained prior to therapy. Since many drugs can precipitate phototoxic reactions resembling severe sunburns, it is also essential to determine all the medications the patient is currently taking. In general, the phototoxicity spectrum is similar to the ultraviolet absorption spectrum of the drug. Most phototoxic drugs absorb predominantly in the UVA region, but concomitant use of phototoxic drugs should be avoided when practical. Common examples of photosensitizing drugs include antibiotics (e.g., sulfonamides, tetracyclines), diuretics (e.g., thiazides), antifungals (e.g., griseofulvin), antidiabetics (e.g., sulfonylureas), tranquilizers (phenothiazines), and antimetabolic or cytostatic drugs (e.g. methotrexate, vinblastine, fluorouracil).

A total skin examination is usually done to substantiate the diagnosis of psoriasis, to establish the severity of the disease, and detect any existent skin cancer. Patients with a "psoriasiform dermatosis", in whom doubt may exist as to the exact diagnosis, may require additional investigation. As indicated, this might include a full blood count, erythrocyte sedimentation rate, an antinuclear antibody assay, and, if necessary, a skin biopsy with immunofluorescence studies. Such an evaluation should be able to exclude a psoriasiform variant of subacute lupus erythematosus that might flare with phototherapy or to detect eczema, pityriasis rubra pilaris, or mycosis fungoides that might be more appropriately treated with other modalities. Patients with very thick plaques tend to be less responsive to UVB, and psoralens with UVA (PUVA) treatment might be necessary to achieve total clearing. UVB has several advantages over PUVA, however, including the absence of nausea, less pruritus, no photosensitivity on the day of the treatment, no need for contraception, shorter irradiation times, and probably fewer long-term side effects.

Before therapy is started, the minimal erythema dose (MED) should be determined. The MED is the minimal amount of energy required to produce a uniform clearly demarcated erythema response, usually at 24 hr. This procedure can be done by irradiating eight 1 cm squares of unexposed skin with UVB (either fluorescent bulbs or a medium-pressure "hot quartz" mercury arc lamp) in a stepwise increasing series of doses. The result can then be used to calculate the first treatment dose, as described below. In some centers, the MED is not determined but is estimated on the basis of the patient's skin type (Table 1).

**Table 1** Skin Types and Range of the Minimal Erythema Dose (MED)

| Skin Type | Reaction to First Sun Exposure of 3 MEDs | MED(mJ/cm$^2$) for UVB |
|---|---|---|
| I | Always burn, never tan | 20–30 |
| II | Usually burn, tan less than average | 25–35 |
| III | Sometimes burn, tan about average | 30–50 |
| IV | Rarely burn, tan more than average | 45–60 |
| V | Brown-skinned persons | 60–100 |
| VI | Dark-skinned persons | 100–200 |

## C. Practical Aspects

### *1. Phototherapy Cabinet*

The ultraviolet lamps used in phototherapy are generally of two types: medium-pressure mercury arc lamps ("hot quartz") with or without metal halides and fluorescent lamps with several types of phosphor coating of the lamp envelope. The different available types of photoirradiation equipment (12), their emission characteristics, and geometrical arrangement have been detailed in Chapter 3.

Since the output from ultraviolet lamps deteriorates gradually, the effective lifetime of most UV lamps is usually limited to 500–1000 hr.

### *2. Safety*

The patient must be adequately protected against electrical hazards and from direct contact with the heated lamps. Hand grasps for support should be present in "stand-up" units. In combination units with optically filtered mercury lamps that allow either UVA or UVA plus UVB radiation, one should ensure that the correct glass filter is in place. Patients as well as operators should always wear protective eye glasses.

### *3. Treatment Regimens*

Determining the patient's MED is the first step in establishing appropriate dosage. The initial dosage may vary and in our center is usually of 0.75–0.9 MED. The genitals and usually the face should be shielded. Other body parts (i.e., breasts) that are at risk for burning may also require shielding during parts of the therapy. For male genitals, an athletic supporter should suffice. A zinc-containing paste, sunscreen, or a towel are usually suitable for other sites.

Several treatment protocols are commonly used with different frequency of treatment and dose increments. As to the frequency of irradiation, UVB is in most cases given three times a week ("B3", usually on Monday, Wednesday, and Friday) or five times a week ("B5" on every weekday). When treatments are given less than three times a week, rates of clearing are much lower.

The dose increments are chosen so that a minimal erythema is produced with each treatment; they usually gradually decrease from 50% to 40%, and so on, until a fixed dose increment of 15–20% is maintained throughout the clearing phase of the treatment. An alternative and more conservative approach, often used in patients in a B-5 regimen or persons at special risk for burning, consists of 20–10–0 protocol (i.e., 20%, 10%, or no increase) with flexible dose increments according to the reaction to the previous treatment.

When a previous treatment resulted in a severe or painful erythema, however, either no treatment is given on the following protocol day or the dose is decreased. When the patient misses a treatment, the last dose is repeated or a lower dose is administered according to the time span between the last and the new irradiation. For total clearing, usually defined as less than 5% total body surface involvement for patients who begin with at least 30% body coverage, an average of 25–30 treatments is usually required in both B3 or B5 regimens, but patients with very thick plaques may need more treatments or a change in therapy.

Often an additional dose is needed to achieve clearing of the lower extremities. This is especially true when fluorescent tube units, which suffer from a relatively reduced ultraviolet emission at both ends of the tubes, are used (13,14).

The endpoint of treatment is the complete flattening of the plaques to a macular nonerythematous state, often with residual postinflammatory hyperpigmentation that gradually resolves.

Erythemogenic doses are usually needed for optimal efficacy with conventional ultraviolet sources (6,15), although some authors suggest that a suberythemogenic UVB schedule, either alone or with adjuvant therapy such as topical tar, can also be suitable for clearing psoriasis (16,17). These data await confirmation, however, in larger patient series with precise definition of the degree of severity of the disease and the use of very accurate dosimetry. The use of topical adjuvants is discussed below.

The remission induced by UVB phototherapy is temporary and its duration displays considerable individual variation. Van Weelden et al. (18) found a similar average remission time in patients cleared with biweekly UVB (5.4 ± 4.3 months) and biweekly PUVA (5.2 ± 2.8 months). The usefulness of maintenance therapy after clearing (e.g., once or twice a week over several months) in prolonging the duration of remission has been controversial for a long time. The interpretation of study results is also complicated by a lack of uniform definitions for the terms “remission” and “flare.”

A multicentered, randomized prospective trial enrolling 165 psoriatic patients compared the duration of remission after UVB clearing with and without UVB maintenance therapy (19). This study demonstrated that postclearing UVB phototherapy, even at the modest average frequency of six treatments per month, significantly and substantially increased the interval during which the disease remained under excellent control. It was estimated that 75% of patients who received maintenance therapy remained clear for at least 133 days after clearing, and more than 50% would still be clear 181 days after initial clearing. Among patients who did not receive maintenance therapy, 75% were still clear after 77 days and one-half remained clear for 151 days; only 28% were still clear after 181 days. Efforts have to be made, therefore, to encourage patient compliance in this regimen especially if the patient cannot rely on enough ambient sunlight exposure to maintain the remission.

Several studies of the phototherapeutic action spectrum support the hypothesis that the longer wavelengths within the UVB region are relatively more therapeutic and less erythemogenic than the shorter wavelengths (1,7). Consequently, ultraviolet sources with peak effective emission between 300 and 320nm, referred to as "selective ultraviolet phototherapy" have been widely used, especially in Europe. Selective ultraviolet therapy has been reported to compare favorably with conventional UVB phototherapy and, in some cases, to be as effective as PUVA (20–23). There is less risk of sunburn reactions but the required treatment times are generally longer. The higher effectiveness of phototherapy at the Dead Sea, which is 1,200 feet below sea level compared to solar phototherapy at other locations is claimed to be the result of, among other factors, a shift towards longer ultraviolet wavelengths at ground level, which would constitute a natural equivalent of selective ultraviolet therapy (6).

Units emitting "low-intensity selective ultraviolet" have been marketed for home therapy of psoriasis, with the possible advantage of lower erythemogenicity and therefore higher safety. In a controlled prospective trial, low intensity selective ultraviolet was found to be less effective in clearing psoriasis when used three times a week than conventional UVB phototherapy (24).

A new development, based on the assumption that the therapeutic optimum for psoriasis is situated around 313 mm, consists of the use of a fluorescent lamp (Philips TL-01) emitting a narrow UV band at 311–312 mm. The results in several studies with a limited number of patients suggest that this treatment can be more effective and safer (fewer burning episodes) than phototherapy with the usual broadband sources that have peak emissions below this apparently optimal wavelength (25–27). There are, however, currently no data available to establish the relative long-term safety of these various approaches.

## D. Adjunctive Topical Therapy

In an attempt to decrease the total UVB exposure required to clear psoriatic plaques and/or to prolong the symptom-free period after the treatment, the combination of UVB with various topical adjunctive agents has received considerable attention. Adjunctive agents that have been advocated include those discussed below.

### *1. Soaks*

Patients are bathed in warm water for half an hour before receiving UVB irradiation. This may slightly enhance clearing and lower the totally required UVB dose (6), but this effect is probably small (28).

### *2. Lubricants*

It has been shown that the application of hydrated petrolatum and other topical lubricants before irradiation alters the optical properties of psoriatic lesions so that less ultraviolet radiation is reflected and scattered from the surface scales (29). This effect is thought to account for the increased effectiveness observed with UVB in combination with topical lubricants (15), and the use of such adjuvants has become a routine element of standard clinical practice.

### *3. Corticosteroids*

Topical corticosteroids are used frequently in conjunction with UVB phototherapy, but there is no consensus, as whether they enhance the therapeutic response to UVB. In a bilateral comparison study, 26 patients were treated with 0.025–0.05% fluocinonide ointment twice daily and erythemogenic UVB 3–5 times weekly (30). There was no significant difference in the number of treatments needed to obtain clearing or in the duration of remission.

In a randomized, double-blind trial of 60 patients, a superpotent corticosteroid, clobetasol propionate, was compared with placebo cream, both three times a week in combination with UVB (31). There was no difference in the number of treatments needed to achieve clearing, but the duration of remission was significantly shorter in the corticosteroid-treated group.

Another prospective, randomized double-blind study of 53 patients compared the effects of fluocinonide cream and bland emollient in conjunction with erythemogenic UVB phototherapy (32). No significant difference was observed in number of treatments or UVB dose needed. However, the patients in the corticosteroid-treated group remained in remission longer than those in the control group. These authors also observed a trend towards a slightly more rapid initial response in the corticosteroid-treated group, which confirms certain earlier findings (33).

On the basis of these partially conflicting data, the beneficial effect of the combination of UVB (in adequate erythemogenic doses) and corticosteroids seems to be limited. We therefore believe that the use of corticosteroids, if any, is best restricted to the initial treatment phase when a quick improvement is desirable, especially in the view of the well-known side effects of prolonged corticosteroid use and their cost.

### *4. Anthralin*

The classic Ingram method for treating psoriasis (34), combining the use of anthralin (cignolin, dithranol) in a paste and ultraviolet irradiation, was initially popularized in Europe. A modified Ingram regimen was introduced in the United States for hospitalized (35) and ambulatory patients (36). Anthralin is not considered photosensitizing, and UVB is administered for its additive effect.

Numerous different treatment regimens exist. In one classic approach, anthralin is usually applied on psoriatic plaques for 6–8 hr in a hard or soft paste formulation with concentrations that increase gradually from 0.05% or 0.1%. Then the paste is removed with mineral oil. A tar bath is given and the patient is exposed to increasing doses of UVB. The Ingram approach has been found superior to a modified Goeckerman therapy in terms of rapidity of response, but it induced comparable remission times (36).

Well-known drawbacks, however, are skin irritation and considerable cosmetic inconveniences such as staining of the skin and clothes. This approach also generally requires hospitalization or day care, with their substantial associated expenses. Several studies have investigated the combination of UVB with the more practical anthralin short-contact therapy (37–39). This combination was significantly less efficient than the classic Ingram regimen and offered no clear advantage over UVB phototherapy alone.

### *5. Tar*

A classic therapeutic approach to widespread psoriasis, the Goeckerman regimen (40) includes the application of crude coal tar for 24 hr, its removal with mineral oil, irradiation with a "hot quartz" mercury lamp, and soaking in warm water with soap.

Crude coal tar, which is the product of the destructive distillation of coal, is a complex mixture of more than 10,000 compounds, most of which have not been identified. Although the efficacy of the Goeckerman regimen in psoriasis is beyond dispute, the therapeutic mechanism, efficacy in terms of rate of clearing and duration of remission when compared to UVB alone, and safety are the subject of debate.

Tar alone has a definite therapeutic effect on psoriasis (15), which is thought to be the result of suppression of DNA synthesis and enzyme inhibition (36). However, the therapeutic mechanism of tar plus ultraviolet light is not known. For many years it was assumed that the phototoxic properties of tar were responsible for the beneficial effect of the Goeckerman regimen. Later it was demonstrated that the action spectrum of tar photosensitization is in the UVA region (41,42) and that conventional light sources are of insufficient UVA irradiance to cause tar–UVA photosensitivity. Furthermore, tar–UVA phototherapy would be limited by an immediate painful burning sensation, referred to as the "smarting reaction", that occurs in many patients and the mechanism of which is poorly understood (43). The same limitation generally holds for selective ultraviolet therapy.

Today, the therapeutic role of each Goeckerman component remains uncertain. One single-blind bilateral comparison study demonstrated that 1% crude coal tar in petrolatum and 5% tar extract in an oil base combined with suberythemogenic UVB doses was superior to combination therapy of UVB plus vehicle alone (44). In this study on a very limited number of patients, the authors compared the different treatment regimens on the basis of the "rate of improvement per week." This parameter, constructed by visual assessment of erythema, scale, and plaque thickness, was found to be 22% for the 1% crude coal tar-UVB treated side vs. 14.7% for the petrolatum–UVB treated side, and 19.6% for the 5% tar extract–UVB treatment vs. 11.4% for the oil–UVB treatment. With erythemogenic UVB, however, no difference was observed. Furthermore, Le Vine and collaborators observed no significant advantage of 5% crude coal tar over the lubricating base when combined with erythemogenic UVB (45).

Anderson et al. conclude in their review (15) that tar offers no benefit to emollients in resolving psoriasis when erythemogenic doses of UVB are used, as in routine phototherapy protocols. On the other hand, tar and suberythemogenic doses of UVB could be more effecacious than either agent alone. These clinical observations are in accordance with the results of DNA synthesis suppressor essays with UVB and tar (44). In a study by Stern and collaborators, the "UVB-sparing effect" of tar in comparison with vehicle alone was found to be of small magnitude (46).

Other considerations should also be included in the discussion. The messy nature of crude coal tar requires that UVB in conjunction with this agent be administered in specialized facilities such as inpatient hospital units or day care centers, which is much more expensive than treatment in an oupatient clinic or physician's office. In the context of sharply rising health care expenses, the cost-effectiveness of this treatment is to be questioned. In recent years, a number of topical tar preparations has become available that are more cosmetically acceptable to outpatients than crude coal tar, but there is so far no well-controlled trial that conclusively demonstrates their additional effectiveness in conjunction with UVB.

Another element of concern is the safety of tar–UVB phototherapy. Crude coal tar contains many polyaromatic hydrocarbons. These compounds can undergo a metabolic conversion by cytochrome P-450 enzymes in the epidermis to well-known tumor-initiating agents (47). An additive tumorigenic effect of combined ultraviolet radiation and application of the polyaromatic hydrocarbon compound benzopyrene has been demonstrated in experimental animals (48).

However, there are no clear data on carcinogenicity related to the combined clinical use of tar and ultraviolet in humans. A follow-up study of 250 patients failed to detect an increased risk of nonmelanoma skin cancer (49) and comparable results were found in a small Swedish study (50). On the other hand, a case control study of 59 patients with nonmelanoma skin cancer with psoriasis and matched controls demonstrated a significant association between skin cancer and high exposure to tar and ultraviolet radiation (51). In another study, mutagenic compounds were isolated from the urine of a number of patients with psoriasis undergoing tar plus ultraviolet treatment (52). The clinical significance of these findings is unclear.

In conclusion, the benefits of tar–ultraviolet phototherapy seem to be quite small considering the problems associated with it. More substantial clinical evidence would be needed of the postulated synergistic effect of both components to consider the Goeckerman regimen or its variants as standard in today's therapy for psoriasis.

## E. Combination Therapy

The idea of combining two or more treatment modalities is not new. It is based on two assumptions. First, the different therapies most likely have different molecular mechanisms and therefore a cumulative therapeutic result can be expected from their combination. Second, they usually have different side effects so that there is less long-term toxicity than if only one therapy had been used. Since psoriasis is a chronic condition and long-term adverse effects of UVB, such as actinic damage and skin cancer risks, are assumed to be dose-dependent, it is an interesting goal to develop effective regimens with reduced exposure to UVB radiation.

### *1. UVA*

Although the addition of UVA to UVB offers no apparent advantages in minimizing the side effects of the latter, some have claimed that a combination of UVB and UVA in a proportion mimicking summer sunlight can be more effective in clearing psoriasis (2). Other studies, however, failed to demonstrate any additional benefit of UVA with UVB compared with UVB alone (3,53).

### *2. PUVA*

In a bilateral comparison study, Momtaz-T and Parrish used both UVB and PUVA three times a week to treat 42 patients with psoriasis that had been difficult to manage with either treatment alone (54). All patients experienced clearing, and a substantial reduction in number of treatments, mean UVB and PUVA dose at clearing, and cumulative irradiation doses was noted. In another bilateral comparison study of 11 patients with moderate or severe psoriasis, treatment with PUVA–UVB twice a week compared favorably to either treatment alone (55).

### *3. Methotrexate*

One study enrolled 26 patients with extensive psoriasis who were treated with a 3 week course of methotrexate followed by a combination of UVB and methotrexate (56). This protocol allowed for clearing of psoriasis at relatively low dosages of UVB (mean of 12 exposures, final dosage 310 $mJ/cm^2$), with a mean total methotrexate dose of 112 mg. In a preliminary study, the authors did not observe an exacerbation of the ultraviolet-induced erythema by methotrexate or phototoxic episodes, as previously reported (57), but this remains a concern.

### 4. *Retinoids*

The efficacy of the aromatic retinoid etretinate is well-established in the treatment of the pustular and erythrodermic variants of psoriasis, while the plaque type is relatively more resistent to etretinate monotherapy (36).

From a theoretical viewpoint, combination therapy with retinoids is particularly appealing because it offers the unique advantage that these agents are not carcinogenic, as opposed to many other psoriasis therapies, and that they might in fact exert even a certain protective action against this hazard.

When etretinate in a dosage of 0.5–1mg/kg/day is combined with UVB or selective ultraviolet therapy, psoriatic patients usually show improvement more quickly than with ultraviolet light alone, and the number of treatments and the cumulative dose can be reduced (58,59).

The introduction of acitretin, the major metabolite of etretinate, into clinical practice is considered an advance due to its more favorable pharmacokinetic properties. As with etretinate, acitretin monotherapy does not lead to complete remission of psoriasis in most patients.

Two studies (60,61) showed that a combination of acitretin and a standard UVB protocol was very effective in clearing psoriasis, with significantly faster response and substantial reductions in the effective cumulative UVB dose. In the larger study (61), the decrease in the psoriasis severity index was 79% in the acitretin and UVB-treated group and 35% in the placebo and UVB-treated group. The response rate, based on a 75% or greater decrease of the psoriasis severity index, was 60% for the combination treatment and only 24% for the control treatment. The median cumulative UVB energy applied to reach 75% clinical improvement was 11.8 $J/cm^2$ vs. 6.9 $J/cm^2$. Isotretinoin, however, has been used with UVB for the treatment of plaque-type psoriasis without success (62).

### 5. *Others*

Oral cyclosporine has been shown to be another effective treatment for psoriasis in several open and two double-blind trials (63). There are, however, many reasons to be concerned about the long-term safety of cyclosporine therapy (64). Its use should therefore be restricted to noncombination regimens until the long-term safety and efficacy data of monotherapy are available.

Gupta et al. treated 20 patients with oral fish oil (3.6 g eicosapentaenoic acid and 2.4 g docosahexaenoic acid daily) vs. placebo for 15 weeks and concomitant suberythemogenic UVB, twice weekly, in weeks 3–11 (65). Patients in the fish oil group had a significantly greater decrease in total body surface area of psoriasis, erythema, plaque thickness, and scaling compared to the control group. On the basis of these results and the presumed role of eicosanoid mediators in the pathogenesis of psoriasis, fish oil was proposed as a valuable adjunctive therapy for psoriasis. Since suboptimal UVB doses were given, however, further studies are needed to assess more accurately the UVB-sparing effect compared to standard erythemogenic UVB treatment regimens.

## II. UVB PHOTOTHERAPY OF OTHER DISORDERS

UVB phototherapy has been advocated in the management of atopic dermatitis (66), although it remains controversial as a primary therapeutic option.

The doses needed to achieve clearing seem to be slightly lower than those given for psoriasis. Several studies indicate that phototherapy can be more effective when UVB is combined with UVA (67,68). Jekler and Larko treated 30 patients with atopic dermatitis in a paired-comparison study with standard UVB and UVA–B equipment (67). Treatment was administered 3 times a week for a maximum of 8 weeks. A significant difference in favor of UVA–B was observed for the total severity score and pruritus score at the end of the treatment. However, no significant difference in healing rate was seen.

In another study, no statistically significant difference between selective ultraviolet therapy and PUVA was found in their efficacy in the treatment of atopic dermatitis (69). UVB has been used to treat other forms of eczema. The effect of UVB and PUVA was investigated in patients with chronic hand eczema of varied origins (contact allergic, irritant, hyperkeratotic) not responding to conventional topical treatment (70,71). It was demonstrated that most patients can benefit from UVB delivered in a standard protocol after a treatment period of 2 or 3 months. Most of the patients did not experience complete clearing, however, and UVB was found to be significantly less effective than PUVA. Moreover, relapses occurred often and early, and maintenance therapy was usually needed.

It is well established that pruritus associated with a variety of causes can be improved by phototherapy (Table 2). The most extensive experience has been accumulated in the treatment of uremic pruritus in patients with chronic renal failure (72–74). Patients are usually treated 1–3 times per week for about 1 month and the improvement rate and length of remission seem to be dependent on the total dose of UVB rather than the particular protocol design. Not all patients experience improvement with this therapy, but a beneficial response seems to be unaffected by the presence of secondary hyperparathyroidism. The therapeutic action mechanism is not known, but it is hypothesized that circulating substances, responsible for the pruritus, are inactivated by UVB. This theory is supported by the observation that bilateral improvement occurs after treatment of one body half. A phototherapeutic trial can also be considered for the relief of pruritus of other internal origins such as primary biliary cirrhosis, either alone or

**Table 2** Therapeutic Indications for UVB

| Established Indications | Other Proposed Uses |
|---|---|
| Psoriasis | Chronic hand eczema |
| Atopic dermatitis | Pruritus of undetermined origin |
| Uremic pruritus | Parapsoriasis |
| Pityriasis rosea | Mycosis fungoides |
| Pityriasis lichenoides acuta and chronica | Chronic urticaria |
| Eosinophilic pustular folliculitis | Polymorphous light eruption |
| | Solar urticaria |
| | Lichen planus |

in combination with cholestyramine (75). Twenty patients with extensive and symptomatic pityriasis rosea were treated with unilateral UVB phototherapy in a bilateral comparison study (76). Erythemogenic exposure on 5 consecutive days resulted in substantially reduced pruritus and extent of disease compared to the untreated side in 50% of the patients. Therapy seemed to be most beneficial to patients receiving treatment within the first week of eruption.

The efficacy of UVB in treating pityriasis lichenoides et varioliformis acuta (PLEVA, Mucha-Habermann's disease) and pityriasis lichenoides chronica has been demonstrated in several small series of patients (77,78). It is unknown, however, whether the natural history of the disease is altered.

Several authors obtained encouraging results with UVB in the treatment of chronic urticaria of different determined or undetermined origins (79,80). The majority of their patients experienced improvement, either permanently or temporarily, and it was suggested that UVB therapy might be worth trying in patients with chronic urticaria.

Furthermore, UVB phototherapy has also been established as an effective treatment of eosinophilic pustular folliculitis in patients infected with human immunodeficiency virus (81).

Phototherapy also has reportedly been beneficial in patients with early mycosis fungoides or parapsoriasis en plaques (82). A complete clinical and histologic remission of disease was achieved in 19 of 34 patients with early mycosis fungoides, with a median duration of about 28 months. Although higher complete response rates are generally achieved with standard therapeutic modalities such as PUVA, topical chemotherapy, or electron beam radiotherapy, UVB might be a therapeutic option for selected patients in early stages of their disease.

Controlled exposures to ultraviolet radiation can be useful in the prevention or improvement of certain photodermatoses. Although PUVA is considered more effective and less likely to provoke clinical symptoms during the course of the desensitization, many patients with polymorphous light eruption and solar urticaria can benefit from repeated UVB exposures (83,84). A variety of immunologic, biochemical, and optical factors can contribute to the raised threshold for triggering symptomatic photosensitivity (6).

There are a number of isolated reports on the efficacy of UVB in other conditions such as reactive perforating collagenosis (85), persistent postlesional erythema in Sweet's syndrome (86), Reiter's disease, lichen planus, and prurigo nodularis (15) that deserve further investigation. In pityriasis rubra pilaris (87) and urticaria pigmentosa (88), the effect of UVB was found to be disappointing. The common belief that ultraviolet irradiation of toes and fingers is an effective prophylactic treatment for chilblains was not supported by the results of a randomized double-blind study (89).

## III. SIDE EFFECTS

Potential acute side effects of UVB irradiation of the skin include "sunburn reactions" (i.e., UVB-induced erythema) and a Koebner exacerbation of psoriasis. Slightly erythemogenic UVB doses are considered necessary to clear psoriasis effectively but severe sunburn reactions should be avoided. Therefore, the therapeutic window in the dose–response diagram can be narrow, and the patient must be carefully questioned each time about his or her tolerance of the previous exposure.

Well-known ocular side effects of UVB include conjunctivitis and keratitis, and therefore eye-protective glasses should always be worn both by the patient and the phototherapy technician.

The most important potential long-term toxicities of UVB phototherapy are accelerated skin aging and increased risk of cutaneous cancer.

Photoaging, in contrast to true chronologic aging, is defined as skin changes as a result of repeated ultraviolet exposure rather than due to passage of time alone. It is characterized clinically by coarseness, wrinkling, laxity, increased fragility, mottled pigmentation, telangiectases, and atrophic or fibrotic areas (90). Although little is known about the precise action spectrum for actinic degeneration of the skin, epidemiologic and experimental data implicate UVB portions of sunlight in photoaging (90). Patients treated with UVB for psoriasis may receive as many as 1,000 MEDs per year, approximately 10 times the typical UVB exposure for persons living in a temperate climate who usually work indoors. Controlled prospective studies focusing on actinic changes in patients treated with today's conventional UVB sources are lacking, precisely because of the slowly progressive nature of these lesions, but a dose-dependent increase can certainly be expected. Since psoriasis often spares the face, shielding of this area can limit the cosmetic significance of actinic damage.

Another element of concern is the carcinogenic risk associated with UVB phototherapy. Ultraviolet radiation is mutagenic and carcinogenic in vitro and in vivo. It also alters immune responses and the distribution of immune competent cells in the body, which may result in suppressed immune responses against tumor cells. There is ample evidence from epidemiological sources and animal studies (90) for a strong corrleation between UVB exposure and nonmelanoma skin cancer incidence. Therefore, it is possible that long-term phototherapy will also increase the risk of nonmelanoma skin cancer. The magnitude of the increase over a patient's lifetime, the relationship of total dose and particular protocol designs to this risk, and the biological behavior of the skin cancers that develop in these patients will require more precise definition.

Until now, no controlled prospective clinical studies such as with PUVA have been undertaken to address these questions. In one case–control study on a limited number of patients with psoriasis treated with UVB (average number of UVB treatments 249, range 101–785; time span of treatments of 0–25 years), no increased risk among the extensively treated patients was found compared to a matched control group (91). The conclusive force of this study, however, is very limited and these findings remain to be confirmed by follow-up studies of a larger and more homogeneous patient population.

Given the difficulty in assembling and following a sufficiently large cohort of UVB-treated patients and the long latency period before nonmelanoma skin cancer development, an alternative approach for estimating the carcinogenic risk associated with phototherapy has mathematical modeling (92). Such models calculate the expected increase in nonmelanoma skin cancer risk as a function of therapeutic UVB exposures, treatment timing, and UVB dose received from sunlight. Treatment timing is an important parameter since exposures early in life may result in a greater carcinogenic burden than comparable exposures later in life.

The results of these calculations indicate that the incidence of nonmelanoma skin cancer among patients receiving phototherapy is expected to be 2.5–7.5 times higher than the risk resulting from natural sunlight exposure alone (92).

This increase in incidence can be justified in many patients with severe psoriasis, especially because there is little reason to believe that the phototherapy-related tumors would be clinically more aggressive than sun-induced tumors. Furthermore, the additional risk can be substantially limited by restricting exposure to affected areas and covering the normally sun-exposed area when possible.

The relationship between UVB phototherapy and melanoma is less clear (93). Although there exist isolated case reports of melanoma occurring in patients who received phototherapy, a clear correlation has not been established. However, childhood UVB exposure (in analogy with childhood sun exposure) and phototherapy of patients with known risk factors for melanoma, such as the dysplastic nevus syndrome, may be associated with an increased incidence of melanoma. These elements should certainly be considered in the decision to undertake therapy.

## ACKNOWLEDGMENT

This work was supported in part by a Fellowship of the Belgian American Educational Foundation Inc., New Haven, Connecticut.

## REFERENCES

1. Parrish JA, Jaenicke KF. Action spectrum for phototherapy of psoriasis. J Invest Dermatol 1981; 76:359–362.
2. Stern WK, Kihiczak G. Photobiology of psoriasis. Arch Dermatol 1974; 109:502–505.
3. Van Weelden H, Young E, Van Der Leun JC. Therapy of psoriasis: comparison of photochemotherapy and several variants of phototherapy. Br J Dermatol 1980; 103:1–9.
4. Parrish JA. The treatment of psoriasis with long-wave ultraviolet light. Arch Dermatol 1977; 113:1525–1528.
5. Young E, Van Der Leun JC. Treatment of psoriasis with longwave ultraviolet light. Dermatologica 1975; 150:352–354.
6. Paul BS, Diette KM, Parrish JA. Therapeutic Photomedicine. In: Fitzpatrick TB, et al., eds. Dermatology in General Medicine. New York: McGraw-Hill 1987: 1522–1533.
7. Fischer T. UV-light treatment of psoriasis. Acta Derm Venerol (Stockh) 1976; 56:473–479.
8. Epstein JH, Fukuyama K, Epstein WL. UVL-induced stimulation of DNA synthesis in hairless mouse epidermis. J Invest Dermatol 1968; 51:445–453.
9. Kramer DM, Pathak MA, Kornhauser A, Wisekmann A. Effect of ultraviolet irradiation on biosynthesis of DNA in guinea pig skin. J Invest Dermatol 1974; 62:388–393.
10. Epstein WL, Fukuyama K, Epstein JH. Early effects of ultraviolet light on DNA synthesis in human skin in vivo. Arch Dermatol 1969; 100:84–89.
11. Friedmann PS. Ultraviolet radiation and the immune system. In: Champion RH, Pye RJ, eds. Recent Advances in Dermatology, 8. New York: Churchill Livingstone, 1990:41–53.
12. Diffey BL. Ultraviolet radiation lamps for the phototherapy and photochemotherapy of skin diseases. In: Champion RH, Pye RJ, eds. Recent Advances in Dermatology, 8. New York: Churchill Livingstone, 1990:21–40.
13. Horwitz SN, Frost P. A phototherapy cabinet for ultraviolet radiation therapy. Arch Dermatol 1981; 117:469–473.
14. Chue B, Borak M, Lowe NJ. Phototherapy units: comparison of fluorescent ultraviolet B and ultraviolet A units with a high-pressure mercury system. J Am Acad Dermatol 1988; 18:641–645.
15. Anderson TF, Waldinger TP, Voorhees JJ. UV-B phototherapy: an overview. Arch Dermatol 1984; 120:1502–1507.
16. Frost P, Horwitz SN, Caputo RV, Berger SM. Tar gel-phototherapy for psoriasis. Arch Dermatol 1979; 115:840–846.
17. Eells LD, Wolff, JM, Garloff J, Eaglstein WH. Comparison of suberythemogenic

and maximally aggressive ultraviolet B therapy for psoriasis. J Am Acad Dermatol 1984; 11:105–110.
18. Van Weelden H, Young E, Van Der Leun JC. Therapy of psoriasis: comparison of photochemotherapy and several variants of phototherapy. Br J Dermatol 1980; 103:1–9.
19. Stern RS, Armstrong RB, Anderson TF, Bickers DR, Lowe NJ, Harber L, Voorhees J, Parrish JA. Effect of continued ultraviolet B phototherapy on the duration of remission of psoriasis: a randomized study. J Am Acad Dermatol 1986; 15:546–552.
20. Tronnier H, Heidbuchel H. Zur Therapie der Psoriasis Vulgaris met ultravioletten Strahlen. Z Hautkr 1976; 51:405–424.
21. Schropl F. Zum heutigen Stand der technischen Entwicklung der selektiven Fototherapie. Dtsch Dermatol 1977; 25:499–504.
22. Tronnier H. Derzeitiger stand der Photochemotherapie fur die dermatologische Praxis. Dtsch Dermatol 1977; 25:262–276.
23. Pullmann, H, Wichmann AC, Steigleder GK. Praktische Enfahrungen mit verschiedenen Phototherapieformen der psoriasis-PUVA, SUP, Teer-UV-Therapie. Z Hautkr 1978; 53:641–647.
24. Paul BS, Stern RS, Parrish JA, Arndt KA. Low-intensity selective UV phototherapy. Arch Dermatol 1983; 119:122–124.
25. Van Weelden H, De La Faille HB, Young E, Van Der Leun JC. A new development in UVB phototherapy of psoriasis. Br J Dermatol 1988; 119:11–19.
26. Green C, Ferguson J, Lakshmipathi T, Johnson BE. 311 nm UVB phototherapy—an effective treatment for psoriasis. Br J Dermatol 1988; 119:691–696.
27. Larko O. Treatment of psoriasis with a new UVB-lamp. Acta Derm Venerol (Stockh) 1989; 69:357–359.
28. Boer J, Schothorst AA, Boom B, Hermans J, Suurmond D. Influence of water and salt solutions on UVB irradiation of normal skin and psoriasis. Arch Dermatol Res 1982; 273:247–259.
29. Anderson RR, Parrish JA. Optical properties of human skin. In: Regan JD, Parrish JA, eds. The Science of Photomedicine. New York: Plenum Press, 1980:147–194.
30. Petrozzi JW. Topical steroids and UV radiation in psoriasis. Arch Dermatol 1983; 119:207–210.
31. Larko O, Swanbeck G, Svartholm H. The effect on psoriasis of clobetasol propionate used alone or in combination with UVB. Acta Derm Venereol (Stockh) 1984; 64:151–154.
32. Dover JS, McEvoy MT, Rosen CF, Arndt KA, Stern RS. Are topical corticosteroids useful in phototherapy for psoriasis? J Am Acad Dermatol 1989; 20:748–754.
33. LeVine MJ, Parrish JA. The effect of topical fluocinonide ointment on phototherapy of psoriasis. J Invest Dermatol 1982; 78:157–159.
34. Ingram JT. The approach of psoriasis. Br Med J 1953; 2:591–594.
35. Farber EM, Harris DH. Hospital treatment of psoriasis: a modified anthralin program. Arch Dermatol 1970; 101:381–389.
36. Farber EM, Abel EA, Charuworn A. Recent advances in the treatment of psoriasis. J Am Acad Dermatol 1983; 8:311–319.

37. Statham BN, Ryatt KS, Rowell NR. Short-contact dithranol therapy—a comparison with the Ingram regimen. Br J Dermatol 1984; 110:703–708.
38. Lebwohl M, Berman B, France DS. Addition of short-contact anthralin therapy to an ultraviolet B phototherapy regimen: assessment of efficacy. J Am Acad Dermatol 1985; 13:780–784.
39. Boer J, Smeenk G. Effect of short-contact anthralin therapy on ultraviolet B irradiation of psoriasis. J Am Acad Dermatol 1986; 15:198–204.
40. Goeckerman WH. Treatment of psoriasis. Northwest Med 1925; 24:229–231.
41. Everett MA, Miller JV. Coal tar and ultraviolet light II. Cumulative effects. Arch Dermatol 1961; 84:937–940.
42. Tanenbaum L, Parrish JA, Pathak MA, Anderson RR, Fitzpatrick TB. Tar phototoxicity and phototherapy for psoriasis. Arch Dermatol 1975; 111:467–470.
43. Diette KM, Gange WR, Stern RS, Arndt KA, Parrish JA. Coal tar phototoxicity: kinetics and exposure parameters. J Invest Dermatol 1983; 81:347–350.
44. Lowe NJ, Wortzman MS, Breeding J, Koudsi H, Taylor L. Coal tar phototherapy for psoriasis reevaluated: erythemogenic versus suberythemogenic ultraviolet with a tar extract of oil and crude coal tar. J Am Acad Dermatol 1983; 8:781–789.
45. Le Vine MJ, White HA, Parrish JA. Components of the Goeckerman regimen. J Invest Dermatol 1979; 73:170–173.
46. Stern RS, Gange RW, Parrish HA, Tang SV, Arndt KA. Contribution of topical tar oil to ultraviolet-B phototherapy for psoriasis. J Am Acad Dermatol 1986; 14:742–747.
47. Bickers DR, Lowy DR. Carcinogenesis: a fifty-year historical perspective. J Invest Dermatol 1989; 92:1215–1315.
48. Gensler HL. Enhancement of chemical carcinogenesis in mice by systemic effects of ultraviolet radiation. Cancer Res 1988; 48:620–623.
49. Pittelkow MR, Perry HO, Muller SA, Maughan WZ, O'Brien PC. Skin cancer in patients with psoriasis treated with coal tar: a 25-year follow-up study. Arch Dermatol 1981; 117:465–468.
50. Larko O, Swanbeck G. Is UVB treatment safe? A study of extensively UVB treated psoriasis patients compared with a matched control group. Acta Derm Venereol (Stockh) 1982; 62:507–512.
51. Stern RS, Zierler S, Parrish JA. Skin carcinoma in patients with psoriasis treated with topical tar and artificial ultraviolet radiation. Lancet 1980; 1:732–735.
52. Wheeler LA, Saperstein MD, Lowe NJ. Mutagenicity of urine from psoriatic patients undergoing treatment with coal tar and ultraviolet light. J Invest Dermatol 1981; 77:181–185.
53. Boer J. Influence of UVA on the erythemogenic and therapeutic effects of UVB irradiation in psoriasis: photoaugmentation effects. J Invest Dermatol 1981; 76:56–58.
54. Momtaz-T K, Parrish JA. Combination of psoralens and ultraviolet A and ultraviolet B in the treatment of psoriasis vulgaris: a bilateral comparison study. J Am Acad Dermatol 1984; 10:481–486.
55. Diette KM, Momtaz-T K, Stern RS, Arndt KA, Parrish JA. Psoralens and UVA and UVB twice weekly for the treatment of psoriasis. Arch Dermatol 1984; 120:1169–1173.
56. Paul BS, Momtaz-T K, Stern RS, Arndt KA, Parrish JA. Combined methotrexate-

ultraviolet B therapy in the treatment of psoriasis. J Am Acad Dermatol 1982; 7:758–762.
57. Korossy KS, Hood AF. Methotrexate reactivation of sunburn reaction. Arch Dermatol 1981; 117:310–311.
58. Marchescu S, Lubach D, Rudolph PO. Die therapie der psoriasis mit Retinoid. Z Hautkr 1982; 57:1410–1420.
59. Orfanos CE, Steigleder GK, Pullmann H, Bloch PH. Oral retinoid and UVB radiation: a new, alternative treatment for psoriasis on an outpatient basis. Acta Derm Venereol (Stockh) 1979; 59:241–244.
60. Lest J, Boer J. Combined treatment of psoriasis with acitretin and UVB phototherapy compared with acitretin alone and UVB alone. Br J Dermatol 1989; 120:665–670.
61. Ruzicka T, Sommerburg C, Braun-Falco O, Koster W, Lengen W, Lensing W, Letzel H, Meigel W, Paul E, Przybilla B, Steinert M, Winzer M, Wiskemann A. Efficiency of acitretin in combination with UVB in the treatment of severe psoriasis. Arch Dermatol 1990; 126:482–486.
62. Roenigk RK, Gibstine C, Roenigk HH. Oral isotretinoin followed by psoralens and ultraviolet A or ultraviolet B for psoriasis. J Am Acad Dermatol 1985; 13:153–155.
63. Gupta AK, Brown MD, Ellis CN, Rocher LI, Fisher GJ, Baadsgaard O, Cooper KD, Voorhees JJ. Cyclosporine in dermatology. J Am Acad Dermatol 1989; 21:1245–1256.
64. Stern RS. Risk assessment of PUVA and cyclosporine: editorial. Arch Dermatol 1989; 125:545–547.
65. Gupta AK, Ellis CN, Tellner DC, Anderson TF, Voorhees JJ. Double-blind, placebo-controlled study to evaluate the efficacy of fish oil and low-dose UVB in the treatment of psoriasis. Br J Dermatol 1989; 120:801–807.
66. Jekler J, Larkö O. UVB phototherapy of atopic dermatitis. Br J Dermatol 1988; 119:697–705.
67. Jekler J, Larkö O. Combined UVA-UVB versus UVB phototherapy for atopic dermatitis: a paired-comparison study. J Am Acad Dermatol 1990; 22:49–53.
68. Falk ES. UV-light therapies in atopic dermatitis. Photodermatology 1985; 2:241–246.
69. Salo O, Lassus A, Juvakoski T. Behandlung der Dermatitis atopica und der Dermatitis seborrhoica mit selektiver UV-Phototherapie und PUVA. Eine vergleichende Studie. Dermatol Monatsschr 1983; 169:371–375.
70. Rosen K, Mobacuen H, Swanbeck G. Chronic eczematous dermatitis of the hands: A comparison of PUVA and UVB treatment. Acta Derm Venereol (Stockh) 1987; 67:48–54.
71. Syovall P, Christensen OB. Local and systemic effect of UVB irradiation in patients with chronic hand eczema. Acta Derm Venerol (Stockh) 1987; 67:538–541.
72. Saltzer EI. Relief from uremic pruritus: a therapeutic approach. Cutis 1975; 16:298–299.
73. Gilchrest BA, Rowe JW, Brown RS, Steinman TI, Arndt KA. Relief of uremic pruritus with ultraviolet phototherapy. N Engl J Med 1977; 237:136–138.
74. Gilchrest BA, Rowe JW, Brown RS, Steinman TI, Arndt KA. Ultraviolet phototherapy of uremic pruritus. Long-term results and possible mechanism of action. Ann Intern Med 1979; 91:17–22.

75. Cerio R, Murphy GM, Sladen GE, MacDonald D. A combination of phototherapy and cholestyramine for the relief of pruritus in primary biliary cirrhosis. Br J Dermatol 1987; 117:265–267.
76. Arndt KA, Paul BS, Stern RS, Parrish JA. Treatment of pityriasis rosea with UV radiation. Arch Dermatol 1983; 119:381–382.
77. LeVine MJ. Phototherapy of pityriasis lichenoides. Arch Dermatol 1983; 119:378–380.
78. MacKinnon JL. Ultraviolet light treatment of a patient with pityriasis lichenoides et varioliformis acuta. A case report. Phys Ther 1986; 66:1542–1543.
79. Hannuksela M, Kokkonen EL. Ultraviolet light therapy in chronic urticaria. Acta Derm Venereol (Stockh) 1985; 65:449–450.
80. Johnsson M, Falk ES, Volden G. UVB treatment of factitious urticaria. Photodermatology 1987; 4:302–304.
81. Buchness MR, Lim HW, Hatcher VA, Sanchez M, Soter NA. Eosinophilic pustular folliculitis in the acquired immunodeficiency syndrome. Treatment with ultraviolet B phototherapy. N Engl J Med 1988; 318:1183–1186.
82. Milstein HJ, Vonderheid EC, Van Scott EJ, Johnson WC. Home ultraviolet phototherapy of early mycosis fungoides: preliminary observations. J Am Acad Dermatol 1982; 6:355–362.
83. Ramsay CA. Solar urticaria treatment by inducing tolerance to artificial radiation and natural light. Arch Dermatol 1977; 113:1222–1225.
84. Addo HA, Sharma SC. UVB phototherapy and photochemotherapy (PUVA) in the treatment of polymorphic light eruption and solar urticaria. Br J Dermatol 1987; 116:539–542.
85. Vion B, Frenk E. Acquired reactive collagen disease in the adult: successful treatment with UVB light. Hautarzt 1989; 40:448–350.
86. Mayerhausen W, Remy W, Borelli S. Acute febrile neutrophilic dermatosis (Sweet's syndrome). Phototherapy of persistent postlesional erythemas. Z Hautkr 1986; 61:205–213.
87. Cohen PR, Prystowsky JH. Pityriasis rubra pilaris: a review of diagnosis and treatment. J Am Acad Dermatol 1989; 20:801–807.
88. Czarnetzki BM, Rosenbach T, Kolde G, Frosch PJ. Phototherapy of urticaria pigmentosa: clinical response and changes of cutaneous reactivity, histamine and chemotactic leukotrienes. Arch Dermatol Res 1985; 277:105–113.
89. Langtry JAA, Diffey BL. A double-blind study of ultraviolet phototherapy in the prophylaxis of chilblains. Acta Derm Venereol (Stockh) 1989; 69:320–322.
90. Taylor, CR, Stern RS, Leyden JT, Gilchrest BA. Photoaging, photodamage and photoprotection. J Am Acad Dermatol 1990; 22:1–15.
91. Larko O, Swanbeck G. Is UVB treatment of psoriasis safe? A study of extensively UVB-treated psoriasis patients compared with a matched control group. Acta Derm Venereol (Stockh) 1982; 62:507–512.
92. Schothorst AA, Slader H, Schouten R, Suurmond D. UVB doses in maintenance psoriasis phototherapy versus solar UVB exposure. Photodermatology 1985; 2:213–220.
93. Stern RS. The carcinogenic risk of UVB phototherapy and PUVA. In: Horsmanheimo M, ed. Proceedings of the Anglo-Nordic Symposium "Long-Term Treatment of Psoriasis." Helsinki, Finland, 1986:26–27.

19

# Photochemotherapy

**Warwick L. Morison**
*Johns Hopkins Medical Institutions, Baltimore, Maryland*

Photochemotherapy involves an interaction between nonionizing radiation and a photosensitizing chemical to produce a beneficial effect in disease. In dermatology it applies to the therapeutic use of psoralens and ultraviolet (UV) A radiation, which is referred to by the acronym PUVA. PUVA therapy is used in several forms with variations in the type of psoralen used, the route of administration, and the nature of radiation source, but, unless otherwise qualified, the term will be used here to refer to oral methoxsalen combined with indoor sources of UVA radiation.

Psoralen photochemotherapy is a very old treatment. Several thousand years ago people in the Middle East and India began using various plant extracts, which contained psoralens, in combination with sunlight to treat vitiligo. The therapy, using purified psoralens, was introduced to the United States 40 years ago, but interest in this treatment was limited since the results obtained in vitiligo were often disappointing. However, the demonstration that PUVA therapy was an effective treatment for psoriasis (1) dramatically changed this: PUVA therapy has become one of the most widely used treatments in dermatology.

## I. PSORALENS

Psoralens belong to the furocoumarin group of compounds. The parent compound psoralen, and many of its derivatives, are widely distributed in plants and fruit such as limes, celery, figs, and lemons. Thus, psoralens are ingested in small quantities in a normal diet; the biological significance of this is unknown. Four psoralens are used for therapeutic purposes (Fig. 1). Methoxsalen, or 8-methoxypsoralen, which is obtained from the seeds of a plant called *Ammi majus* that grows in the Nile Valley, is the most commonly used psoralen.

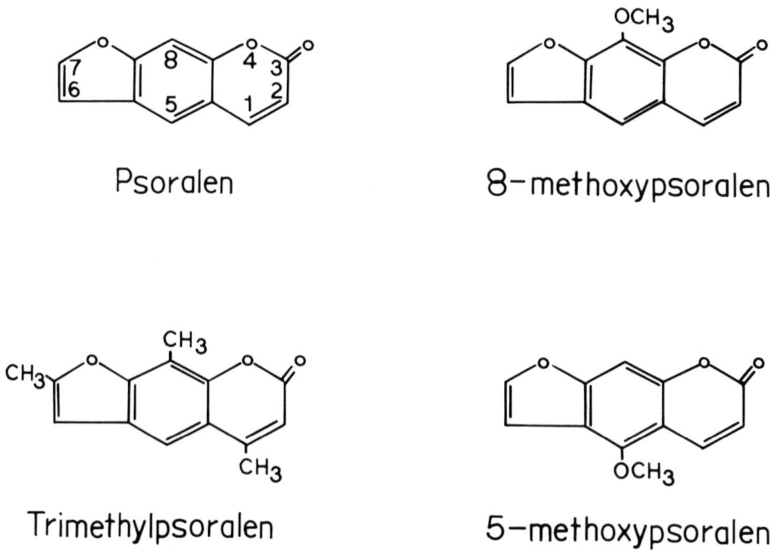


**Figure 1** Psoralens used in therapy. (From Morison WL. Phototherapy and Photochemotherapy of Skin Disease, 1990.)

Trioxsalen, or 4, 5'-8-trimethylpsoralen, is a synthetic compound used for the treatment of vitiligo. Bergapten, or 5-methoxypsoralen, is available in Europe and is now undergoing trials in the United States to confirm its efficacy in psoriasis. Psoralen is used in some Mediterranean countries for the treatment of psoriasis and vitiligo.

After oral administration, psoralens are distributed to all tissues of the body; in the absence of photochemical binding, they are rapidly metabolized in the liver and excreted in urine and feces. Several features of the pharmacology of these compounds are important in therapy (2). Psoralens are poorly soluble in water and this limits their absorption from the gastrointestinal tract; liquid formulations partially overcome this problem. Second, there is a very significant but saturable first-pass effect in the liver so that a low dose of psoralen may be without effect. Finally, there is a very large interindividual, and a smaller but still significant, intraindividual variation in absorption of psoralens reflected in both peak blood levels and time of peak levels after administration.

## II. UVA RADIATION

The absorption spectrum of psoralen in vitro is maximal in the UVC and UVB region, with a small amount of absorption in the UVA portion of the UV spectrum. However, in vivo determinations of the action spectra for psoralen photosensitization have revealed most activity > 320 nm. Early studies found a peak for the action spectrum for delayed erythema between 340 and 380 nm, and this led to the use of fluorescent bulbs with a peak emission at 350 nm for indoor treatment (Fig. 2). A recent study in mice suggests that maximal photosensitization occurs in the 320–340 nm waveband (3), but a radiation source with peak emission in this region has not been tested for clinical use.

Fluorescent bulbs having the designation "PUVA bulb" all have the same emission spectrum and are mounted in various configurations in square or cylindrical stand-up units, hand-and-foot units, and lie-down bed units. These are the most commonly used sources of radiation for PUVA therapy, but other options are available. Sunlight can be used with oral trioxsalen but it is unsafe for use with methoxsalen because the therapeutic dose is too close to the phototoxic dose. Of course sunlight has some additional problems such as a variable content of UVA and UVB radiation, its unpredictable nature, and associated hazards of insects and peeking neighbors. Metal halide lamps can also be used as a source of UVA radiation, their main advantage being a high irradiance and main disadvantage being high cost.

## III. BIOLOGICAL EFFECTS

### A. Cellular

Psoralens interact photochemically with DNA, RNA, proteins, and cell membranes but the most important interaction is thought to be with DNA. Psoralens

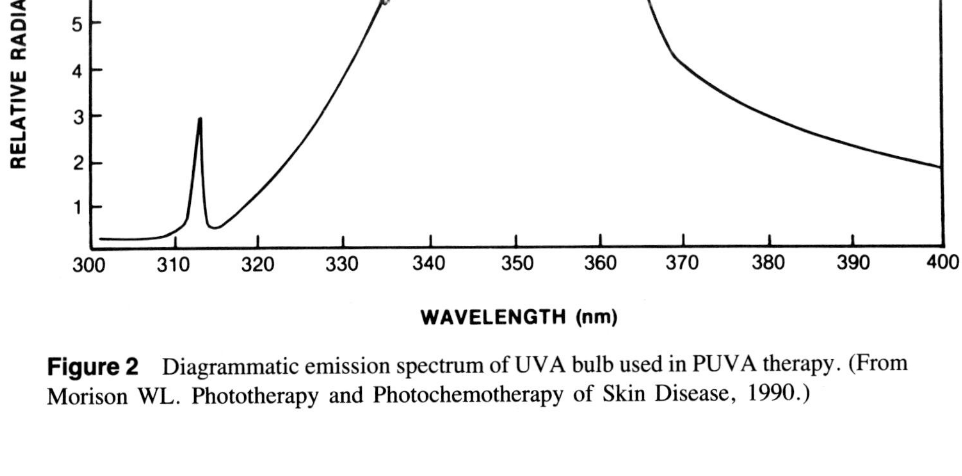


**Figure 2** Diagrammatic emission spectrum of UVA bulb used in PUVA therapy. (From Morison WL. Phototherapy and Photochemotherapy of Skin Disease, 1990.)

intercalate between the strands of DNA and, upon absorption of a photon, form a monofunctional adduct with a pyrimidine molecule on one strand. The psoralens used in therapy can absorb a second photon and react with a pyrimidine molecule on the sister strand to cross-link DNA. These lesions may be lethal to a cell or be repaired, but such repair is slower and less complete than it is with lesions caused by UV radiation alone.

## B. Cutaneous

Psoralens and UVA radiation induce erythema and pigmentation in skin that resembles the responses to UVB radiation. However, there are several important differences. Erythema from PUVA treatment appears later and lasts longer (Fig. 3); it usually first appears 36–48 hr after exposure but may be delayed to 72–96 hr and persists for up to 7–14 days. The dose–response curve for PUVA-induced erythema is also much steeper so that the dose required to produce an intense red erythema is only a few multiples of that producing a faint erthema (Fig. 4). Erythema is associated with pruritus in the form of a deep, burning itch that may persist for weeks and is probably due to direct damage to cutaneous nerves. Pigmentation is deeper and persists longer than following exposure to UVB radiation. Pigmentation following treatment with topical psoralen and exposure to UVA radiation can persist for months or a year or more.

## C. Ocular

Psoralens can enter the ocular media and lens. Because UVA radiation is mainly absorbed by the lens, psoralens can thus form photoproducts with lens proteins

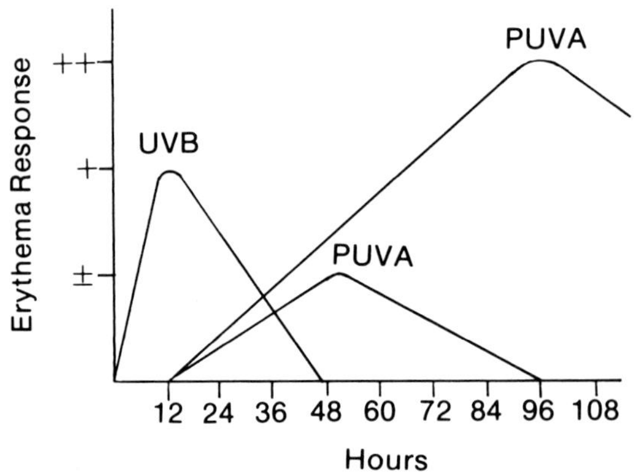


**Figure 3** Time course of erythemal responses induced by ultraviolet (UV) A, UVB, and PUVA. Note that a more intense erythema develops later with PUVA. (From Morison WL. Phototherapy and Photochemotherapy of Skin Disease, 1990.)

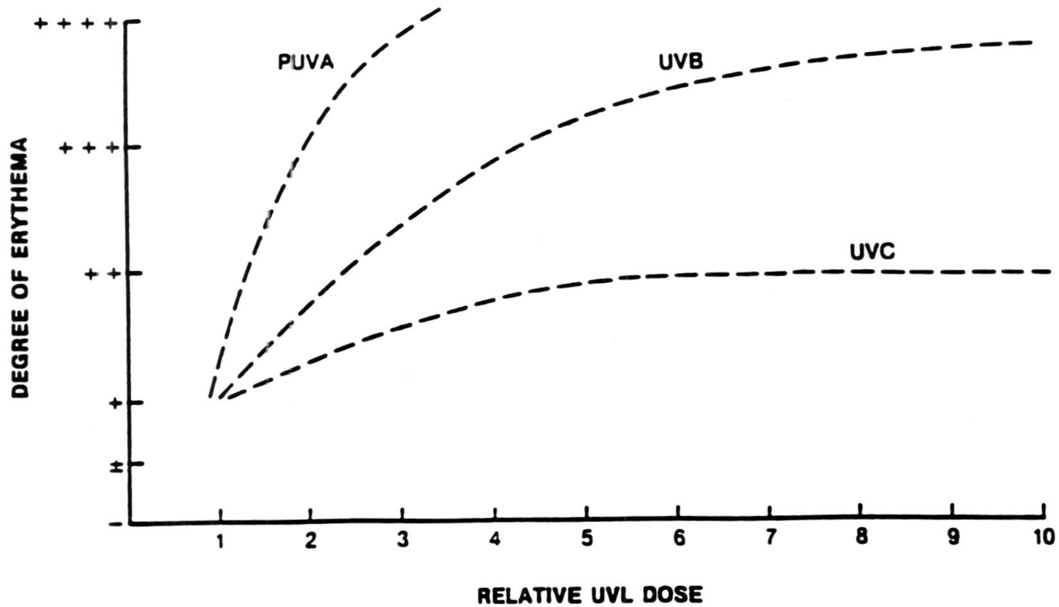


**Figure 4** Dose–response curve for erythema induced by PUVA, UVB, and UVC radiation. (From Morison WL. Phototherapy and Photochemotherapy of Skin Disease, 2nd edition. Raven Press, 1991. New York.)

and their accumulation results in cataracts. These cataracts are readily induced in mice and rats but not in rabbits (4, 5). This variability in susceptibility between species, coupled with the routine use of eye protection in human subjects, may explain why there have been only a few anecdotal reports of cataracts in humans after PUVA therapy (6–8) and no increased incidence in prospective studies (9, 10).

### D. Immunologic

Treatment with psoralens and UVA radiation can induce specific immunologic alterations in mice and other rodents, such as local and systemic suppression of contact hypersensitivity, but the changes are selective and panimmunosuppression does not occur (11). PUVA therapy in humans can alter the function and distribution of circulating lymphocytes and the expression of contact allergy; the significant of these changes is unknown.

## IV. MECHANISM OF ACTION OF PUVA

PUVA therapy is beneficial in more than 30 skin diseases; while the pathogeneses of these conditions are largely unknown, they probably are very diverse. In these circumstances it is not too surprising that the mechanism of action of PUVA therapy is also largely unknown. There are three main theories, and it is possible that more than one mechanism may apply to a given disease.

### A. Suppression of DNA Synthesis

PUVA therapy suppresses synthesis of DNA in the epidermis. This is considered to be a possible mechanism for its effect in psoriasis, which is characterized by hyperproliferation of the epidermis.

### B. Selective Cytotoxicity

In cutaneous T-cell lymphoma, PUVA therapy appears to be selectively toxic for the neoplastic cells in the skin. If inflammatory or other cells responsible for mediating a disease have a similar susceptibility to the treatment, this might explain its effect in conditions such as eczema and vitiligo.

### C. Photoimmunologic Alterations

Most of the diseases that respond to PUVA therapy are thought to have an immunologic pathogenesis; if PUVA therapy can selectively alter immune pathways, this could explain some of its effects.

## V. THERAPY OF PSORIASIS VULGARIS

Psoriasis is the most common indication for PUVA therapy and will be used as a model for describing the treatment.

### A. Evaluation of the Patient

PUVA theraepy is often a long-term treatment that is time-consuming for the patient and associated with certain risks. Therefore, it is important to recognize the risk/benefit ratio for each patient.

#### *1. The Patient*

Age and gender are important since PUVA therapy is relatively contraindicated in the young and women of reproductive age. Motivation should be evaluated because treatment is unlikely to be successful if the patient is unwilling to expend the time and effort required. PUVA therapy is also unlikely to be successful in patients who are unable to follow instructions since it is a relatively complicated treatment.

#### *2. The Disease*

PUVA therapy is indicated in patients with extensive disease on exposed areas that is causing significant disability. The term extensive is of course difficult to quantify, and, for example, marked disease of the palms and soles is extensive in terms of disability. Several features may suggest that PUVA therapy is not the ideal treatment. Associated severe arthritis may point to use of a systemic agent. Involvement of the face or a history of worsening on exposure to sunlight suggests a photosensitive element and a poor response to PUVA therapy.

#### *3. Previous Treatment*

A failure to respond to UVB phototherapy is often an indication to try PUVA therapy. Past exposure to ionizing radiation or arsenic is usually a contraindication to PUVA therapy because of a greatly increased risk of skin cancer.

#### *4. Other Conditions*

Xeroderma pigmentosum, lupus erythematosus, and lactation are all absolute contraindications to PUVA therapy. Relative contraindications include a history or family history of melanoma and a history of nonmelanoma skin cancer and pregnancy. Uremia and severe hepatic failure may interfere with drug metabolism and excretion. Medication with photosensitizing drugs is a warning to be watchful for unexpected phototoxicity. Aphakia requires strict adherence to eye protection.

#### *5. Social and Economic Factors*

The patient must be socially and geographically available for treatment; deep-sea fishermen and traveling salespeople are poor candidates for PUVA therapy. The treatment is expensive and the patient must be able to afford it or have adequate insurance.

#### 6. Physical Examination

A complete skin examination is essential before commencing treatment to assess extent and severity of disease, detect skin cancer, and evaluate any photoaging.

#### 7. Laboratory Investigations

A test for serum antinuclear antibody is the only test currently recommended prior to PUVA therapy, but its value is questionable (12–14). A preferable approach is to do a full serologic evaluation for lupus erythematosus in any patient with photosensitivity or other features that suggest associated connective tissue disease.

#### 8. Ophthalmologic Examination

All patients require a complete eye examination when initiating PUVA therapy, and this should be repeated yearly.

### B. Treatment Schedules

Oxsoralen ultra (ICN Pharmaceuticals) brand of methoxsalen is given in a dosage of 0.4 mg/kg body weight 1 hr prior to exposure to UVA radiation (15). The initial dosage of UVA radiation is usually based on the skin type of the patient (1 J/cm$^2$ for skin type I, 2 J/cm$^2$ for type II, etc.) but may be based on a determination of the minimum phototoxic dose (16). Various schedules of treatment have been established to be effective (17).

#### 1. BIW and TIW

Treatments twice (BIW) or three (TIW) times each week are effective clearance schedules for most patients; 20–25 treatments are required. The dosage of UVA radiation is increased by 0.5 J/cm$^2$ (skin types I–III) or 1 J/cm$^2$ (skin type IV–VI) at each treatment.

#### 2. 11011

Treatment on Monday, Tuesday, Thursday, and Friday provides faster clearing and is more successful in skin types V and VI. The dosage is increased by 0.5–1 J/cm$^2$ every third treatment and about 15 treatments are necessary for clearing.

#### 3. Combination Therapies

Patients with very thick plaques, a high skin type, or a failure to respond to PUVA alone may experience clearing with a combination therapy. A preliminary course of methotrexate in usual dosage for 3 weeks followed by PUVA plus methotrexate is one approach (18). Etretinate in a dosage of 1 mg/kg body weight for 2 weeks followed by PUVA plus etretinate is also successful (19–21).

Finally, UVB plus PUVA therapy is useful in patients for whom methotrexate and etretinate are contraindicated (22).

#### 4. Maintenance Treatment

Once satisfactory control of disease has been achieved the frequency of treatment can be reduced, first to weekly exposure and then to every other week or even monthly treatment. During this phase of maintenance treatment the dosage of UVA radiation is held constant.

### C. Response to Therapy

The first change observed in the psoriatic lesions is a decrease in the amount of scaling. This is followed by a flattening of the plaques to the level of the surrounding skin. The erythema then fades to leave normally pigmented skin or, particularly in skin type IV–VI, hyperpigmented skin. This hyperpigmentation may persist for months. The most persistent lesions are usually on the elbows and knees; clearing of these sites may take twice as much treatment as lesions elsewhere. The aim of treatment in most patients should be control of disease to $< 5\%$ of its original extent rather than total clearing, so as to keep the overall exposure as low as possible and reduce the risk of long-term adverse effects. All clinical varieties of psoriasis of the fingernails respond to treatment, but this usually requires 3–4 months of treatment.

### D. Protection During Therapy

Two important aspects of protection are reviewed here.

#### 1. During Therapeutic Exposure

The eyes must be shielded with UVA-opaque goggles unless the periorbital skin is affected, in which case the eyes must be kept closed. The face, which is the site of actinic damage, and the genitalia in men should be shielded unless affected by a significant amount of psoriasis.

#### 2. Before and After Treatment

From the time of ingestion of psoralen until sunset on the days of treatment, patients must protect exposed skin with clothing or UVA-protective sunscreen and their eyes with wraparound, UV-opaque sunglasses whenever they are exposed to sunlight. In addition, they must avoid deliberate exposure to sunlight on these days. Some centers advise patients to wear protective sunglasses for a full 24 hr after ingestion of methoxsalen, although compliance on the day of treatment is the most important consideration. Exposure to sunlight should be discouraged on nontreatment days since sun-induced pigmentation interferes with therapy.

## E. Adjunctive Therapy

Liberal use of moisturizers should be encouraged since PUVA therapy tends to cause dryness of the skin. Adjunctive treatment with topical antipsoriatic agents such as corticosteroids, tar, and anthralin probably contributes very little to the success of therapy except in relatively nonexposed sites such as the scalp and intertriginous areas (23).

## F. Psoralen Bath Therapy

Topical exposure to psoralen as a very dilute solution in a bath prior to UVA radiation has been extensively used in Europe but appears to have gained limited acceptance elsewhere. One study in the United States found a methoxsalen bath treatment to be successful (24), although the high cost of the methoxsalen solution is likely to deter use of such a treatment. The main advantage claimed for this topical therapy is the reduced systemic exposure to psoralen, but the medication can be detected in serum, and therefore eye protection is still essential.

# VI. THERAPY OF OTHER DISEASES

PUVA therapy has been used in the treatment of many diseases with varying degrees of success. This experience is summarized in Table 1. A few of these conditions will be discussed.

## A. Diseases of the Palms and Soles

Plaque-type psoriasis, pustular psoriasis, endogenous eczema, and palmar–plantar pustulosis all respond well to PUVA therapy (25), and are important indications for this treatment since other therapies are often ineffective or have greater toxicity. A specialized hand-and-foot irradiator provides optimal therapy; in a small office, the treatment can be done using the door panel of a stand-up unit. Disease on the soles is usually slow to respond, presumably due to the thickness of the stratum corneum at that site; higher increments of UVA radiation and more treatments are required compared to those for psoriasis on the general body surface. If disease fails to respond, a fungal infection or contact dermatitis must be considered.

## B. Erythrodermic and Generalized Pustular Psoriasis

These conditions are difficult to treat with PUVA therapy alone (26). Recurrent phototoxicity and difficulty in differentiating between disease and phototoxicity complicate management of erythrodermic psoriasis. A low rate of success and a slow response are features of the therapy of generalized pustular psoriasis. For these reasons, combined treatment with etretinate or methotrexate plus PUVA therapy is preferable and successful in achieving clearing in most patients with

**Table 1** Response of Various Dermatoses to PUVA Therapy

| | Controlled trial | Open trial | Anecdotal |
|---|---|---|---|
| Psoriasis | | | |
| Vulgaris | +++ | | |
| Pustular | | ++ | |
| Erythrodermic | | + | |
| Palms and soles | | | |
| Psoriasis | +++ | | |
| Pustulosis | ++ | | |
| Eczema | +++ | | |
| Mycosis fungoides | ++ | | |
| Parapsoriasis en plaques | | +++ | |
| Pityriasis lichenoides chronica | | +++ | |
| Pityriasis lichenoides acuta | | +++ | |
| Vitiligo | ++ | | |
| Atopic eczema | ++ | | |
| Other endogenous eczemas | | ++ | |
| Photodermatoses | | | |
| PMLE | +++ | | |
| Solar urticaria | +++ | | |
| Chronic photosensitive eczema | | ++ | |
| Erythropoietic protoporphyria | | | ++ |
| Lichen planus | | +++ | |
| Graft-versus-host disease | ++ | | |
| Urticaria pigmentosa | ++ | | |
| Granuloma annulare | | +++ | |
| Alopecia areata | | +++ | |
| Dermatitis herpetiformis | | ++ | |
| Acne vulgaris | + | | |
| Chronic urticaria | | | |
| Physical | | | +++ |
| Dermographism | + | | |
| Idiopathic | | | ++ |
| Pruritus | | | |
| Polycythemia vera | | | ++ |
| Idiopathic | | | ++ |
| Eosinophilic pustular folliculitis | | | +++ |
| Erythrokeratoderma | | | +++ |
| Darier's disease | | | ++ |
| Ichthyosis linearis circumflexa | | | +++ |
| Histiocytosis X | | | +++ |
| Pityriasis rubra pilaris | | | + |
| Scleromyxedema | | | ++ |
| Transient acantholytic dermatosis | | | +++ |
| Xeroderma pigmentosum | | − | |

+++, Complete response; ++, partial response; +, minimal response; −, no response.
*Source:* Morison WL. Phototherapy and Photochemotherapy of Skin Disease, 2nd edition. Raven Press. 1991. New York.

these conditions. Once the disease is cleared, maintenance treatment with PUVA therapy alone is usually adequate.

## C. Cutaneous T-Cell Lymphoma

The eczematous and plaques forms of cutaneous T-cell lymphoma (CTCL), poikiloderma atrophicans vasculare, and premycotic forms of parapsoriasis all respond to PUVA therapy in most instances. The exception is the patient with a brief history and rapid progression of disease; this small subgroup of patients is difficult to treat with any modality. Patients with the tumor phase of CTCL can be treated with PUVA therapy, but adjunctive local treatment with X-ray is required for the tumors. As an alternative, it may be preferable to use another treatment such as electron beam therapy or extracorporeal photopheresis. The latter treatment is also preferable for patients with erythrodermic CTCL or Sézary's syndrome (see Chap. 20).

Analysis of the results of various studies (27–32) suggests that CTCL should be treated in a manner different from the approach to psoriasis. When clearing of disease is achieved on clinical grounds, abnormal cells are still present in the dermis, and treatment should be continued until the skin is clear of disease on histologic examination; this usually requires about 10 additional exposures. Maintenance treatment, initially weekly and later every other week, should probably be continued forever since suspension of treatment is usually followed by return of disease that is more difficult to control with resumption of PUVA therapy.

## D. Atopic Eczema

Generalized atopic eczema responds to PUVA therapy; while most patients can be cleared of disease, only about 75% can be controlled on a long-term basis (33). Eczema is more difficult to treat than psoriasis for several reasons. Patients are more introspective, questioning and demanding, often refusing to believe that their condition is improved despite obvious clearing of their disease. Eczema usually continues to fluctuate in severity until an adequately high dosage of UVA radiation is reached to achieve suppression of the disease. Of course, the patient often attributes an exacerbation to the treatment during this period, and the clinician must be insistent upon continuing therapy. These problems can be circumvented to some extent by using a short course of prednisone as adjunctive therapy to suppress disease while raising the dosage of UVA radiation as rapidly as possible. It is also necessary to use antibiotics as supplementary treatment to control secondary infection.

## E. Vitiligo

Repigmentation of vitiligo is achieved in about 70% of patients after an adequate course of PUVA therapy (34); however, an adequate course may consist of up to

150–300 exposures. Because of the amount of treatment involved, patients should be carefully selected and closely monitored.

### 1. Selection of Patients

Three points should be considered before suggesting PUVA therapy.

1. Site of disease: lesions on the face respond well, the trunk less well and distal limbs not at all.
2. Motivation of patient: a prolonged course of therapy requires a strongly motivated patient.
3. Age of patient: children usually should not be treated until they, and not just their parents, are concerned about their disease.

### 2. Treatment Schedules

Two approaches to treatment are commonly used.

1. Indoor therapy: Oral methoxsalen and exposure to UVA radiation in an office or clinic give the best results. A three times per week or twice per week schedule is used with the aim of producing and maintaining a faint pink color in the areas of vitiligo. Topical application of methoxsalen as a solution is used in some centers and good results have been reported. This technique is associated with a high frequency of unexpected blistering erythemas, which can Koebnerize to produce more vitiligo.
2. Outdoor therapy: Trioxsalen and exposure to sunlight achieve reasonable results in a sunny climate. Trioxsalen is taken orally in a dosage of 0.6 mg/kg body weight and exposure to sunlight is commenced 2 hr later, starting with 5 min and increasing gradually up to 1 hr on each side. The exposure should ideally be done in the middle of the day and of course the treatment is restricted to the summer months in colder climates.

## F. Photodermatoses

Patients who experience significant limitation of normal activities due to photosensitivity can be desensitized by PUVA therapy. Polymorphous light eruption can be prevented in about 90% of patients by three times per week treatment for 4 weeks in spring followed by 1 hr of exposure to noonday sunlight each week during the summer (35). Solar urticaria shows a similar response (36). Patients with photosensitive eczemas can also be desensitized but, since they are usually very sensitive to UV radiation, suppression with high dosages of prednisone is required in the early phase of treatment, and this usually necessitates a brief period of hospitalization (37).

## VII. PROBLEMS AND RISKS

Every treatment is associated with some problems and adverse effects, and PUVA therapy is not an exception.

### A. Problem Cases

#### 1. *The Nonresponder*

Patients with psoriasis should be responding to treatment by the tenth exposure, and if not, a reason must be sought. The most common cause is missed treatments so that a patient who is on a twice a week clearance schedule for example, is really only keeping appointments on a weekly basis, which is a maintenance schedule. A low cutaneous methoxsalen concentration is the next most common cause of a slow response; this is suggested by an absence of pigmentation of normal skin. The patient may not be taking the mediation, or not absorbing it very well, or it may be inactivated rapidly by liver enzymes induced by other medication such as carbamazepine (Tegretol) or phenytoin (Dilantin).

#### 2. *Development of a New Rash*

New eruptions may appear during the therapy. Appearance of lesions of polymorphous light eruption, guttate psoriasis, and transient acantholytic dermatosis are indications for continuing treatment. Lupus erythematosus and bullous pemphigoid may very rarely be induced by treatment and require cessation of treatment. Seborrheic dermatitis of the face may occur and does not respond to the therapy, so topical treatment is required.

#### 3. *Photosensitive Psoriasis*

A history of exacerbations in summer and involvement of the face are two warnings that a patient may have photosensitive psoriasis. Associated polymorphous light eruption and Koebnerization of this rash are the most common causes of photosensitivity (38). These patients can be treated with PUVA therapy, but combination treatment with etretinate or methotrexate is usually required.

#### 4. *Associated Human Immunodeficiency Virus Infection*

Human immunodeficiency virus (HIV) infection can precipitate or aggravate psoriasis. This must be considered in any patient with sudden appearance of erythrodermic or pustular psoriasis or very active, inflammatory disease. The best treatment has not been determined but PUVA therapy and etretinate appear to be safe and effective, while methotrexate is contraindicated because it exacerbates the HIV infection (39, 40).

### B. Short-Term Adverse Effects

These can be conveniently classified into three groups, according to cause.

### 1. *Methoxsalen*

Nausea and, much less commonly, other gastrointestinal symptoms are the most frequently occurring problems due to the medication, but they are readily solved by taking the drug with some food, treating later in the day, or, occasionally, by reducing the dosage. Methoxsalen has an effect on the brain and headache, dizziness, insomnia, and depression are common adverse effects but rarely require any intervention. Bronchoconstriction and xanthemata have been reported but are very rare.

### 2. *Phototoxic Events*

Symptomatic erythema will occur in at least 10% of patients regardless of how carefully the treatment is given; the main reason is the unpredictable absorption of methoxsalen (17). Erythema is most common in patients with skin types I and II and women over the age of 60 years, while the most common sites are breasts and buttocks. Pruritus is a warning signal of an impending erythema and usually persists for a week or two. A psoriasisform eruption on maximally exposed areas associated with intense pruritus (so-called subacute phototoxicity) is a manifestation of phototoxicity and usually develops slowly over 1–2 weeks. Friction blisters on the margins of the hands and feet and photo-onycholysis are rarer forms of phototoxicity.

### 3. *Nonphototoxic Events*

Treatment is stressful since it usually involves standing in a very hot chamber for a prolonged period of time. Tachycardia occurs, although more serious arrhythmias have not been reported. Hypertrichosis is common, particularly on the face, and although it occurs in both men and women, it is usually only noted by women. The frequency of recurrent herpes simplex on the face is increased but primary infection is probably not more common.

## C. Long-Term Adverse Effects

### 1. *Photoaging*

Chronic exposure to PUVA therapy produces changes in the skin that resemble the premature aging changes due to sunlight. Early in the course of treatment the skin appears dry and wrinkled but this is reversible once treatment is stopped. Later in therapy there is freckling, telangiectases, and disturbance of melanization in the form of macules of hypopigmentation. These changes are only partially reversible (41–43). The development of photoaging is related to the total dose of treatment, being more marked at higher doses, and the intensity of treatment, so that a high doses over a short period is more likely to produce changes. Individual susceptibility varies; people with fair skin are more prone and those with dark skin are relatively immune to the changes.

### 2. Nonmelanoma Skin Cancer

PUVA treatment is mutagenic in a variety of systems and carcinogenic in mice. Thus, not suprisingly, it is also carcinogenic in humans and results in an increased frequency of squamous cell carcinomas and, to a lesser degree, basal cell carcinomas (44–50). PUVA therapy acts as a cocarcinogen with X-rays and arsenic so that patients who have been previously exposed to these agents have a greatly increased risk of developing skin cancer. Sunlight and UVB radiation may also be cocarcinogens. PUVA therapy can be a carcinogen by itself, and a dose–response relationship has been demonstrated. The precise risk of skin cancer from PUVA therapy is not known since the prospective studies examining the question had a skewed selection towards patients with severe disease and exposure to other carcinogens and immunosuppressive agents, but it is probably 10–30 times higher than the risk in the general population (44, 45). The biological behavior of PUVA-induced neoplasms appears to be similar to or perhaps less aggressive than sunlight-induced skin cancer, although this is hard to establish because lesions are usually quickly diagnosed and excised.

### 3. Cataracts

Prospective studies have not found an increased frequency of cataracts in patients treated with PUVA therapy (9, 10) although there are reports of cataracts occurring in some patients who did not use eye protection (6–8).

## VIII. CONCLUSION

PUVA therapy has revolutionized photomedicine as it applies to dermatology since it has brought many practitioners into contact with photons who never previously had great interest in photobiology. It has proven to be an effective treatment for psoriasis and many other diseases and is associated with only moderate toxicity. Nonmelanoma skin cancer is the major adverse effect but this is readily treated if it occurs and is preventable by restricting treatment in susceptible people whenever possible. The potential adverse effects of cataracts, melanoma, and immunosuppression have not developed but should be monitored for in prospective studies of exposed patients now underway.

## REFERENCES


1. Parrish JA, Fitzpatrick TB, Tanenbaum L, Pathak MA. Photochemotherapy of psoriasis with oral methoxsalen and longwave ultraviolet light. N Engl J Med 1974; 291:1207–1211.
2. Brickl R, Schmid J, Koss FW. Clinical pharmacology of oral psoralen drugs. Photodermatology 1984; 1:174–186.
3. Cripps DJ, Lowe NJ, Lerner AB. Action spectra of topical psoralens: a re-evaluation. Br J Dermatol 1982; 107:77–82.

4. Cloud TM, Hakim R, Griffin AC. Photosensitization of the eye with methoxsalen. Arch Ophthalmol 1961; 66:689–694.
5. Parrish JA, Chylack LT, Woehler ME, Cheng H-M, Pathak MA, Morison WL, Krugler J, Nelson WF. Dermatological and ocular examinations in rabbits chronically photosensitized with methoxsalen. J Invest Dermatol 1979; 73:256–258.
6. Pedvis-Leftick A, Cyrlin MN, Solomon LM. Cataracts in a patient with vitiligo who received phtochemotherapy. Arch Dermatol 1979; 115:1253–1255.
7. Cyrlin MN, Pedvis-Leftick A, Suger J. Cataract formation in association with ultraviolet photosensitivity. Ann Dermatol 1980; 12:786–790.
8. Woo TY, Wong RC, Wong JM, Anderson TF, Lerman S. Lenticular psoralen photoproducts and cataracts of a PUVA-treated psoriatic patient. Arch Dermatol 1985; 121:1307–1308.
9. Stern RS, Parrish JA, Fitzpatrick TB. Ocular findings in patients treated with PUVA. J Invest Dermatol 1985, 85:269–273.
10. Boukes RJ, Bruynzeel DP. Ocular findings in 340 long-term treated PUVA patients. Photodermatology 1985; 2:178–180.
11. Morison WL. Photoimmunology: study of the effects of nonionizing radiation on the immune system. Ann NY Acad Sci 1984; 453:105–113.
12. Stern RS, Morison WL, Thibodeau LA, Kleinerman RA, Parrish JA, Geer DE, Fitzpatrick TB. Antinuclear antibodies and oral methoxsalen photochemotherapy (PUVA) for psoriasis. Arch Dermatol 1979; 115:1320–1324.
13. Picascia DD, Rothe M, Goldberg NS, Roenigk HH. Antinuclear antibodies during psoralens plus ultraviolet A (PUVA) therapy—are they worthwhile? J Am Acad Dermatol 1987; 16:574–577.
14. Tuffanelli DL. Antinuclear antibodies and photosensitivity in lupus erythematosus—relevant in PUVA therapy? J Am Acad Dermatol 1987; 16:614–616.
15. Levins PC, Gange RW, Momtaz-T K, Parrish JA, Fitzpatrick TB. A new liquid formation of 8-methoxypsoralen: bioactivity and effect of diet. J Invest Dermatol 1984; 82:185–187.
16. Wolff K, Gschnait F, Hönigsmann H, Konrad K, Parrish JA, Fitzpatrick TB. Phototesting and dosimetry for photochemotherapy. Br J Dermatol 1977; 96:1–10.
17. Melski JW, Tanenbaum L, Parrish JA, Fitzpatrick TB, Bleich HL, 28 participating investigators. Oral methoxsalen photochemotherapy for the treatment of psoriasis: a cooperative clinical trial. J Invest Dermatol 1977; 68:328–335.
18. Morison WL, Momtaz K, Parrish JA, Fitzpatrick TB. Combined methotrexate-PUVA therapy in the treatment of psoriasis. J Am Acad Dermatol 1982; 6:46–51.
19. Fritsch PO, Hönigsmann H, Jaschke E, Wolff K. Augmentation of oral methoxsalen-photochemotherapy with an oral retinoic acid derivative. J Invest Dermatol 1978; 70:178–182.
20. Lauharanta J, Juvakoski T, Lassus A. A clinical evaluation of the effects of an aromatic retinoid (Tigason), combination of retinoid and PUVA, and PUVA alone in severe psoriasis. Br J Dermatol 1981; 104:325–332.
21. Parker S, Coburn P, Lawrence C, Marks J, Shuster S. A randomized double-blind comparison of PUVA-etretinate and PUVA-placebo in the treatment of chronic plaque psoriasis. Br J Dermatol 1984; 110:215–220.
22. Momtaz-T K, Parrish JA. Combination of psoralens and ultraviolet A and ul-

traviolet B in the treatment of psoriasis vulgaris: a bilateral comparison study. J Am Acad Dermatol 1984; 10:481–486.
23. Morison WL, Parrish JA, Fitzpatrick TB. Controlled study of PUVA and adjunctive topical therapy in the management of psoriasis. Br J Dermatol 1978; 98:125–132.
24. Lowe NJ, Weingarten D, Bourget T, Moy LS. PUVA therapy for psoriasis: comparison of oral and bath-water delivery of 8-methoxypsoralen. J Am Acad Dermatol 1986; 14:754–760.
25. Morison WL, Parrish JA, Fitzpatrick TB. Oral methoxsalen photochemotherapy of recalcitrant dermatoses of the palms and soles. Br J Dermatol 1978; 99:297–302.
26. Hönigsmann H, Gschnait F, Konrad K, Wolff K. Photochemotherapy for pustular psoriasis (von Zumbusch). Br J Dermatol 1977; 97:119–126.
27. Gilchrest BA, Parrish JA, Tanenbaum L, Haynes HA, Fitzpatrick TB. Oral methoxsalen photochemotherapy of mycosis fungoides. Cancer 1976; 38:683–689.
28. Lowe NJ, Cripps DJ, Dufton PA, Vickers CFH. Photochemotherapy for mycosis fungoides. Arch Dermatol 1979; 115:50–53.
29. MacKie RM, Foulds IS, McMillan EM, Nelson HM. Histological changes observed in the skin of patients with mycosis fungoides receiving photochemotherapy. Clin Exp Dermatol 1980; 5:405–413.
30. Briffa DV, Warin AP, Harrington CI, Bleehen SS. Photochemotherapy in mycosis fungoides. Lancet 1980; 2:49–53.
31. Rosenbaum MM, Roenigk HH, Caro WA, Esker A. Photochemotherapy in cutaneous T cell lymphoma and parapsoriasis en plaques. J Am Acad Dermatol 1985; 13:613–622.
32. Hönigsmann H, Brenner W, Rauschmeier W, Konrad K, Wolff K. Photochemotherapy for cutaneous T cell lymphoma. J Am Acad Dermatol 1984; 10:238–245.
33. Morison WL, Parrish JA, Fitzpatrick TB. Oral psoralen photochemotherapy of atopic eczema. Br J Dermatol 1978; 98:25–32.
34. Ortonne J-P, Mosher DB, Fitzpatrick TB. Vitiligo and Other Hypomelanoses of Hair and Skin. New York: Plenum, 1982:267–282.
35. Parrish JA, LeVine MJ, Morison WL, Gonzalez E, Fitzpatrick TB. Comparison of PUVA and beta-carotene in the treatment of polymorphous light eruption. Br J Dermatol 1979; 100:187–191.
36. Parrish JA, Jaenicke KF, Morison WL, Momtaz K, Shea C. Solar urticaria: treatment with PUVA and mediator inhibitors. Br J Dermatol 1982; 106:575–580.
37. Morison WL, White HAD, Gonzalez E, Parrish JA, Fitzpatrick TB. Oral methoxsalen photochemotherapy of uncommon photodermatoses. Acta Dermatol 1979; 59:366–368.
38. Ros A-M, Eklund G. Photosensitive psoriasis. J Am Acad Dermatol 1987; 17:752–758.
39. Duvic M, Johnson TM, Rapini RP, Freese T, Brewton G, Rios A. Acquired immunodeficiency syndrome-associated psoriasis and Reiter's syndrome. Arch Dermatol 1987; 123:1622–1632.
40. Ranki A, Puska P, Mattinen S, Lagerstedt A, Krohn K. Effect of PUVA on immunologic and virologic findings in HIV-infected patients. J Am Acad Dermatol 1991; 24:404–410.

41. Stern RS, Parrish JA, Fitzpatrick TB, Bleich HL. Actinic degeneration in association with long-term use of PUVA. J Invest Dermatol 1985; 84:135–138.
42. Rhodes AR, Harrist TJ, Momtaz-T K. The PUVA-induced pigmented macule: a lentiginous proliferation of large, sometimes cytologically atypical, melanocytes. J Am Acad Dermatol 1983; 9:47–58.
43. Gschnait F, Wolff K, Hönigsmann H, Stingl G, Brenner W, Jaschke E, Konrad K. Long-term photochemotherapy: histopathological and immunofluorescence observations in 243 patients. Br J Dermatol 1980; 103:11–22.
44. Stern RS, Thibodeau LA, Kleinerman RA, Parrish JA, Fitzpatrick TB, and 22 participating investigators. Risk of cutaneous carcinoma in patients treated with oral methoxsalen photochemotherapy for psoriasis. N Engl J Med 1979; 300:809–813.
45. Stern RS, Laird N, Melski J, Parrish JA, Fitzpatrick TB, Bleich HL. Cutaneous squamous-cell carcinoma in patients treated with PUVA. N Engl J Med 1984; 310:1156–1161.
46. Stern RS, Lange R. Non-melanoma skin cancer occurring in patients treated with PUVA five to ten years after first treatment. J Invest Dermatol 1988; 91:120–124.
47. Henseler T, Christophers E, Hönigsmann H, Wolff K, and 19 other investigators. Skin tumors in the European PUVA study. J Am Acad Dermatol 1987; 16:108–116.
48. Ros A-M, Wennersten G, Lagerholm B. Long-term photochemotherapy for psoriasis: a histopathological and clinical follow-up study with special emphasis on tumour incidence and behavior of pigmented lesions. Acta Derm Venereol (Stockh) 1983; 63:215–221.
49. Eskelinen A, Halme K, Lassus A, Idanpaan-Heikkila J. Risk of cutaneous carcinoma in psoriatic patients treated with PUVA. Photodermatology 1985; 2:10–14.
50. Stern RS, members of the photochemotherapy follow-up study. Genital tumors among men with psoriasis exposed to psoralens and ultraviolet A radiation (PUVA) and ultraviolet B radiation. N Engl J Med 1990; 322:1093 1097.

# 20

# Extracorporeal Photopheresis: Principles and Practice

**Alain H. Rook, Michael H. Berkson, and Benjamin R. Vowels**
*University of Pennsylvania, Philadelphia, Pennsylvania*

Extracorporeal photopheresis is a new form of immunotherapy that involves the use of a leukapheresis device that permits the extracorporeal exposure of potentially pathogenic peripheral blood leukocytes to 8-methoxypsoralen (8-MOP) and ultraviolet A radiation followed by the reinfusion of the treated cells (1, 2). Successful clinical trials have resulted in the addition of photopheresis to the clinician's therapeutic armamentarium for the treatment of advanced forms of cutaneous T-cell lymphoma. Long-term follow-up of patients with Sézary syndrome who participated in an initial multicenter clinical trial has indicated the potential for the significant prolongation of life as well as the potential for cure of this devastating malignancy (3). Moreover, current clinical trials suggest the usefulness of photopheresis in the management of a variety of autoimmune diseases including systemic sclerosis, pemphigus vulgaris, and rheumatoid arthritis (4–6).

## I. PROCEDURE

The treatment is accomplished using the UVAR photopheresis system (Therakos, West Chester, PA) as previously described for cutaneous T-cell lymphoma (1). Two hours after the ingestion of 8-MOP (0.5 mg/kg Oxsoralen Ultra), patients undergo a discontinuous leukapheresis procedure with exposure of removed leukocytes to ultraviolet A radiation. During the procedure, approximately 240 ml of leukocyte-enriched blood is mixed with 300 ml of the patients plasma and 200 ml of sterile normal saline plus approximately 10,000 units of heparin. We have occasionally found the need for greater concentrations of heparin when treating patients with systemic sclerosis due to heightened coagulability. The final buffy coat preparation contains an estimated 25–50% of the total peripheral blood mononuclear cell compartment and has a hematocrit from 2.5 to 7%. The buffy coat is then passed as a 1 mm film through a sterile cassette surrounded by ultraviolet A-emitting bulbs, permitting a 180 min exposure to ultraviolet A light, yielding an average exposure per lymphocyte of 2 $J/cm^2$. After exposure of the cells to ultraviolet A, the buffy coat is returned to the patient. The entire procedure requires approximately 3.5 hr.

Throughout the course of treatment, psoralen levels require intermittent assessment. Plasma levels at or above 100 ng/ml or levels within the photopheresis buffy coat bag above 50 ng/ml are recommended. Substantial intraindividual variation in psoralen absorption can occur depending upon gastric contents and disease state. For example, patients with systemic sclerosis who have bowel involvement may have extremely poor absorption of the 8-MOP. If this aspect of their disease worsens, absorptive capacity may diminish, thus requiring an increase in dosage. It is not unusual for some patients to have marked variations in plasma psoralen levels from treatment to treatment without a discernible cause. Thus, measurements are taken several times annually to ensure the maintenance of therapeutic levels of 8-MOP.

Obtaining vascular access prior to the leukapheresis phase of treatment may sometimes be difficult. This may not be an uncommon problem among patients with systemic sclerosis. Since reasonably high venous flow rates are required for the treatment, intermittent femoral vein cannulation is recommended in such instances. The regular use of indwelling Hickman catheters has resulted in frequent infection at our center, therefore vena porta catheters are preferred.

## II. CLINICAL INDICATIONS

### A. Cutaneous T-Cell Lymphoma

Cutaneous T-cell lymphoma (CTCL) represents the most common type of adult T-cell lymphoma. It consists of a proliferation of a clonal population of malignant CD4+ helper T lymphocytes, which apparently develops initially within the skin. The Sézary syndrome represents an advanced form of CTCL associated with the dissemination of the malignant T lymphocytes beyond the skin to the lymph nodes and to the peripheral blood, with a concurrent cutaneous erythroderma (7). The clinical course of the Sézary form of CTCL is almost invariably one of relentless progression to death, with a median survival of approximately 30 months from disease onset.

Administration of conventional psoralen and ultraviolet A photochemotherapy (PUVA) to involved areas of the skin produces a high rate of initial clinical response in plaque stage CTCL (8). A study from Scandinavia has indicated that nearly 75% of patients with plaque-stage CTCL will initially experience a complete response to PUVA therapy. However, extracutaneous disease appears to be little affected by this treatment. In contrast, an initial multicenter trial conducted by Edelson and colleagues, employing photopheresis in the treatment of advanced CTCL, produced promising results (1). Twenty-nine patients with exfoliative erythroderma, resistant to systemic chemotherapy, electron beam radiotherapy, topical mechlorethamine, and PUVA photochemotherapy, and an additional eight patients with localized plaque disease, underwent photopheresis on 2 successive days on a monthly basis. Twenty-four of 29 (83%) erythrodermic and 3 of 8 (38%) plaque patients had improvement in their cutaneous disease, with a mean time to development of the positive response of 22.4 weeks. Nine patients (24%) experienced a marked response with better than a 75% improvement in extent of their skin lesions. Thirteen patients (35%) had a 50–75% improvement in their skin lesions. Those with the most prominent degree of improvement had generally experienced noticeable clearing of their skin by the third or fourth month of treatment. Eight of 10 patients who had lymph node involvement, as demonstrated histologically at the initiation of therapy, responded to treatment. It is the authors' bias, described more fully below, that patients with extensive nodal involvement should receive concomitant adjuvant therapy, such as with interferon alpha, in addition to photopheresis.

In addition to improvement in the cutaneous manifestations of CTCL, an apparent decrease in the extent of peripheral blood involvement was also observed. Five of six patients with elevated numbers of CD4+ peripheral blood cells who were studied experienced a decrease in the absolute numbers of these cells during their therapy, presumably reflecting a decrease in the numbers of circulating malignant cells that typically bear CD4 on their membrane. The number of atypical cells within the peripheral blood with a cerebriform nuclear morphology (Sézary cells) also decreased. In addition, Southern blot analysis has indicated that some individuals have had an apparent disappearance of the malignant clone from their peripheral blood in response to photopheresis therapy (9). At the authors' institution, four patients treated with photopheresis who originally had the Sézary syndrome are now in complete clinical remission without evidence of a peripheral blood clone on Southern blot analysis. Future use of more sensitive techniques of molecular diagnosis, such as the polymerase chain reaction, with patient specific oligonucleotide probes (10), will permit us to define more precisely the nature of possible minimal reservoirs of residual disease.

Follow-up evaluation of erythrodermic patients who participated in the initial multicenter trial of photopheresis has suggested that this therapy may prolong survival beyond that expected with conventional therapies (3). Median survival of the original 27 erythrodermic patients is currently approaching 55 months, whereas previous studies of a comparable patient population, employing conventional therapies, have revealed survival rates of 30–40 months. During this period of extended observation, few adverse effects of photopheresis have been recorded. These results are remarkable not only in view of recent data demonstrating a failure to prolong survival of CTCL patients with the use of intensive multidrug chemotherapeutic regimens combined with electron beam irradiation, but also in light of the high degree of morbidity associated with such treatment (11).

These findings regarding the well-tolerated nature of photopheresis and the potential for the prolongation of life serve as the basis for our early consideration of the use of photopheresis in the treatment of certain stages of CTCL. We routinely recommend this treatment for patients who have demonstrable evidence of circulating malignant cells. Thus, patients with the Sézary syndrome and individuals with extensive plaque disease, in whom atypical peripheral blood cells, characterized by cerebriform nuclear morphology are present, are considered to be excellent candidates for photopheresis. In contrast, those without evidence of malignant peripheral blood cells have faired less well at our institution. None of three patients with plaque disease with greater than 10% skin involvement but no peripheral blood involvement has responded to photopheresis, whereas 3 of 3 patients with a similar degree of skin disease, in addition to the presence of atypical cells within the peripheral blood, have had substantial clinical improvement. Likewise, erythrodermic patients who lack atypical blood cells are less likely to respond than are those erythrodermic individuals with Sézary syndrome. However, those with an especially large tumor burden (white blood count $>15,000/mm^3$, widespread bulky lymphadenopathy, visceral disease, or numerous cutaneous tumors) are unlikely to respond to photopheresis used as a single-treatment modality. Although the authors are aware of several cases of tumor stage disease that have improved while undergoing photopheresis treatment, this form of CTCL is a good example of the need for concurrent therapy with electron beam irradiation, interferon alpha, or low-dosage methotrexate (see below).

A number of additional parameters have been identified that may predict responsiveness to photopheresis. In a study of 10 patients with Sézary syndrome, we determined that increased numbers of CD8+ peripheral blood T cells correlated with a favorable response to therapy (Table 1). Conversely, an elevated number of CD4+ T cells was associated with a lack of response to photopheresis. Heald and colleagues have made similar observations in regard to T-cell numbers (12). In studies of their patient population, the mean percentage of CD4+ and CD8+ peripheral blood T cells among responders was 55.4 and 15.6%, respectively, while the nonresponders had a mean of 74.8 and 7.8%, respectively. Since the malignant T cells bear CD4 on their surface, an increase in the numbers of CD4+ cells would appear to indicate an increase in the tumor

**Table 1** Peripheral Blood Lymphocytes in Patients with CTCL Undergoing Photopheresis

| | %Lymphocyte Phenotype[a] | | | | | |
|---|---|---|---|---|---|---|
| Patients | CD3 | CD4 | CD8 | CD29 | CD45RA | CD16 |
| Responders: | 77 | 42 | 37 | 15 | 17 | 4 |
| n=5 | 49 | 35 | 28 | 32 | 23 | 4 |
| | 74 | 29 | 48 | 22 | 25 | 1 |
| | 62 | 56 | 35 | 19 | 34 | 18 |
| | 56 | 6 | 19 | 34 | 8 | 0 |
| Mean ± SD | 64 ± 12 | 34 ± 18 | 33 ± 11 | 24 ± 8 | 21 ± 10 | 5 ± 7 |
| Nonresponders: | 98 | 83 | 3 | 78 | 1 | 1 |
| n=5 | 92 | 91 | 2 | 51 | 3 | 0 |
| | 97 | 95 | 2 | 95 | 1 | 0 |
| | 98 | 93 | 2 | 89 | 1 | 1 |
| | 99 | 93 | 0 | 96 | 2 | 0 |
| Mean ± SD | 97 ± 3 | 91 ± 5 | 2 ± 1 | 82 ± 19 | 2 ± 1 | 0 ± 1 |

[a]Percentage of cells expressing indicated surface molecule.
CD3, pan T cell; CD4, helper T cell; CD8, suppressor/cytotoxic T cell; CD4/CD29, helper/inducer T cell; CD4/CD45RA, suppressor/inducer T cell; CD16, natural killer cell.

cell burden, which, as previously stated, correlates with a poor prognosis. We have also observed a correlation between normal levels of peripheral blood natural killer cell activity and response to therapy. Overall, these results imply that CD8+ cytolytic and suppressor T cells and natural killer (NK) cells may be important components of the antitumor immune response. Moreover, these data may reflect the requirement for a greater degree of immune integrity so as to be able to generate an immune response against the photoaltered malignant cells.

In addition to these immunologic parameters, Heald and colleagues have determined that duration of disease prior to the initiation of photopheresis, as well as the extent of prior treatment, also correlated with response to therapy. The majority of individuals with the Sézary syndrome who responded to photopheresis had their disease for less than 1 year, whereas nonresponders generally had a disease duration of greater than 2 years. Furthermore, patients who received extensive pretreatment with systemic chemotherapy responded less frequently to photopheresis. Since an intact host immune system appears to be necesssary to obtain a response to photopheresis, substantial prior use of immunosuppressive agents may impair the response to this therapy. These observations, along with those indicating a low frequency of adverse effects of photopheresis, suggest that this therapy should be considered in the early phases of the clinical management of patients with CTCL who have the malignant cells within the peripheral blood.

For patients who manifest disease parameters suggestive of a lower probability of a therapeutic response to photopheresis (extensive tumor burden, low numbers of peripheral blood CD8+ T cells, depressed NK cell activity), adding certain adjunctive therapies in combination with photopheresis may improve the clinical course. Judicious use of electron beam therapy as a procedure to reduce tumor load for patients with cutaneous tumors is a common practice at our institution. Although there are anecdotal reports of patients with tumors who responded to photopheresis alone, in our experience, concomitant electron beam therapy is usually required to produce tumor flattening. For patients with a poor prognosis, we also favor the early use of interferon alpha in low dosages. Patients are started on 2–2.5 million units subcutaneously three times weekly to minimize flulike symptoms. Dosages are gradually increased, as tolerated, to 5–7.5 million units four to five times weekly. The majority of patients treated in this manner have experienced a substantial decrease in the numbers of circulating atypical cells. At our center, a patient with Sézary syndrome with a white blood cell count of 15,500/mm$^3$, 46% of which were atypical, and a Southern blot analysis demonstrating a clonal rearrangement of the T-cell receptor $\beta$ chain gene, experienced complete clearing of his erythroderma, return of his white blood count to normal levels, and disappearance of the malignant clone while on a combined regimen of photopheresis and interferon alpha (9) (Fig. 1, see color plate).

Use of methotrexate in dosages of 15–25 mg weekly or azathioprine daily have also been suggested for patients with elevated white blood cell counts. Although these drugs may provide benefit by impeding the in vivo proliferation of the malignant T cells, one must also consider their potential for suppression of the host response to the photoaltered tumor cells induced by photopheresis. As an example of a photopheresis-induced immune response that may be impaired by drugs, we have recently determined that photopheresis induces the marked production of tumor necrosis factor alpha (TNF$\alpha$) by treated monocytes (13). The treatment-induced TNF$\alpha$ may be an important component of the antitumor response. Since corticosteroids are known to inhibit TNF$\alpha$ production in addition to other immune responses, we attempt to avoid the use of prednisone in our patients. For reasons such as these we advocate, whenever possible, the use of immune-enhancing agents such as interferon alpha rather than potentially immunosuppressive drugs. In addition to interferon alpha, the immune-augmenting biological response modifiers interferon gamma and interleukin 2 may have some therapeutic activity against CTCL, although they have not yet been used in combination with photopheresis.

## B. Autoimmune Disease

### *1. Systemic Sclerosis*

Systemic sclerosis is an autoimmune disease characterized by the increased deposition of collagen within the skin and often within certain visceral organs including the lungs, heart, gastrointestinal tract, and kidneys. Recent observations have implicated the immune system as a leading factor in the genesis of the increased collagen production (14, 15). Elevated serum levels of interleukin 2 and soluble interleukin 2 receptors in patients with active clinical disease support the association of T-cell activation with disease progression. Furthermore, early biopsy of involved tissues has revealed the infiltration of activated helper T cells which may be releasing certain cytokines, particularly transforming growth factor $\beta$, which is a potent stimulator of collagen synthesis (16).

In view of evidence of peripheral blood T-cell activation, a pilot study was conducted that involved the use of photopheresis in the treatment of two patients with rapidly progressive disease (17). In the first case, a 60-year-old man had a 6 month history of disease with skin thickening involving greater than 80% of the body surface, mild pulmonary involvement, proteinuria (1300 mg/24 hr), severe Raynaud's phenomenon with ulcerated digits, and evidence of microangiopathic hemolysis. Marked clinical improvement was observed within the first several months of treatment with the photopheresis being administered on 2 consecutive days every 4 weeks. By the third month of therapy, the proteinuria, hemolysis, and Raynaud's phenomenon were no longer evident. After 1 year of therapy, the only residual disease was mild skin thickening of the hands with acrosclerosis. Treatment of a second patient with extensive skin thickening also resulted in skin softening and a decrease in the area of cutaneous involvement.

These initial results prompted the development of a multicenter randomized trial to examine the efficacy and safety of photopheresis in the treatment of

**Table 2** Change in Severity Scores During Treatment of Patients with Scleroderma

| | Evaluation Interval | | | |
|---|---|---|---|---|
| | 6 Month | | 10 Month | |
| Treatment Group | Improved ⩾15% | Worse⩾15% | Improved ⩾15% | Worse⩾15% |
| Photochemotherapy | 21/31(68%)[a] | 3/31(10%) | 20/29(69%) | 3/29(10%) |
| D-Penicillamine | 8/25(32%) | 8/25(32%) | 9/18(50%) | 3/18(16%)[b,c] |

[a]The proportion of patients responding to therapy at 6 months was significantly greater ($p=0.02$) in the photochemotherapy arm than in the D-penicillamine arm.

[b]Five of seven patients who dropped out between 6 and 10 months of therapy within the penicillamine arm had more than 15% worsening from baseline in their severity score at the time of drop-out.

[c]Patients not shown had less than 15% change in their skin severity score during therapy

systemic sclerosis (Table 2). Seventy-nine patients with systemic sclerosis of recent onset (mean duration of symptoms, 1.8 years) and progressive skin involvement entered a randomized, parallel group, single-blinded clinical trial comparing photopheresis given on 2 consecutive days every 4 weeks to treatment with D-penicillamine. Skin severity scores (skin hardening), percentage skin surface area involvement, hand closure, and oral aperture were evaluated monthly by examiners blinded to the nature of the treatment. In addition, pulmonary function tests, skin biopsies, and serologic values were obtained at baseline and after 6 and 12 months of treatment. During this trial, 56 patients received 6 months of therapy (31 received photopheresis) while 47 received 10 months of therapy (29 on photopheresis). By 6 months of treatment, 21 of 31 (68%) patients who received photopheresis had experienced softening of the skin in comparison to 8 of 25 (32%) who received D-penicillamine (Table 2). It is noteworthy that whereas only 3 of 31 (10%) of those who received photopheresis had experienced significant worsening of their skin severity score after 6 months of treatment, 8 of 25 (32%) who received D-penicillamine had significant worsening. Thus, in the early phases of treatment, a significantly higher response rate was obtained with photopheresis ($p=0.02$). At both the 6 and 10 month evaluation point, the mean skin severity score, mean percentage involvement, and mean oral aperture measurements were significantly improved from baseline among those who received photopheresis. Mean right and left hand closure measurements had also improved significantly by 10 months of therapy. Skin biopsy studies demonstrated an association between clinical improvement and decreased thickness of the dermal layer. No significant change in pulmonary function was observed in either treatment group after 12 months of therapy.

It should be emphasized that adverse effects of photopheresis were minimal and did not require discontinuation of treatment by any of the patients. In contrast, six patients who received D-penicillamine were required to discontinue this drug permanently due to its adverse effects. Thus, photopheresis appears to produce early improvement with few side effects when used for aggressive cases of recent-onset systemic sclerosis.

### 2. *Pemphigus Vulgaris*

A pilot study was undertaken to determine the efficacy of photopheresis in the therapy of cases of pemphigus vulgaris that were resistant to treatment with corticosteroids and immunosuppressive drugs (4). Four patients with persistent disease despite a lengthy history of treatment with prednisone in combination with azathioprine or cyclophosphamide were started on photopheresis on 2 consecutive days every 4 weeks. All patients initially had improvement in the extent of their skin disease, which allowed for tapering of other medications. Three of the four patients eventually experienced long-term remissions that permitted discontinuation of all treatment. Reductions in serum levels of anti-epidermal cell autoantibodies occurred in association with clinical improvement.

Although relapses occurred in all three, remission was rapidly reinduced following three to four additional monthly cycles of photopheresis. Photopheresis produced no adverse effects in any of the four patients during several years of follow-up evaluation.

During this trial, frequent treatment cycles (2 consecutive days every 2 weeks) were useful for the rapid induction of remission, whereas treatment cycles every 3–4 weeks were usually satisfactory for the maintenance of remission. The authors' experience indicates that once clinical improvement occurs, gradual tapering of corticosteroids and immunosuppressive medications can proceed. However, simultaneous abrupt tapering of photopheresis along with the tapering of other medications may result in the early recurrence of skin lesions.

#### 3. *Rheumatoid Arthritis*

Malawista and colleagues conducted a pilot study of the use of photopheresis for recent-onset rheumatoid arthritis (6). Seven patients received photopheresis on 2 successive days every 4 weeks. By week 16, four of the seven had experienced significant improvement in the numbers and severity of clinically involved joints. Two of the patients had complete cessation of disease activity. A slow recurrence of clinical symptoms was observed approximately 3 months after discontinuation of treatment. As with the other autoimmune diseases, treatment was extremely well tolerated, with only one episode of mild hypotension noted in over 100 treatments. The substantial response rate in this small study and the minimal adverse effects of photopheresis in comparison to other drugs commonly used in the treatment of rheumatoid arthritis have resulted in the development of a multicenter, randomized controlled trial to evaluate the efficacy of photopheresis in the treatment of rheumatoid arthritis.

#### 4. *Psoriasis*

Vonderheid et al. treated four patients with a history of refractory psoriasis with photopheresis administered on alternate weeks (18). Two patients received methotrexate at a dosage of 25 mg every other week alternating with the photopheresis. Although all four patients initially improved in response to photopheresis, exacerbations eventually occurred upon discontinuation of the methotrexate in the two receiving this drug and in the other two, apparently in association with a seasonal change. When methotrexate was readministered to the two individuals, but at a previously ineffective dosage (12.5 mg every other week), steady improvement was again noted in one. These results suggest that photopheresis, particularly when used in combination with other therapies, may provide benefit to patients with refractory extensive psoriasis.

### C. Miscellaneous

A number of additional clinical indications for photopheresis are currently being examined in pilot studies or randomized trials, including a double-blind trial in the treatment of multiple sclerosis (19). The rationale for this study is based upon the salutary effects of a photopheresis-like procedure in preventing experimental allergic encephalitis in a murine model (see below). Patients are randomized to receive either photopheresis or sham photopheresis on an every-4-week schedule.

A pilot study of the treatment of chronic lymphocytic leukemia (CLL) indicated that photopheresis was not likely to provide benefit for patients with advanced disease (20). Since the majority of patients with CLL who were treated in this study had white blood cell counts greater than 100,000/$mm^3$, it is not known whether individuals with lower white counts would respond to photopheresis. By analogy with the therapeutic experience of the Sézary syndrome, patients with a smaller tumor burden and a more intact immune system respond more often to photopheresis than do those with an extensive tumor burden. Therefore, those individuals with CLL with a lower white blood count would be more likely to respond.

Recently, the use of photopheresis to treat five patients with acquired immunodeficiency syndrome (AIDS)-related complex was reported (21). Following 6 months of treatment, 4 of the patients had a decrease in disease-related symptoms and an improvement in delayed-type hypersensitivity skin tests. Enlarged lymph nodes disappeared in all five individuals. A larger trial with longer follow-up is currently underway in an effort to confirm these results and to determine whether photopheresis can prolong survival of patients with human immunodeficiency virus infection and depressed CD4+ T-cell numbers.

Other disorders under clinical investigation to determine their responsiveness to photopheresis include systemic lupus erythematosus, graft-versus-host disease, and allograft rejection. Results of studies using animal models for each of these conditions suggest that photopheresis may be of benefit. In addition, the authors have successfully treated a patient with severe scleromyxedema, with a reduction in the degree of cutaneous sclerosis.

## III. ADVERSE EFFECTS

Despite being used for a variety of medically complex conditions, photopheresis has been extremely well tolerated, with minimal adverse effects. The most frequent adverse effect has been the sporadic occurrence of psoralen-induced nausea. This is usually mild and short-lived but occasionally can be associated with vomiting and diaphoresis. Psoralen levels may be warranted to determine if the dosage is excessive.

Hypotension, although uncommon, may occur during the leukapheresis phase, particularly among those taking antihypertensive agents or diuretics that can lead to intravascular volume compromise. We frequently wait until the conclusion of the treatment to administer vasodilators or diuretics. As an alternative, small volumes of normal saline may be infused just prior to the initiation of the treatment as a preventive measure. Nevertheless, our observation that patients with advanced cardiomyopathy and aortic stenosis have tolerated photopheresis without difficulty indicates that substantial homeostatic derangements of vascular volume are often required for the development of hypotension.

A low-grade fever occurring 4–12 hr after reinfusion of the treated cells is not unusual during the early phases of therapy of patients with CTCL. This appears to be unrelated to bacteremia and is associated with the most marked clinical responses to treatment. Recent observations indicate that photopheresis is responsible for the induction of proinflammatory and pyrogenic cytokines from monocytes including TNF $\alpha$ and interleukin 1 $\alpha$ and $\beta$ and interleukin 6 (13). This immunologic effect is the likely explanation for the posttreatment febrile response. However, since this therapy involves venous cannulation of the frequently immunosuppressed CTCL patient, it is recommended that blood cultures be obtained and that all potential sources of infection be sought.

Although normal leukocytes are exposed to 8-MOP and ultraviolet A during the photopheresis procedure, depletion of these blood elements has not been noted. Furthermore, clinical evidence of photopheresis-induced immunosuppression, such as the development of neoplasia or opportunistic infections has not been observed. Thus, the fact that photopheresis produces minimal adverse effects and provides substantial benefit for CTCL, as well as for certain autoimmune diseases, indicates that this therapy has clear advantages in comparison to chemotherapeutic and immunosuppressive agents in the treatment of these diseases.

## IV. MECHANISM OF ACTION

Ultraviolet A irradiation activates psoralen contained within leukocytes leading to the cross-linking of DNA, which ultimately results in the proliferative arrest of treated cells (22). Large or activated mononuclear cells appear to be particularly sensitive to the antiproliferative effects of 8-MOP and ultraviolet A, possibly due to an increased number of psoralen-binding sites. This observation may account for the apparent heightened susceptibility to photopheresis of the often large, malignant T cells that are typical of CTCL. Activated T cells within the peripheral blood of patients with systemic sclerosis are also likely targets for psoralen and ultraviolet A-mediated damage. Thus, the treated cells are rapidly eliminated from the body simply by the process of "photodestruction."

It should be emphasized that photodestruction of pathogenic T cells probably represents only a minor mode of action of photopheresis, since removal of malignant cells by leukapheresis frequently has little effect on disease progression. More important mechanisms of action of photopheresis have been suggested by recent observations using a number of different experimental animal models. Study of the murine model of experimental allergic encephalitis has been especially useful in our understanding the mechanisms of photopheresis

(23). In this model, rats injected with myelin basic protein develop a paralytic illness associated with T cell destruction of the nervous system. The pathogenic T cells that mediate this destruction can be isolated and cloned in vitro. When naive syngeneic rats are inoculated with the cloned T cells, all of the features of experimental allergic encephalitis are reproduced. However, if the pathogenic clones are first treated with psoralen and ultraviolet A and then infused, the animals are protected from the development of disease upon subsequent challenge with the pathogenic T cells. Protection from disease appears to be mediated by the generation of clone-specific suppressor T cells that have developed in response to the psoralen and ultraviolet A-modified pathogenic cells. Perez and colleagues, employing a model of cutaneous allograft rejection, have similarly obtained evidence of stimulation of an antigen-specific suppressor T cell response when alloreactive effector T cells are treated with psoralen and ultraviolet A and infused into syngeneic animals (24). These results indicate that, at least in the setting of these animal models, an active immunization process can occur following the administration of photoinactivated syngeneic T-cell clones.

As cited above, new evidence that suggests an additional mode of action of photopheresis has emerged from the observation that photopheresis induces the release of inflammatory mediators from mononuclear cells. In this regard, monocytes isolated subsequent to the photopheresis treatment are found to be producing large quantities of TNF$\alpha$ compared to untreated cells (13). This finding has important implications for the treatment of CTCL as well as for systemic sclerosis. In regard to CTCL, TNF$\alpha$ can augment a number of antitumor immune responses and may also exert a direct cytotoxic effect on lymphoma cells (25). In the case of systemic sclerosis, TNF$\alpha$ has been shown to inhibit collagen synthesis by skin fibroblasts (14). In addition to the stimulation of TNF$\alpha$ release, photopheresis also induces the release of other lymphokines including interleukin 1 and interleukin 6. Current studies are underway to define the full scope of the lymphokine-inductive effects of photopheresis.

The appropriate future clinical applications of photopheresis will depend upon a thorough understanding of the mechanisms of action of this new therapy. As our comprehension evolves, we will have the capacity to use additional pharmacologic agents together with photopheresis to enhance its therapeutic benefit further. One example is the combined use of interferon alpha with photopheresis in the treatment of CTCL. Photopheresis is in its early phases of use and a great deal remains to be learned.

## REFERENCES

1. Edelson R, Berger C, Gasparro F, et al. Treatment of cutaneous T cell lymphoma by extracorporeal photochemotherapy: preliminary results. N Engl J Med 1987; 316:297–303.

2. Edelson R. Light activated drugs. Sci Am 1988; 259:68–75.
3. Edelson R. Photopheresis—a novel and meaningful biologic response modification. In: Edelson R, Ashwell JA, eds. Antigen and Clone-Specific Immunoregulation. New York: New York Academy of Sciences, 1990.
4. Rook AH, Heald PW, Nahass GT, et al. Treatment of autoimmune disease with extracorporeal photochemotherapy: pemphigus vulgaris—preliminary report. Yale J Bio Med 1989; 62:647–652.
5. Rook AH, Freundlich B, Jegasothy BV, et al. Treatment of systemic sclerosis with extracorporeal photochemotherapy. Results of a multicenter trial. Arch Dermatol 1992; 128:337–346..
6. Malawista S, Trock D, Edelson R. Treatment of rheumatoid arthritis by extracorporeal photochemotherapy: a pilot study. Arthritis Rheum 1991; 34:646–654.
7. Edelson RL. Cutaneous T cell lymphoma: mycosis fungoides, Sézary syndrome, and other variants. J Am Acad Dermatol 1980; 2:89–106.
8. Thomsen K, Hammar H, Molin L, Volden G. Retinoids plus PUVA (RePUVA) and PUVA in mycosis fungoides, plaque stage. A report from the Scandinavian mycosis fungoides group. Acta Derm Venereol (Stockh) 1989; 69:536–538.
9. Rook AH, Prystowsky MB, Cassin M, Boufal M, Lessin SR. Combined therapy of the Sézary syndrome with extracorporeal photochemotherapy and low dose interferon alpha. Clinical, molecular, and immunologic observations. Arch Dermatol 1991; 127:1535–1540.
10. Lessin SR, Rook AH, Rovera G. Molecular diagnosis of cutaneous T-cell lymphoma: polymerase chain reaction amplification of T-cell antigen receptor $\beta$-chain gene rearrangements. J Invest Dermatol 1991; 96:299–302.
11. Kaye FJ, Bunn PA, Steinberg SM, et al. A randomized trial comparing combination electron beam radiation and chemotherapy with topical therapy in the initial treatment of mycosis fungoides. N Engl J Med 1989; 321:1784–1790.
12. Heald P, Perez M, Christensen I, Dobbs N, McKiernan G, Edelson R. Photopheresis therapy of cutaneous T-cell lymphoma: the Yale–New Haven Hospital experience. Yale J Biol Med 1989; 62:629–638.
13. Vowels BR, Cassin M, Boufal M, Walsh L, Rook AH. Extracorporeal photochemotherapy induces production of tumor necrosis factor and IL-6 by adherent peripheral blood mononuclear cells. J Invest Dermatol 1991; 96:585.
14. Kahari V, Chen YQ, Su MW, Ramirez F, Uitto J. Tumor necrosis factor $\alpha$ and interferon-$\gamma$ suppress the activation of human type I collagen gene expression by transforming growth factor-$\beta$. J Clin Invest 1990; 86:1489–1495.
15. Kahaleh MB, LeRoy EC. Interleukin-2 in scleroderma: correlation of serum level with extent of skin involvement and disease duration. Ann Intern Med 1989; 110:446–450.
16. Kulozik M, Hogg A, Lankat-Buttgereit B, Krieg T. Co-localization of transforming growth factor $\beta 2$ with $\alpha 1$(I) procollagen mRNA in tissue sections of patients with systemic sclerosis. J Clin Invest 1990; 86:917–922.
17. Rook AH, Freundlich B, Nahass GT, et al. Treatment of autoimmune disease with extracorporeal photochemotherapy: progressive systemic sclerosis. Yale J Biol Med 1989; 62:639–645.
18. Vonderheid EC, Kang CK, Kadin M, Bigler RD, Griffin TD, Rogers. TJ. Ex-

tracorporeal photopheresis in psoriasis vulgaris: clinical and immunologic observations. J Am Acad Dermatol 1990; 23:703–712.
19. Rostami AM, Galetta S, Farber RE, et al. A double-blind clinical trial of extracorporeal photopheresis in chronic progressive multiple sclerosis. Neurology 1990; 40:393–394.
20. Wieselthier J, Yu T, Rothstein T, Koh H. Extracorporeal photopheresis for treatment of lymphoproliferative malignancies. J Invest Dermatol 1990; 94:591.
21. Bisaccia E, Berger C, Klainer AS. Extracorporeal photopheresis in the treatment of AIDS related complex: a pilot study. Ann Intern Med 1990; 113:270–275.
22. Song PS, Tapley KJ. Photochemistry and photobiology of psoralens. Photochem Photobiol 1979; 29:1177–1197.
23. Khavari PA, Edelson RL, Lider O, Gasparro FP, Weiner HL, Cohen IR. Specific vaccination against photoinactivated cloned T cells. Clin Res 1988; 36:662.
24. Perez M, Edelson R, LaRoche L, Berger C. Specific suppression of antiallograft immunity by immunization with syngeneic photoinactivated effector lymphocytes. J Invest Dermatol 1989; 92:669–676.
25. Grunfeld C, Palladino MA. Tumor necrosis factor: immunologic, antitumor, metabolic, and cardiovascular activities. Adv Intern Med 1990; 35:45–72.

# 21

# Lasers in Photomedicine

**Roy G. Geronemus and Robin Ashinoff**
*New York University School of Medicine, New York, New York*

The laser as a surgical and medical tool has become an increasingly important asset in the dermatologic armamentarium. New developments and improvements on older ideas have expanded and perfected the use of lasers in cutaneous diseases significantly. The term "laser" is an acronym for *l*ight *a*mplification by *s*timulated *e*mission of *r*adiation. Laser light is unique in that it is of one uniform wavelength, its photons of energy are synchronized in time and space or spatially coherent, and it is collimated, which means that the beams of light are parallel to one another. Lasers also produce a very intense beam of light that has been harnessed to produce desired clinical effects.

Goldman and his co-workers were the first to use laser light on human skin (1, 2). Over the last 30 years, physicians, scientists, and engineers have expanded our knowledge and clinical use of lasers so that now argon, carbon dioxide, and dye lasers are used routinely in most medical and surgical specialties.

The skin contains two chromophores that absorb various wavelengths of light selectively: oxyhemoglobin and melanin. Oxyhemoglobin has three absorption peaks: 418, 542, and 577 nm (3–5). Melanin absorbs broadly with a maximum absorption in the range from 300 to 500 nm, which then decreases steadily in the visible range.

Visible laser light is either transmitted down a fiberoptic cable or an articulated arm with multiple mirrors. An optical system of lenses enables the beam to be focused at a specific distance. The spot size of the laser can be constant or variable, depending on the laser. Other parameters important in understanding lasers are the duration of laser exposure, which is measured in seconds, milliseconds, or nanoseconds; whether the laser is pulsed or continuous or can be used in a gated or shuttered manner; the power as measured in watts or joules/sec; the power density or irradiance as measured in watts/cm$^2$; and the energy and energy density (fluence) as measured in joules and joules/cm$^2$, respectively. The fluence or energy density is probably the most clinically useful term as it relates the power density (watts/cm$^2$ on joules/sec/cm$^2$) to the time of laser exposure of the skin (3).

The gain medium is the substance, which may be a liquid, solid, or gas, that is energized within the laser cavity, amplified by reflecting mirrors in the laser cavity, and which then emits a specific wavelength of laser light. In the argon, carbon dioxide ($CO_2$), and excimer lasers, the gain medium is a gas; in the dye lasers, it is a liquid; and in the Q-switched ruby, neodymium–yttrium-aluminum-garnet (Nd-YAG), and copper vapor lasers, it is a solid (Table 1).

The different types of lasers also have varying modes of emission, which has a profound effect on the type of injury induced by each laser (Table 1). Continuous-wave lasers produce a continuous beam of laser light, and because of this feature, often produce nonspecific or unwanted injury to structures near those being treated. Continuous-wave lasers can be "shuttered" or "gated" by the use of mechanical shutters that can be open or closed intermittently to obstruct or allow the laser light to exit from the laser cavity. When the laser is used in a shuttered pattern, one can limit the amount of laser light exposure and, therefore, energy exposure to differing degrees over the continuous mode.

Pulsed lasers, on the other hand, produce extremely short pulses of light (in the nanosecond range) with relatively high peak power, often 100 times that of a

**Table 1** Lasers Used in Photomedicine

| Type | Gain Medium | Wavelength (nm) | Emission Modes |
|---|---|---|---|
| Argon-pumped tunable wave dye | Liquid: organic fluorescent dye in methanol | 488–630 | Continuous, shuttered |
| Flashlamp-pumped pulsed dye | Liquid: organic fluorescent dye in methanol | 577 or 585 | Pulsed |
| Q-switched ruby | Solid: ruby crystal | 694 | Q-switched |
| Carbon dioxide | Gas: $CO_2$ + nitrogen + helium | 10,600 | Continuous, shuttered, superpulsed |
| Argon | Gas: argon | 488–514 | Continuous, shuttered |
| Nd-YAG | Solid: yttrium aluminum crystal | 1,060 | Continuous, shuttered, pulsed |
| Copper vapor | Solid: copper | 511, 578 | Continuous, shuttered |
| Excimer | Gas: variable halide gases | 193, 248, 308, 350 | Pulsed |

continuous-wave laser. The short pulse duration accounts for the rather specific limited injury caused by these lasers (Table 1).

Superpulsing refers to $CO_2$ lasers that have the ability to produce a train of short pulses with high peak powers. This laser cannot produce a single pulse, however. This design was initiated to attempt to limit the amount of thermal damage caused by the $CO_2$ laser (Table 1).

The argon laser in the blue–green portion of the visible light spectrum has 80% of its emission between 488 and 514 nm and is a continuous type of laser. It can be used to treat vascular lesions as well as melanocytic lesions (Table 2). The $CO_2$ laser is a continuous-wave laser with an emission wave-length of 10,600 nm, has its energy absorbed by the water in skin, and can be used to vaporize or incise tissue (Table 2). The dye lasers with yellow light emissions of 577–585 nm can be used to treat a variety of vascular lesions. The flashlamp-pumped pulsed dye laser has become the treatment of choice for port wine stains, especially those occurring in children (Table 2). The continuous wave argon-pumped dye laser can also be used to treat various vascular lesions of the skin as well as numerous skin cancers, when utilized as part of photodynamic therapy (Table 2). Unlike the pulsed dye laser, the argon-pumped continuous dye laser has a greater propensity for injuring nearby tissue by virtue of its continuous emission mode (Table 1).

The copper vapor laser has been reported to be useful in the treatment of vascular lesions, and the excimer lasers using halide gases may have usefulness in the treatment of melanocytic processes and in performing fine surgery. The Nd-YAG laser is used for conditions in which a deeper level of destruction is necessary, since it penetrates to a depth of approximately 4–6 mm with wide diffusion (Table 2). Although it is similar to the carbon dioxide laser, the Nd-YAG laser has a wavelength of 1060 nm and is not specifically absorbed by a single cutaneous chromophore (Table 1).

The newest laser in clinical use is the Q-switched ruby laser. The "Q" stands for "quality factor," which means the rate of discharge of energy. Q-switched lasers have the ability to allow excessive energy to build up in the laser cavity and then, by the use of an electromagnetic or chemical switch, to allow very short, single, extremely high power pulses to be emitted (Table 1). The Q-switched ruby laser has already shown good to excellent results in the treatment of tattoos and shows promise in the treatment of pigmented lesions (Table 2).

With some familiarity with the terms used in speaking about lasers, one can more fully understand the capacities and shortcomings of our more commonly used lasers.

## I. DYE LASER

The theory of selective photothermolysis proposed by Anderson and Parrish (6) brought forth the idea that by carefully selecting the appropriate wavelength of

**Table 2** Clinical Uses and Value of Lasers Used in Photomedicine

| Laser | Clinical Applications | Advantages | Disadvantages |
|---|---|---|---|
| Argon-pumped tunable wave dye | Vascular lesions; photodynamic therapy | Photocoagulation of larger vascular lesions without scarring | Hypertrophic scarring in PWS in children; dermal fibrosis |
| Flashlamp-pumped pulsed dye | Vascular lesions, especially port wine stains (PWS) | Treatment of choice for PWS in children | Treated area turns grey–black; multiple treatments required |
| Q-switched ruby | Tattoos; pigmented lesions | Removal of tattoos without scarring; promising for pigmented lesions | Multiple treatments; incomplete removal; color dependent |
| Carbon dioxide | Vaporization; cutting; coagulation | Hemostasis; lymphatics and nerves sealed | Nonspecific thermal injury; decreased wound tensile strength |
| Argon | Vascular and pigmented lesions | Similar to argon-pumped tunable wave dye laser | Similar to argon-pumped tunable wave dye laser |
| Nd-YAG | Deeper coagulation | Can be used for deeper lesions | Cutaneous scarring |
| Copper vapor | Vascular lesions | Data preliminary on PWS and telangiectases; area does not turn black | More operator dependent; skill required |
| Excimer | Superficial cutting | Extremely delicate surgery | Ultraviolet emission range: mutagenic potential |

light, pulse duration, and energy density, one can achieve selective destruction of multicellular organs (such as blood vessels) or subcellular organelles, such as melanosomes. For cutaneous blood vessels, oxyhemoglobin is the chromophore to be targeted and it has three distinct absorption peaks: 418 nm, 542 nm, and 577 nm (3, 4, 5). At shorter wavelengths of light, epidermal melanin tends to compete with the underlying dermal vessels for the absorption of laser light. However, even at 577 nm epidermal melanin still absorbs laser light and can be the cause of epidermal damage, especially in more darkly pigmented skin (7).

Continuous-wave lasers, like the argon and carbon dioxide lasers, are likely to cause dermal scarring because the duration of laser exposure is not less than the thermal relaxation time of cutaneous vessels (8, 9). The thermal relaxation time of a target vessel is the time it takes for that vessel to cool to 50% of its normal temperature after laser exposure. If the duration of laser exposure (pulsewidth) is kept below the thermal relaxation time of a cutaneous vessel, the amount of heat conducted to surrounding structures (such a dermal collagen) is kept to a minimum. In this way, very selective vascular injury can be induced with minimal dermal and epidermal disruption, resulting in less clinical scarring. Depending on their size, the thermal relaxation time for superficial cutaneous vessels is in the tens of $\mu$sec for capillaries, and is about 1 msec for arterioles and venules (6, 10).

A wavelength of 577 nm allows for a depth of injury of about 0.5 mm from the dermal–epidermal junction and less competition by melanin than at the lower wavelengths of oxyhemoglobin absorption. Increasing the laser wavelength beyond 577 nm greatly increases the exposure dose necessary for vascular injury. However, it has been found more recently that light of a wavelength of 585 nm and a pulsewidth of 360 $\mu$sec can produce vascular injury to a depth of 1.2 mm without losing specificity for cutaneous blood vessels at usual exposure doses (4).

Purpura is the clinical response immediately following pulsed dye laser threshold doses regardless of the pulsewidth of laser exposure (3, 8, 11–19). At very short pulsewidths of 1 $\mu$sec, one sees microvascular rupture and hemorrhage, while at longer durations (up to 450 $\mu$sec), an intravascular coagulum is produced without extravasation (4, 19, 20). In addition, the exposure dose necessary for damage increases with increasing pulsewidth. However, perivascular collagen is histologically altered by very short pulsewidths, as opposed to longer laser pulses. Several mechanisms have been suggested to explain these phenomena. At very short pulsewidths of 1 $\mu$sec, the laser energy is still mostly confined to erythrocytes within the vessel. Red blood cells have been shown to be altered by pulsed dye laser irradiation in several ways. One of these alterations includes the appearance of electronlucent areas within red blood cells, which possibly represent steam formation within the cell. Endothelial cells and pericytes also display considerable damage, while surrounding cells such as mast cells, neutrophils, macrophages, fibroblasts, as well as collagen bundles appear unaffected by laser irradiation at pulsewidths of 1.5 $\mu$sec (21). It has been proposed that vessel rupture and subsequent hemorrhage may be due to microvaporization of erythrocytes. This mechanism is also suggested by studies that have shown that the skin temperature has an effect on the clinical purpura threshold (22, 23). Another possible mechanism to explain the hemorrhage seen with very short pulsewidths is rapid thermal expansion inducing pressure waves capable of causing rupture of the vessel (24).

Glassberg et al. (13) utilized the pulsed dye laser in vitro to irradiate endothelial cell monolayers with and without overlying red blood cells to discern the target of pulsed dye laser energy. In the cocultures, laser irradiation impaired endothelial cell DNA replication when cells were not covered by the red blood cells, but the covered endothelial cells' DNA replication was not affected. Erythrocytes are clearly one target of pulsed dye laser irradiation, as evidenced by lysis of these cells and the subsequent release of hemoglobin. Endothelial cells could also be a target for pulsed dye laser energy in vivo, however. In vitro, the coculture of erythrocytes and endothelial cells is an open system. This open system allows heat and vaporized steam to escape, while in vivo the dissipation of heat within a vessel might subject vascular endothelial cells to direct damage by heat conduction and mechanical trauma.

It appears that platelet thrombi also may play an important role in the formation of an intravascular coagulum necessary to the clinical response of vascular lesions treated with the pulsed dye laser. Ashinoff and Geronemus reported a case of thrombocytopenia–absent radii syndrome in an infant with a port wine stain of his face who did not respond to the pulsed dye laser after three attempts at varied energy densities (25). The patient's severe thrombocytopenia likely prevented the formation of platelet thrombi and therefore no clinical lightening was ever noted.

As the pulsewidth is increased beyond the thermal relaxation time of a cutaneous vessel, the heating of that vessel is less efficient and heat is lost to the surrounding tissues. Therefore, a longer pulse duration is necessary to achieve the clinical purpura threshold. In addition, a longer pulsewidth of 360 μsec allows for a slower heating and the formation of an intravascular coagulum with less mechanical vessel damage.

Histopathologic studies of port wine stains treated with the pulsed dye laser (17, 21) have shown that immediately following laser exposure there are multiple areas of agglutinated erythrocytes, fibrin, and platelet thrombi that extend to the midreticular dermis. At 24 hr following laser irradiation, there is persistent fibrin, platelet thrombi, and polymorphonuclear leukocytes within affected vessels and no discernible endothelial cells. Unaffected vessels also show edema and fibrillary degeneration of their walls. By 1 month after dye laser irradiation, no abnormally dilated port wine stain blood vessels are seen. The blood vessels in laser-irradiated sites appear to have thicker walls, smaller diameters, and more prominent endothelial cells and pericytes than normal adjacent vessels (8). Endothelial cells appear to be normal ultrastructurally. Both intra- and extracellularly, hemosiderin is found around some vessels in the papillary and superficial reticular dermis. However, the epidermis appears normal by both light and electron microscopic examination.

These results are very unlike those seen with either argon or carbon dioxide laser treatment of port wine stains (8). In both the argon and carbon dioxide laser-treated port wine stains, there is marked epidermal necrosis, denatured collagen, necrotic appendages and, finally, scar formation with fibrosis in the papillary and upper reticular dermis. This correlates with what one sees clinically. The pulsed dye laser does not cause clinical scarring when appropriate energy densities are used, while carbon dioxide and argon laser treatment of port wine stains do and are not recommended for the treatment of macular port wine stains.

Port wine stains have been shown to have few perivascular mast cells (26). Posttreatment biopsies of port wine stains, however, reveal that by 1 year after pulsed dye laer treatment the abnormal ectatic vessels are not only replaced by small arterioles and venules but there are also numerous activated mast cells seen perivascularly. Since mast cells make heparin and possibly other angiogenic factors, it seems that mast cells probably play an important role in the formation of normal vasculature in port wine stains treated with the pulsed dye laser.

The dye utilized in the dye laser is an alcohol in a fluorescent base that can be excited by a flashlamp or an argon laser. The argon-pumped dye laser is a continuous-wave laser that does not produce optimal vascular damage. However, some clinicians have found that by using a very small spot size and low energy, one can achieve good clinical results in treating port wine stains with the argon-pumped tunable dye laser (27).

The flashlamp-pumped dye laser is the treatment of choice for port wine stains, especially in children (Figs. 1 and 3; Fig. 2, see color plate). It is also an appropriate treatment modality for vascular spiders, actinically induced telangiectases, erythema and ectases due to rosacea, and poikiloderma of Civatte (16). By choosing an appropriate energy dose, one can achieve excellent lesion lightening with a very low risk of hyper- or hypopigmentation or scarring as evidenced by changes in skin texture. Immediately following treatment, the area turns grey–black with surrounding erythema. Over the next 6–8 weeks, the treated areas gradually lighten. Often, multiple treatment sessions are required to lighten certain lesions significantly, but even multiple retreatments do not result in scarring (Table 2).

Numerous clinical trials of the pulsed dye laser in the treatment of port wine stains have shown that it is the treatment of choice for macular lesions, especially those in children (9, 11–14, 17–19). Garden et al. (19) found that 73% of 52 patients with port wine stains experienced 50% or greater lightening of their lesions. Tan et al. (12) reported on 35 children aged less than 14 years with pink to red macular port wine stains. After an average of 6.5 treatments, lesions in all patients reportedly cleared. Reyes and Geronemus (28) reported on 73 patients aged less than 14 years with macular port wine stains treated with the pulsed dye laser. The average number of treatments was two and 87% of the patients had a 50% or greater lightening of their lesions.

In all these clinical studies, complications have been minor and transitory. These included cutaneous depressions and hyper- and hypopigmentation.

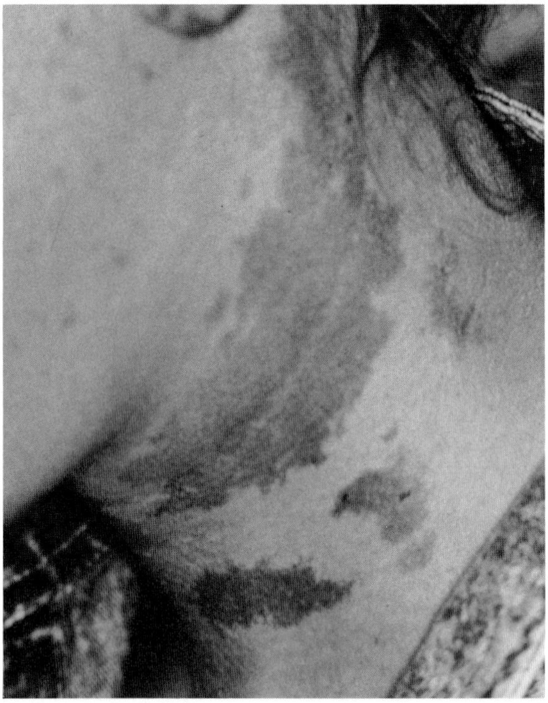

(A)

**Figure 1** A. A patchy flat port wine stain of the left side of the neck prior to treatment. B. Two years later, after a total of 10 treatment sessions, using a 585 nm pulsed dye laser and 7.0 J/cm$^2$, the port wine stain is almost completely gone with no textural or pigmentary change.

The pulsed dye laser has even shown promise in the treatment of capillary hemangiomas (29), but its limited depth of penetration will likely limit its effectiveness in the treatment of cutaneous hemangiomas (Fig. 4, see color plate).

The greatest clinical impact of the pulsed dye laser has been the ability to treat very young children with port wine stains with great efficacy and minimal morbidity. Fewer treatments and a more complete response are likely if treatments are begun earlier, before the lesion grows in surface area and depth and becomes hypertrophic. The earliest possible treatment also enables the removal of the port wine stain before the child reaches school age and suffers the psychological burden associated with having a port wine stain (30).

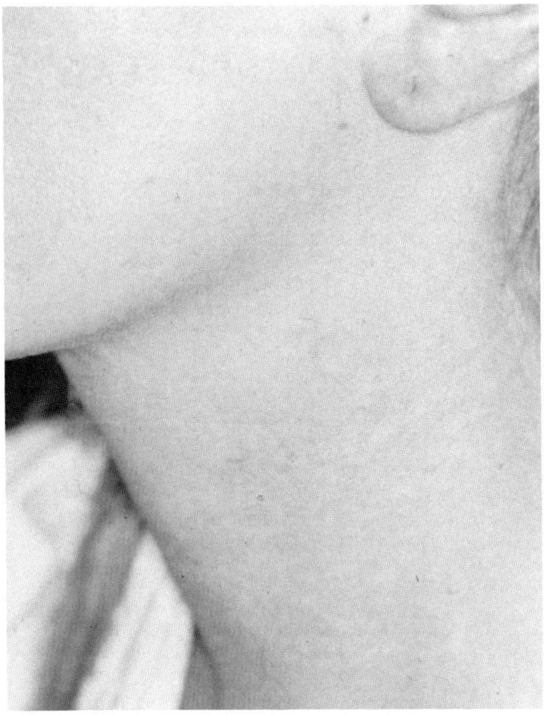

(B)

**Figure 1** *(continued)*

## II. Q-SWITCHED RUBY LASER

The first laser ever developed was the ruby laser, introduced by T. H. Maiman in 1960 (31). Soon this laser was adopted into the medical profession. In 1960 Goldman first demonstrated that a ruby laser with a pulse duration of 500 μsec could selectively affect melanocytes and melanosomes (32). However, it was soon learned that unless pulse duration was limited, excessive cutaneous damage would occur. Melanin has a broad spectrum of absorption throughout the neuroltraviolet, visible, and near-infrared wavelengths. At a wavelength of 694 nm, ruby laser light causes injury to a depth of about 0.1 mm into skin, has minimal absorption by hemoglobin, and so allows melanocytes to be selectively damaged without direct vascular injury. Melanosomes are the primary target of Q-switched ruby laser energy (33, 34). The calculated thermal relaxation time of a melanosome is approximately 1 μsec. Melanosomal disruption is induced by 10–12 nsec Q-switched Nd-YAG, and 40 nsec Q-switched ruby lasers because

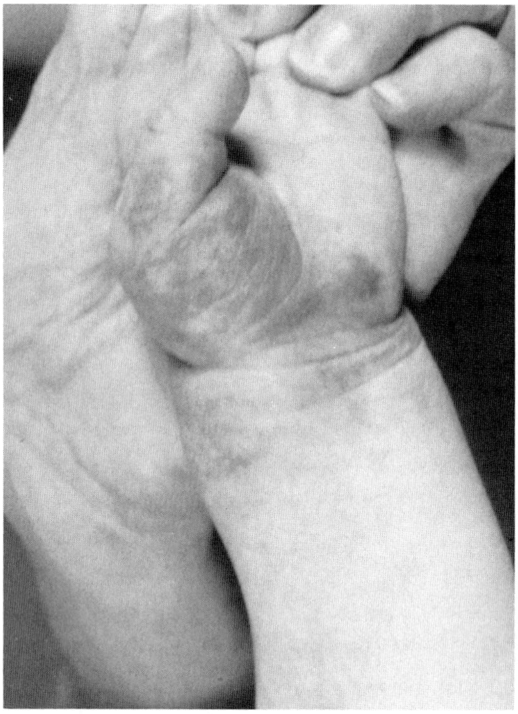

(A)

**Figure 3** A. An immature, flat, red port wine stain on the arm and hand of a young child prior to treatment. B. There is significant lightening after 7 months and 4 treatment sessions using a 585 nm pulsed dye laser and 6.25 $J/cm^2$.

these lasers have a pulse duration less than the thermal relaxation time of a melanosome (35).

The earliest studies of the pulsed ruby laser by Goldman utilized a laser with a pulse duration of 1 μsec, which produced nonspecific superficial and deep necrosis and a broad zone of degenerated collagen when used to treat tattoos (32). Q-switching is a technique that allows storage of laser energy in the laser cavity before it is released, so that high peak power outputs with a pulse duration in the order of nanoseconds can be produced. Yutes et al. (36) were able to show as early as 1967 that by using a Q-switched ruby laser rather than a long pulse ruby laser (1.8 msec pulse duration), treatment of black tattoos could be effective in pigment removal without scarring. Others have substantiated these results (37–43).

More recent studies have focused on the mechanisms, nature, and extent of

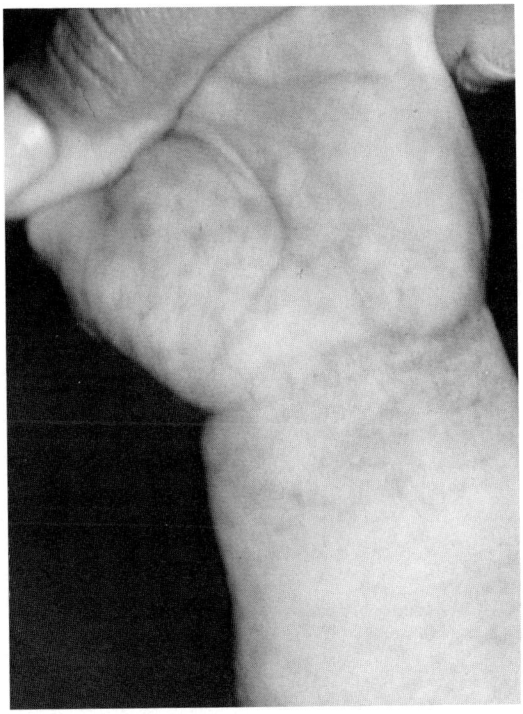

(B)

**Figure 3** *(continued)*

pigmented cell injury caused by the Q-switched ruby laser (44). Studies in guinea pigs have shown that the laser energy is selectively absorbed by pigment in the skin, since albino animals fail to develop clinical or histopathologic evidence of injury after irradiation at any exposure doses. Brown or black pigs develop an immediate white macule at the site of laser irradiation that gradually fades over 30 min after treatment with dosages greater than or equal to the clinical threshold dosage, which is the dosage that causes a visible, immediate skin change. Depigmentation may develop and is transient, lasting 4–8 weeks in pigmented pigs irradiated with dosages equal to the immediate clinical threshold dosage. Those pigmented animals irradiated with greater than the clinical threshold dosage develop permanent leukotrichia: hairs in the treated site remain white for at least 4 months after irradiation. The pigmented pigs receiving less than clinical threshold dosages develop hyperpigmentation at 24 and 48 hr after irradiation.

Histologic correlations show that in albino guinea pigs the only immediate change is intercellular edema and a slight disarray of keratinocytes only seen at the higher exposure dosages. By 24 hr after exposure, there are no changes noticeable, and all internal organelles, such as nuclei, melanosomes, and mitochondria, are present with normal internal structure as confirmed by electron microscopic examination.

In contrast, the epidermis of pigmented guinea pigs shows reproducible changes after exposure, and these are more marked in black than in the brown animals. The changes are also more extreme, as the incident exposure increases. Subepidermal vesicles are seen on electron microscopic examination and are found to be at the level of the lamina lucida. Pigment-laden cells in the basal layer are vacuolated, and there is condensation of pigment and nuclear material at the periphery of both keratinocytes and melanocytes. These cells have been called "ring cells." At the threshold dosage, the melanosome is the only organelle showing changes; however, at higher dosages, nuclear damage is also seen. With increasing exposures, melanosomal changes begin as internal electron lucency and focal disruptions of the outer melanosomal membrane ("popcorn melanosomes") to complete melanosomal disruption. At 4 months after irradiation, biopsies of those areas treated with above-threshold dosages reveal a normally pigmented epidermis, but there are decreased or no melanocytes in the follicular epithelium, which correlates with the clinical finding of white hairs in these areas.

Another interesting finding is the fact that at 48 hr after exposure in those pigmented animals receiving subthreshold doses, there is an increase in keratinocytic pigmentation, and melanocytic dendrites are longer, wider, and more highly pigmented. This increase in pigmentation appears to be transient and is not apparent by 1 week after irradiation. This finding may indicate that low levels of ruby laser light, which are incapable of destroying melanosomes, may, in fact, stimulate melanocytes.

These studies in guinea pigs demonstrate that selective photothermolysis is possible with the Q-switched ruby laser. Based on the fact that only pigmented guinea pigs show clinical and microscopic evidence of disruption of melanosomes, melanin has been shown to be the selective chromophore of ruby laser light absorption, and the melanosome is now known to be the principal organelle affected. Nuclear injury is postulated to be caused by thermomechanical injury initiated by nearby melanosomes. This mechanism of thermomechanical damage is also postulated to account for the separation at the lamina lucida seen in the area of melanosomal damage. Since the dermoepidermal junction is strongly affected by changes in temperature, the thermal theory of damage could account for the separations seen.

The transient whitening response seen in skin treated with threshold and above-threshold dosages is thought to be the result of empty spaces within ruptured melanosomes, which represent small vapor cavities induced by steam and other thermal products. These cavities could scatter incident light enough to account for the transient whitening of skin that is seen.

Preliminary studies of the Q-switched ruby laser in the treatment of pigmented lesions such as epidermal nevi, nevus of Ota, lentigines, Becker's nevus, café-au-lait spots, and tattoos have shown very encouraging results (43) (Table 2).

A recent study of patients with 57 tattoos revealed significant lightening or total clearing in 78% of the amateur tattoos and 23% of the professional tattoos (42) (Table 2). The patients were treated at a mean interval of 3 weeks, and the optimal energy density was found to be between 4 and 8 J/cm$^2$. Responses were related to the energy density used, with better responses seen at higher densities. Blue–black tattoos responded the best, while green and yellow sites responded less well, and red areas responded poorly or not at all. The clinical threshold response is whitening, as seen in guinea pig studies.

The authors did perform histologic and electron microscopic correlations. However, there was extreme intersubject and even intrasubject variability. Even within the same tattoo, there was great variation in the density and depth of tattoo pigment. Before treatment, tattoo pigment was seen as electron dense clumps in discrete intracellular granules in fibroblasts, macrophages, and mast cells, especially perivascularly. The professional tattoos showed most of the pigment located superficially, while the amateur tattoos displayed pigment throughout the dermis and even in the subcutaneous fat.

After Q-switched ruby laser treatment, there were dosage-related changes to pigment-containing cells similar to changes seen in guinea pig skin. Vacuolization of keratinocytes and melanocytes with melanosomal rupture was also seen. Immediately following treatment, pigment particles appeared changed, were noticeably smaller, and became extracellular. Adjacent cells without pigment were unaltered, and most of the dermal collagen appeared normal as well. However, there was little correlation between clinical response and the amount of residual tattoo pigment. No fibrosis was seen in specimens taken after 5 months. Hypopigmentation was the most common side effect of treatment and was transient in most subjects, returning to normal within 4–12 months. However, 1 year after Q-switched ruby laser treatment, 4 of 10 patients had persistent confettilike hypopigmentation.

The authors speculate that perhaps a decrease in tattoo particle size or alteration of pigment granule structure is enough to make the pigment less noticeable. Or perhaps the pigment is eliminated in part by rephagocytosis or transepidermal loss. These, however, are all speculations, and the exact mechanisms of Q-switched ruby laser removal of tatoo pigment are unknown.

The ruby laser has also been used in the past to treat pigmented lesions such as melanoma and benign nevi, with variable results (45–47).

## III. CARBON DIOXIDE LASER

Carbon dioxide lasers are commonly used by dermatologists and other specialists for the treatment of a variety of cutaneous disorders. The carbon dioxide laser is a continuous-wave laser that emits infrared radiation at 10,600 nm (Table 1). Since the infrared segment of the electromagnetic spectrum is invisible, a helium–neon laser at 632 nm of very low intensity is utilized as an aiming beam.

The light of the carbon dioxide laser is absorbed primarily by water, thus there is little thermal scatter within the skin. The depth of penetration is variable, depending upon the energy fluence delivered by this laser. Unlike that with the flashlamp-pumped pulsed dye laser and the Q-switched ruby laser, the thermal damage created by the carbon dioxide laser is nonspecific; however, this damage can be limited by the degree of thermal damage produced at various energy fluences. Limiting thermal damage between 50° and 60°C will minimize the amount of irreversible denaturation of tissue proteins. Temperatures above 50–60°C lead to irreversible denaturation, increasing the possibility of fibrosis and subsequent clinical scarring. Vaporization of tissue occurs when the tissue water reaches over 100°C, resulting in conversion of the tissue into a plume consisting of steam and charred tissue. The extent of thermal damage to the surrounding tissue can be limited to 600 $\mu$m or less, unless very high energy fluences are utilized. The effect of the carbon dioxide laser upon tissue is dependent upon the power density of laser light delivered to the skin surface. Lower power densities can be obtained by defocusing the beam and, thus, tissue can be superficially vaporized with a power output ranging from 250 to 750 watts/cm$^2$. Tissue vaporization in this range can lead to a depth of penetration of less than 0.1 mm. This same laser can be utilized at much higher power densities of 100,000–300,000 watts/cm$^2$ with a spot size of 0.1 mm to create a focused beam, which can allow one to cut tissue as with a scalpel.

Although the carbon dioxide laser is a continuous-wave laser, its light can be shuttered electronically to emit individual pulses ranging from 0.01 sec to several seconds. The carbon dioxide laser light can also be superpulsed, which is a method of minimizing thermal damage by delivering a train of high-energy pulses with pulsewidth ranging from 0.1 to 100 msec with a repetition rate of 10–1,000 pulses per second. The very high peak power can be obtained up to several thousand watts per pulse, which, in conjunction with the short pulsewidth, allows for minimal thermal injury. Thermal damage is minimized to such a degree that the hemostatic properties usually found with the carbon dioxide laser are not noted when the super pulsed mode is utilized. The long-term clinical benefits of super-pulsed application of the carbon dioxide laser have also not yet been demonstrated.

The formation of steam that results upon the heating of tissue beyond 100°C requires the use of a suction or smoke evacuation system to remove this laser-produced plume. A report by Garden et al. demonstrated the presence of DNA from the papilloma virus following the treatment of warts (48). The recovery of these viral particles within the laser plume has reinforced the need for appropriate suction and protection of the surgeon and ancillary personnel with masks during removal of viral particles. This may limit the therapeutic application of this laser.

The carbon dioxide laser has the following unique therapeutic characteristics that affect the treatment of cutaneous disease: the laser allows for hemostasis by sealing small blood vessels (0.5 mm in diameter or less); since the laser does not contact the skin surface, there is minimal risk of postoperative infection as the heat of the laser sterilizes the wound; the minimal degree of the thermal damage diminishes the incidence of scarring, and histologic margins are preserved to a greater degree than with thermoelectrosurgery and other thermal mechanisms of tissue removal; excision with the carbon dioxide laser seals lymphatic vessels, which may minimize postoperative edema and the lymphatic spread of tumor cells; and nerve endings are sealed, which minimizes postoperative discomfort (Table 2). In the focused or cutting mode with its high-power densities and small focal point, the laser is used as a cutting tool when hemostasis is required, if electrosurgery is contraindicated, in the presence of infected surgical wounds, and for the excision of highly vascular tissue.

The carbon dioxide laser is used extensively for the excision of malignant neoplasms. Thermal damage, however, is noted in the tissue margins of the excisions, which limits the ability to evaluate the histologic margins (Table 2). This may occasionally lead to false-positive histologic interpretations of the tumor margins. Mohs' micrographic surgery has been performed following the excision of tissue with the carbon dioxide laser. The limitation of this technique is that since the laser is absorbed in water, there is often difficulty with the adherence of the tissue to the slide during the preparation of frozen sections. The concomitant use of albumin to enhance the adherence of the tissue to the slide may obviate this problem. Carbon dioxide laser excision has not been established to be a preferred method of skin cancer excision, as opposed to routine scalpel excision. Comparable cure rates with carbon dioxide and laser scalpel excision of malignant melanoma have been found by Bandieramonte et al. (49). However, Bilik et al. reported that malignant melanomas excised with the carbon dioxide laser actually had a higher recurrence rate than those excised with the scalpel (50).

Initial reports utilizing the carbon dioxide laser for the treatment of keloids were favorable (51); however, long-term follow-up and additional reports have refuted this theory (52). In our experience, the carbon dioxide laser has been useful to excise extensive keloids in a bloodless fashion and to remove them when they are infected; however, it has only been of benefit when used in conjunction with postoperative injection of corticosteroids into the wound and postoperative pressure dressings.

Tensile strength following excision using the carbon dioxide laser is less than in scalpel-excised wounds and, thus, sutures must remain in place longer (Table 2). The successful reconstruction of post-Mohs surgery wounds, in which the carbon dioxide laser was utilized to excise the tissue, has been reported by Bailin (53). However, one study of full-thickness skin grafting following carbon dioxide laser excision of malignant melanomas found a higher incidence of graft failure (54).

Vaporization of tissue in the defocused mode has provided the treatment of a wide variety of cutaneous disorders. A common use for the carbon dioxide laser is the treatment of warts of all types. The effectiveness of this laser for this purpose has not been well established; however, successes ranging from 80 to 95% have been noted for treatment of common warts (55). The carbon dioxide laser remains particularly helpful in the treatment of recalcitrant, hyperkeratotic, or painful warts. In the treatment of periungual warts, particular care must be taken to avoid thermal damage to the nail matrix, which could result in a permanent dystrophy of the nail. Similar to the other surgical modalities, treatment of warts with this technique in patients with diffuse involvement of the papilloma virus, with or without immunosuppression, is often unsuccessful.

Treatment of disorders involving the genitalia has been successful: balanitis xerotica obliterans (56), Zoon's balanitis (57), erythroplasia of Queyrat (58), and lichen planus (59). Because these mucosal surfaces heal well with minimal postoperative pain and morbidity, the carbon dioxide laser should be considered a reasonable treatment for these entities.

Use of the carbon dioxide laser for the treatment of actinic cheilitis has been particularly successful (60). This approach results in significant success with minimal morbidity compared with other therapeutic approaches. The laser vaporizes abnormal tissue at very low power densities and healing takes place within 2 or 3 weeks.

Vaporization of various aspects of the nail unit and periungual tissue has been performed successfully with the carbon dioxide laser. Precise matrixectomies and removal of digital mucous cysts have been reported (61, 62). Podiatrists have commonly used this laser for the treatment of onychomycosis; however, its efficacy has not been documented.

Epidermal nevi of the verrucous type have responded well to the carbon dioxide laser; however, scarring is possible if vaporization extends into the deeper papillary and reticular dermis (63). Recurrences of extensive epidermal nevi have also been noted. Various tumors of the skin have responded well to vaporization with the carbon dioxide laser, including rhinophyma (64), adenoma sebaceum (65), multiple syringomas (66), superficial basal cell carcinomas, trichoepitheliomas, hydrocystomas, lymphangiomas, actinic keratoses, ephelides, seborrheic keratoses, neurofibromas, and xanthlomas (64–68). Controlled studies do not currently exist to demonstrate the efficacy of this technique compared to other surgical modalities.

Tattoos have been removed with the carbon dioxide laser with some success; however, tatoo pigment is routinely replaced with a scar (69), although the scarring may be minimized by the use of chemical exfoliation postoperatively (70).

Although the carbon dioxide laser has been reported to be of benefit in the treatment of port wine stains, it is not considered the treatment of choice (71). It is a reasonable therapeutic alternative for very hypertrophic port wine stains in which the pulsed dye or other lasers capable of treating vascular lesions cannot provide an adequate depth of vascular injury. Since the vascular tissue is routinely replaced with the scar when the carbon dioxide laser is utilized, its use for the treatment of vascular lesions must be considered carefully.

Dover et al. have reported on the use of low-power-density carbon dioxide laser vaporization for treatment of solar lentigines (72). This technique allows one to limit thermal damage to the epidermis. Low-power vaporization may be of benefit for superficial lesions, such as flat warts, for which extensive thermal damage is not required or desired.

The carbon dioxide laser can sometimes be used in both the focused and defocused mode to treat the same lesions. For example, extensive rhinophyma might be debulked while utilizing the carbon dioxide laser to excise the excess tissues and then the beam is defocused to vaporize the nose to the desired shape (Fig. 5). The carbon dioxide laser could likewise excise full-thickness tumors and then be utilized to vaporize the peripheral margins where in situ histologic changes have been found.

## IV. ARGON LASER

The argon laser is a continuous-wave laser emitting light at 6 major different wavelengths from 457 to 514.5 nm. This light is emitted in the visible blue–green portion of the electromagnetic spectrum. This light can be "gated" or "shuttered" to deliver individual pulses or the beam of light can be emitted continuously (Table 1). There is significant absorption by oxyhemoglobin with the argon laser; however, there is competitive absorption by melanin, which has a broad absorption spectrum ranging from 300 to 500 nm. The presence of melanin can sometimes prevent absorption of the argon laser light by oxyhemoglobin, particularly in persons with darkly pigmented skin. The effect upon tissue in the treatment of vascular lesions is that of photocoagulation. Thermal injury occurs in the upper 1 mm of the dermis. Unlike the precise vascular injury found with the flashlamp-pumped pulsed dye laser, there is epidermal and papillary dermal damage surrounding the blood vessel, which results in the presence of

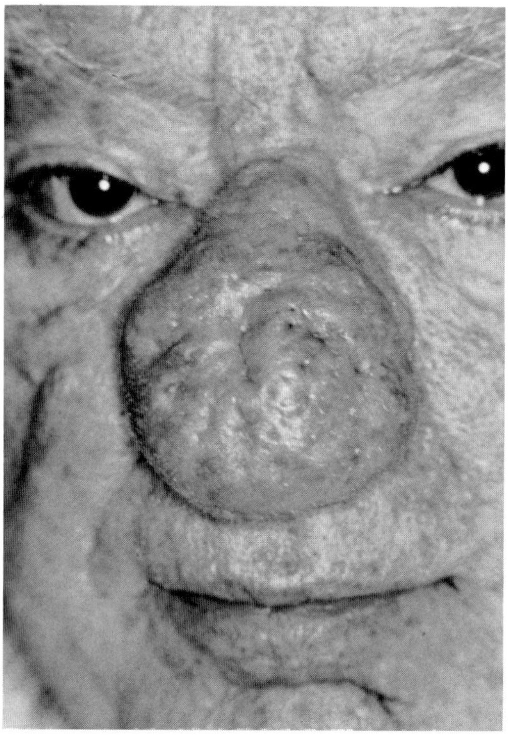
(A)

**Figure 5** A. Disfiguring rhinophyma prior to treatment. The nodular areas were excised using the $CO_2$ laser in the focused mode with a power density of 1911 W/cm$^2$. The final remodeling was done with a defocused beam and a power density of 478 W/cm$^2$. B. An excellent cosmetic result 3 months later, but some scarring is noted. (from Ref. 74.)

fibrosis histologically (Table 2). This laser has been utilized clinically for the treatment of port wine stains, telangiectases, low-flow vascular malformations, as well as tumors that contain a vascular component. It has also been utilized to treat some pigmented lesions, in view of its absorption by melanin.

Prior to the advent of the flashlamp-pumped pulsed dye laser, the argon laser was utilized as the treatment of choice for port wine stains (73). While beneficial lightening was noted, permanent loss of pigmentation was found in a minimum of 20% of patients and incidences of hypertrophic scarring have been reported to be at 5% (74) (Table 2). Certain anatomical areas are more prone to hypertrophic scarring: the upper lip is the area of greatest concern for abnormal healing. In view of the relatively low safety threshold in the treatment of port wine stains in children, the use of the argon laser is presently limited mostly to the treatment of port wine stains in adults with very hypertrophic lesions. It is still considered the

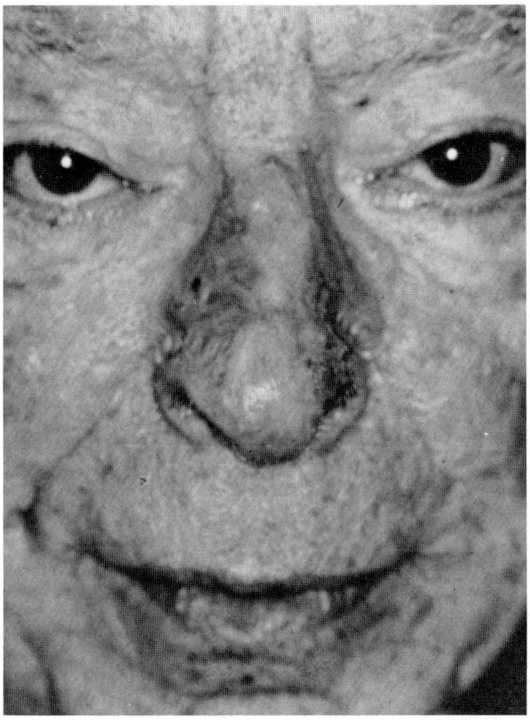

(B)

**Figure 5** *(continued)*

treatment of choice by many physicians for very hypertrophic port wine stains and for nodular lesions that occur within otherwise macular vascular birthmarks.

Prior to treatment of a port wine stain in its entirety, a test treatment is usually performed with different energy densities to help determine the best therapeutic dosage. The clinical response to the argon laser is dependent upon the fibrosis that develops around the blood vessel wall. Thus it is important to wait approximately 4 months following the test treatment to determine the presence or absence of textural change of the skin.

Various types of telangiectases have been successfully treated with the argon laser. Large-diameter vessels on the nose and cheeks are good examples of the type of lesion that will respond well to argon laser photocoagulation. Other vascular disorders that respond well to the argon laser include postrhinoplasty erythema, cherry hemangiomas, venous lakes, pyogenic granulomas, and angiokeratomas. Various tumors of the vascular component, including Kaposi's sarcoma, angiolymphoid hyperplasia, glomus tumors, and some adenoma sebaceum, contain a large enough vascular component within the lesion to be a

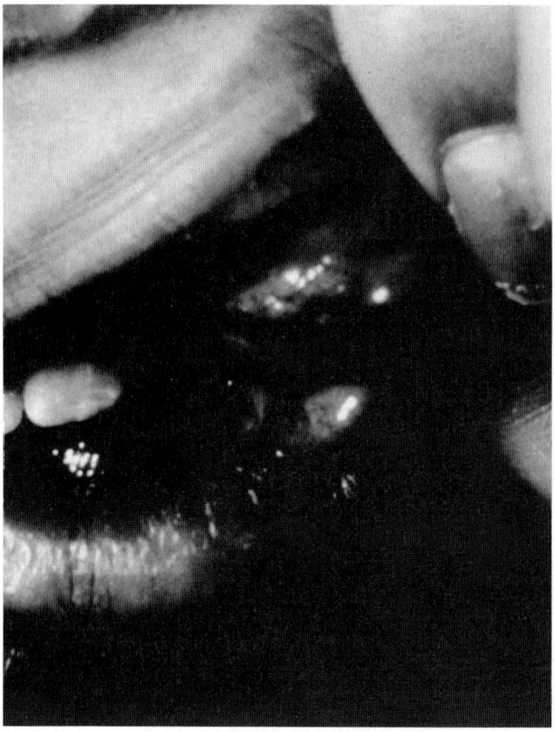
(A)

**Figure 6** A. A venous malformation involving the upper lip before treatment with the argon laser with a continuous beam and a power density of 64 W/cm$^2$. B. Eight weeks after treatment there is marked resolution of the lesion.

clinical target for the argon laser (75–77). The treatment of Kaposi's sarcoma with this laser is best suited in lesions that are isolated and are either macular or slightly nodular in their morphology.

The argon laser works particularly well for "low-flow" vascular malformations, particularly of the venous type. Unless there is very deep subcutaneous involvement of the venous malformation, these superficial lesions should respond quite dramatically with immediate blanching of the lesion. The mucosal surfaces have a tendency to respond quite well to this technique when treatment is indicated (Fig. 6).

Use of the argon laser for pigmented lesions has been well demonstrated for the treatment of lentigines and other epidermal and dermal processes. As with the reports from treatment of vascular disorders of the skin, permanent pigmentary change is possible with this therapeutic technique. Arndt reported the successful treatment of malignant melanoma in situ on the nose with the argon laser, only to

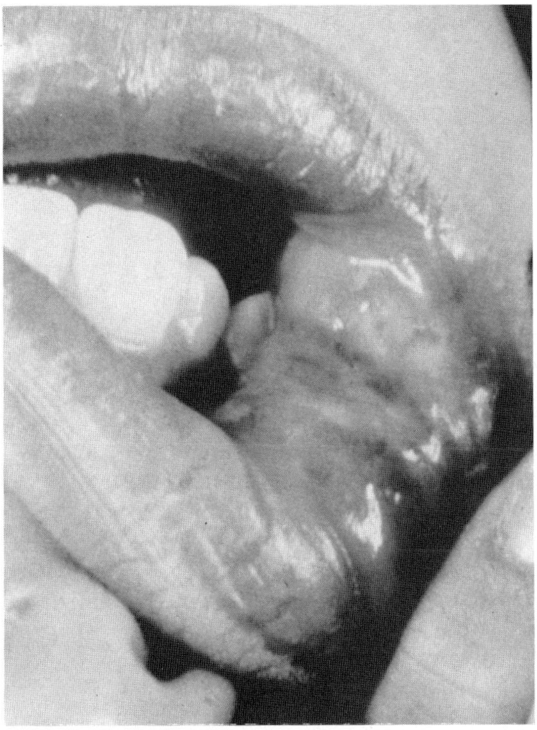

(B)

**Figure 6** *(continued)*

find a recurrence several years later (78, 79). The argon laser should not be considered an effective treatment for a malignant melanoma in which surgical margins are required.

Other pigmented lesions have been treated with the argon laser including nevus of Ota, labial lentigines, melasma, and Becker's nevus (80). The mechanism of pigment destruction in these cases has not been proven to be chromophore-specific and probably represents a nonspecific photothermal effect.

## V. SUMMARY

Research efforts on multiple fronts have allowed the field of laser photobiology to soar over the past 30 years. Clinical lasers are now available that can treat various skin conditions that previously could not be treated or not treated effectively.

The theory of selective photothermolysis has been the basis for the development of lasers targeted to treat specific skin conditions. The development of the pulsed dye laser has probably been the single most important, clinically applicable accomplishment in the last decade of laser advancements. For the first time, port wine stains in children and adults can be treated without cutaneous scarring or permanent pigmentary changes.

In the future, additional concerted basic and clinical research efforts will undoubtedly increase our knowledge of cutaneous biological processes and enable us to utilize leaser to treat additional cutaneous disorders safely and effectively.

## REFERENCES

1. Goldman L, Blanry DJ, Kindel DJ, et al. Pathology of the effect of the laser beam on the skin. Nature 1963; 197:192.
2. Goldman L, Blaney DJ, Kindel DJ, et al. Effect of the laser beam on the skin. J Invest Dermatol 1963; 40:121–122.
3. Dover, JS, Arndt KA, Geronemus RG, et al., eds. Understanding lasers. In: Illustrated Cutaneous Laser Surgery, A Practitioner's Guide. Norwalk, CT: Appleton and Lange, 1990: 1–19.
4. Tan OT, Murray S, Kurban AK. Action spectrum of vascular specific injury using pulsed irradiation. J Invest Dermatol 1989; 92:868–871.
5. Anderson RR, Parrish JA. The optics of human skin. J Invest Dermatol 1981; 77:13–19.
6. Anderson RR, Parrish JA. Selective photothermolysis: precise microsurgery by selective absorption of pulsed radiation. Science 1983; 220:524–527.
7. Tan OT, Kerschmann R, Parrish JA. The effect of epidermal pigmentation on selective vascular effects of pulsed laser. Lasers Surg Med 1985; 4:365–374.
8. Tan OT, Carney M, Margolis R, et al. Histologic responses of portwine stains treated by argon, carbon dioxide and dye lasers: a preliminary report. Arch Dermatol 1986; 122:1016–1022.
9. van Gemert MJ, Welch AJ, Amin AP. Is there an optimal laser treatment for portwine stains? Lasers Surg Med 1986; 6:76–83.
10. Anderson RR, Parrish JA. Microvasculature can be selectively damaged using dye lasers: a basic theory and experimental evidence in human skin. Lasers Surg Med 1981; 1:263–276.
11. Garden JM, Tan OT, Parrish JA. The pulsed dye laser use at 577 nm wavelength. J Dermatol Surg Oncol 1987; 13:134–138.
12. Tan OT, Sherwood K, Gilchrest BA. Treatment of children with portwine stains using the flashlamp-pumped tunable dye laser. N Engl J Med 1989; 320:416–421.
13. Glassberg E, Lask GP, Tan EML, Uitto J. The flashlamp-pumped 577-nm pulsed tunable dye laser: clinical efficacy and in vitro studies. J Dermatol Surg Oncol 1988; 14:1200–1208.
14. Tan OT, Polla LL, Garden JM. The treatment of port-wine stains with the pulsed dye laser. Arch Dermatol 1988; 124:889–896.
15. Polla LL, Tan OT, Garden JM, Parrish JA. Tunable pulsed dye laser for the treatment of benign cutaneous vascular ectasia. Dermatologica 1987; 144:11–17.
16. Geronemus RG. Poikiloderma of Civatte. Arch Dermatol 1990; 126:547–548.

17. Morelli JG, Tan OT, Garden JM, et al. Tunable dye laser (577 nm) treatment of port wine stains. Lasers Surg Med 1986; 6:94–99.
18. Tan OT, Stafford TJ. Treatment of portwine stains at 577 nm. Med Instrum 1987; 21:218–221.
19. Garden JM, Polla LL, Tan OT. The treatment of portwine stains by the pulsed dye laser. Arch Dermatol 1988; 124:889–896.
20. Garden JM, Tan OT, Kerschmann R, et al. Effect of dye laser pulse duration on selective cutaneous vascular injury. J Invest Dermatol 1986; 87:653–657.
21. Nakagawa H, Tan OT, Parrish JA. Ultrastructural changes in human skin after exposure to pulsed laser. J Invest Dermatol 1985; 84:396–400.
22. Paul BS, Anderson RR, Jarve J, Parrish JA. The effect of temperature and other factors on selective microvascular damage caused by pulsed dye laser. J Invest Dermatol 1983; 81:333–336.
23. Tan OT, Kerschmann R, Parrish JA. Effect of skin temperature on selective vascular injury caused by pulsed dye laser irradiation. J Invest Dermatol 1985; 85:441–444.
24. Anderson RR, Jaenicke KF, Parrish JA. Mechanisms of selective vascular changes caused by dye lasers. Lasers Surg Med 1983; 3:211–215.
25. Ashinoff R, Geronemus RG. Thrombocytopenia-absent radii syndrome and lack of response to the pulsed dye laser. Arch Dermatol. 1990; 126:1520–1521.
26. Tan OT, Whitaker D, Garden JM, Murphy G. Pulsed dye laser (577 nm) treatment of portwine staines: ultrastructural evidence of neovascularization and mast cell degranulation in healed lesions. J Invest Dermatol 1988; 90:395–398.
27. Scheibner A, Wheeland RG. Argon-pumped tunable dye laser therapy for facial port-wine stain hemangiomas in adults—a new technique using small spot size and minimal power. J Dermatol Surg Oncol 1989; 15:277–289.
28. Reyes BA, Geronemus R. Treatment of port wine stains with the flashlamp-pumped dye laser during childhood. J Am Acad Dermatol 1990, 23.1142–1148.
29. Ashinoff R, Geronemus RG. Capillary hemangiomas and treatment with the flash-lamp-pumped pulsed dye laser. Arch Dermatol 1991; 127:202–205.
30. Lanigan SW, Cotterill JA. Psychological disabilities amongst patients with port wine stains. Br J Dermatol 1989; 121:209–215.
31. Maiman TH. Stimulated optical radiation in ruby. Nature 1960; 187:493–494.
32. Goldman L, Wilson RG, Hornby P, Meyer RG. Radiation from a Q-switched ruby laser. Effect of repeated impacts of power output of 10 megawatts on a tattoo of man. J Invest Dermatol 1965; 44:69–71.
33. Polla LL, Margolis RJ, Dover JS, et al. Melanosomes are a primary target of Q-switched ruby laser irradiation in guinea pig skin. J Invest Dermatol 1987; 89:281–286.
34. Murphy GF, Shepard RS, Paul BS, et al. Organelle-specific injury to melanin-containing cells in human skin by pulsed laser irradiation. Lab Invest 1983; 49:680–685.
35. Anderson RR, Margolis RJ, Watenabe S, et al. Selective photothermolysis of cutaneous pigmentation by Q-switched Nd:YAG laser pulses at 1064, 532, 355 nm. J Invest Dermatol 1989; 93:28–32.
36. Yules RB, Laub DR, Honey R, et al. The effect of Q-switched ruby laser radiation on dermal tattoo pigment in man. Arch Surg 1967; 95:179–180.

37. Laub DR, Yules RB, Arras M, et al. Preliminary histopathological observation of Q-switched ruby laser radiation on dermal tattoo pigment in man. J Surg Res 1968; 8:220–224.
38. Reid WH, McLeod PJ, Ritchie A, Ferguson-Pell M. Q-switched ruby laser treatment of black tattoos. J Plast Surg 1983; 36:455–459.
39. Vance CA, McLeod PJ, Reid WH, et al. Q-switched ruby laser treatment of tattoos: a further study. Lasers Surg Med 1985; 5:179.
40. Reid WH, Murphy MJ, Miller ID, Paul JP. Experiences with the pulsed ruby laser treatment of tattoos. Lasers Surg Med (Suppl) 1990; 2:50.
41. Dover JS, Arndt KA. Dermatology. JAMA 1990; 263:2633–2635.
42. Taylor CR, Gange RW, Dover JS, et al. Treatment of tattoos by Q-switched ruby lasers: a dose–response study. Arch Dermatol 1990; 126:893–899.
43. Scheibner A. Removal of tattoos and benign pigmented lesions using the ruby laser. Lasers Surg Med (Suppl) 1990; 2:51.
44. Dover JS, Margolis RJ, Polla LL, et al. Pigmented guinea pig skin irradiated with Q-switched ruby laser pulses. Arch Dermatol 1989; 125:43–49.
45. Goldman L, Wilson R, Hornby P, Meyer R. Laser radiation of malignancy in man. Cancer 1965; 10:533–545.
46. Ohsiri T, Maruyama Y. The ruby and argon lasers in the treatment of naevi. Ann NY Acad Med 1983; 12:388–395.
47. Goldman L. Optical radiation hazards to the skin. In: Sliney D, Wolbarsht, eds. Safety with Lasers and Other Optical Sources: A Comprehensive Handbook. New York: Plenum Press, 1983: 167–169.
48. Garden JM, O'Banion K, Shelnitz LS, et al. Papillomavirus in the vapor of carbon dioxide laser-treated verrucae. JAMA 1988; 259:1199–1202.
49. Bandieramonte G, Chiesa F, Lupi M. Lasers microsurgery oncology: indications, techniques and results of 5-year experience. Lasers Surg Med 1987; 7:478–486.
50. Bilik R, Kahanov CL, Rubin S. Morbidity and recurrence rate after surgical treatment of malignant melanoma by scalpel versus $CO_2$ laser beam. Surg Gynecol Obstet 1987; 1:333–338.
51. Bailin P. Use of the $CO_2$ laser for non-port-wine stain cutaneous lesions. In: Arndt K, et al., eds. Cutaneous Laser Therapy: Principles and Methods. New York: John Wiley and Sons, 1983: 187–199.
52. Apfelberg DB, Maser M, White D, Lash H. Failure of carbon dioxide laser excision of keloids. Lasers Surg Med 1989; 9:382–389.
53. Bailin PL, Ratz JL, Lutz-Nagey L. $CO_2$ laser modification of Mohs surgery. J Dermatol Surg Oncol 1981; 7:621–623.
54. Lejeune FJ, Van Hoof G, Gerard A. Impairment of skin graft after $CO_2$ laser surgery in melanoma patients. Br J Surg 1980; 67:318–320.
55. McBurney EI, Rosen DA. Carbon dioxide laser treatment of verrucae vulgares. J Dermatol Surg Oncol 1984; 10:45–48.
56. Ratz JL. Carbon dioxide laser treatment of balanitis xerotica obliterans. J Am Acad Dermatol 1984; 10:925–928.
57. Baldwin H, Geronemus R. Carbon dioxide laser vaporization of Zoon's balanitis: a case report. J Dermatol Surg Oncol 1989; 15:491–494.
58. Greenbaum S, Glogau R, Stegman S, Tromovitch T. Carbon dioxide laser treatment of erythroplasia of Queyrat. J Dermatol Surg Oncol 1989; 15:747–754.

59. Bain L, Geronemus R. The association of lichen planus of the penis with squamous cell carcinoma in situ and verrucous squamous cell carcinoma. J Dermatol Surg Oncol 1989; 15:413–417.
60. Stanley RJ, Roenigk RK. Actinic cheilitis: treatment with the carbon dioxide laser. Mayo Clin Proc 1988; 63:230–235.
61. Leshin B, Whitaker DL. Carbon dioxide laser matrixectomy. J Dermatol Surg Oncol 1988; 14:608–611.
62. Huener CJ, Wheeland RL, Bailin PL, Ratz JL. Lasers: treatment of myxoid cysts with carbon dioxide laser vaporization. J Dermatol Surg Oncol 1988; 29:357–369.
63. Ratz JL, Bailin PL, Lakeland RF. Carbon dioxide laser treatment of epidermal nevi. J Dermatol Surg Oncol 1989; 15:413–417.
64. Bohigegian RK, Shapshy SR, Hybey RL. Management of rhinophyma with carbon dioxide laser: Lahey Clinic experience. Lasers Surg Med 1988; 8:397–401.
65. Garden JM, Geronemus RG. Dermatologic laser surgery. J Dermatol Surg Oncol 1990; 16:156–168.
66. Wheeland RG, Bailin PL, Reynolds O, Ratz JL. Carbon dioxide ($CO_2$) laser vaporization of multiple facial syringomas. J Dermatol Surg Oncol 1986; 12:223–228.
67. Wheeland RG, Bailin PL, Kronberg E. Carbon dioxide ($CO_2$) laser vaporization for the treatment of trichoepitheliomata. J Dermatol Surg Oncol 1984; 10:470–474.
68. Roenigk RK, Ratz JL. $CO_2$ laser treatment of cutaneous neurofibromas. J Dermatol Surg Oncol 1987; 13:187–190.
69. Bailin PL, Ratz JL, Levine HL. Removal of tattoos by $CO_2$ laser. J Dermatol Surg Oncol 1980; 6:997–1001.
70. Ruiz-Esparza J, Goldman M, Fitzpatrick R. Tattoo removal with minimal scarring: the chemo-laser technique. J Dermatol Surg Oncol 1988; 14:1372–1376.
71. Ratz J, Bailin P. The case for use of the carbon dioxide laser in the treatment of port-wine stains. Arch Dermatol 1987; 123:74–75.
72. Dover JS, Smoller BR, Stern RS, et al. Low-fluence $CO_2$ laser irradiation of lentigines. Arch Dermatol 1988; 124:1219–1224.
73. Cosman B. Experience in the argon laser therapy of port wine stains. Plast Reconstr Surg 1980; 65:119–129.
74. Dover J, Arndt K, Geronemus R, et al., eds. Illustrated Cutaneous Laser Surgery: A Practitioner's Guide. Norwalk, CT: Appleton & Lange, 1990: 81.
75. Achauer BM, VanderKam VM. Argon laser treatment of telangiectasia of the face and neck: 5 years' experience. Lasers Surg Med 1987; 7:495–498.
76. Arndt KA. Adenoma sebaceum: successful treatment with the argon laser. Plast Reconst Surg 1982; 70:91–93.
77. Wheeland RG, Bailin PL, Norris MG. Argon laser photocoagulation therapy of Kaposi's sarcoma: a clinical and histologic evaluation. J Dermatol Surg Oncol 1985; 11:1180–1184.
78. Arndt KA. Argon laser treatment of lentigo maligna. J Am Acad Dermatol 1984; 10:953–957.
79. Arndt KA. New pigmented macule appearing 4 years after argon laser. J Am Acad Dermatol 1986; 14:1092.
80. Apfelberg DB, Maser M, Lash H, et al. The argon laser for cutaneous lesions. JAMA 1981; 245:2073–2075.

22

# Photodynamic Therapy

**Allan R. Oseroff**
*Roswell Park Cancer Institute, Buffalo, New York*

## I. INTRODUCTION

A primary goal of all therapies is to affect targeted cells while sparing normal cells and tissues; the challenge is to maximize this selectivity. Photochemotherapy provides a means for achieving high selectivity. Photochemotherapy, (called photodynamic therapy or PDT when singlet oxygen is involved in the damage process), involves two components: light and light-activated compounds (photosensitizers), which are relatively innocuous in the dark but initiate toxic photochemical reactions upon absorbing light of the appropriate wavelength. Since damage requires the simultaneous presence of the photosensitizer and the light, this approach permits two ways to achieve selectivity: by spatial localization of the illumination and by preferential targeting of the photosensitizer to the desired sites. Photochemotherapy actually permits *double selectivity*. The amount of selectivity depends on both the degree of preferential accumulation of the photosensitizer in the targeted cells and the extent that the illumination is spatially localized to the target.

In some cases, one can use minimally selective photochemotherapy. For example, in the case of whole-body psoralen and ultraviolet A therapy (PUVA), illumination is not limited to the lesions, and the photosensitizer (psoralen) is not localized to specific tissue sites. Therapeutic effects are confined to skin because of the limited penetration of the UVA radiation, but both lesional and normal skin is affected. For a single PUVA treatment the dosages of photosensitizer and light are limited by what is an acceptable amount of phototoxic damage to normal skin. Multiple treatments are necessary to attain the desired therapeutic effect.

In contrast to PUVA, if we wish to cause enough photodamage to destroy lesions with a single treatment, it is necessary to increase selectivity through localization of the light or the photosensitizer. Either approach can be effective. Spatial control of the illumination is acceptable when the treated lesions are discrete and readily visualized, but is problematic when there are small, diffuse, or poorly delineated lesions interspersed with normal tissues. In general, it is more useful to rely on localizing the photosensitizer to the lesions, since it is then possible to treat without precisely knowing or defining the lesional boundaries. If photosensitizer localization is sufficiently precise it should be possible to use widespread illumination but still achieve microscopically selective photodamage: killing the targeted cells but sparing adjacent tissues that are also illuminated.

Photochemotherapy can be used for lesions accessible to light that is transmitted through the body surface or directed internally, via an optical fiber. Fibers may be inserted within tumors, permitting treatment of large masses. Thus, cutaneous, endobronchial, intra-abdominal, bladder, and central nervous system malignancies are amenable to this technique. The prototype, first-generation photosensitizer is hematoporphyrin derivative and the partially purified fraction of hematoporphyrin derivative called Photofrin II. Because hematoporphyrin derivative/Photofrin II produces singlet oxygen upon illumination (1), treatment with these compounds is appropriately called photodynamic therapy. The tumor-localizing properties of hematoporphyrin derivative were first described by Lipson (2), and subsequently by Dougherty (3). Photodynamic therapy with hematoporphyrin derivative or Photofrin II has been used on a variety of tumor types with significant clinical benefit (4–22, 35). Since 1977 about 5,000 patients have been treated with hematoporphyrin derivative or Photofrin II. Results are very encouraging with palliation of advanced tumors and cures of early disease. Phase III trials for bladder, lung, and esophageal cancers are currently underway. At our institution we have had excellent results in the treatment of several different types of cutaneous lesions including basal- and squamous-cell carcinomas (4, 14, 22, 35), recurrent or metastatic breast cancer (10), Kaposi's sarcoma, and cutaneous T-cell lymphoma.

Although it is clinically useful as a sensitizer, Photofrin II has several limitations. It is not a pure compound, so that pharmacokinetic studies and other analyses are difficult. Photofrin II has high uptake and prolonged retention in the reticuloendothelial system and in the skin, where it can cause delayed phototoxicity. Its optical properties are also not ideal; it has a relatively small absorbance near the lower end of the "therapeutic window" (see below). Many second-generation photosensitizer with potentially improved properties have been proposed, and some are entering clinical trials.

## II. UNDERLYING PRINCIPLES

### A. Photochemical Processes

Photosensitizer-induced reactions may involve formation of active oxygen species such as singlet oxygen ($^1O_2$*) or superoxide radical anion, cationic radicals, photoredox reactions of the photosensitizer with biomolecules, or light-induced breakdown products that are toxic to cells. As noted above, in the case of Photofrin II, the principal reaction seems to involve production of singlet oxygen ($^1O_2$*). In this multistep reaction, the photosensitizer absorbs a photon and initially goes from the ground state to an excited singlet state, and then, through intersystem crossing, to an excited triplet state. The energy in this triplet state is transferred to ground (triplet) state oxygen, resulting in high reactive, excited (singlet) state oxygen and a ground state photosensitizer. This process can be repeated when the photosensitizer absorbs another photon. Thus, as long as the illumination persists, a single photosensitizer molecule can produce multiple molecules of reactive oxygen in a quasi-"enzymatic" fashion. The amount of singlet oxygen formed depends on the product of photosensitizer concentration and the amount of illumination (fluence). Experiments with a photosensitizer linked to monoclonal antibodies, in which the quantity of sensitizer at the cell surface is known, suggest that on the order of $10^{10}$ single oxygen molecules are necessary to lethally damage a single cell (23). In the absence of photobleaching (see below), a simple reciprocal relationship occurs; lower concentrations of photosensitizer necessitate higher doses of light, and conversely.

#### 1. *Photodynamic Threshold Dose*

Both in culture and in animal models, photodynamic therapy-induced injury appears to exhibit a "threshold" for lethal damage. Below this threshold, the injury is reversible; above it, the damage is fatal. In culture this property is evidenced as a shoulder in the survival curve. *In vivo*, the threshold is apparent in histologic sections as a sharp boundary between living and dead cells. The phototoxic threshold ($PT_{th}$) can be expressed as a combination of the photosensitizer and light doses, which corresponds to a threshold level of photoxic damage, or a threshold *photodynamic dose*.

#### 2. *Photobleaching*

In addition to interacting with molecular oxygen to produce single oxygen, the excited-state photosensitizer molecule can undergo a side reaction leading to loss of absorbance and photosensitizing ability. This process of *photobleaching* can modify the reciprocity between photosensitizer level and light, since with irradiation there will be a progressive loss of sensitizer. Photobleaching therefore places a limit on the total amount of phototoxicity ($PT_{max}$) that can be produced by a given level of photosensitizer for a given light dose. While this can be a problem, it can also be an advantage. Photosensitizer near the surface of the tissue will absorb light and decrease the illumination at a depth. As photosensitizer near the surface bleaches, more light will penetrate more deeply into the tissue.

The combination of photobleaching and a threshold dosage for irreversible injury can also increase selectivity. For example, in the absence of photobleaching and threshold effects if there is an n-fold difference in photosensitizer accumulation between lesions and surrounding skin, there would be an identical n-fold difference in the resulting photodamage. However, this difference can be amplified by photobleaching. Because of photobleaching, there will be maximal phototoxicities $PT(lesion)_{max}$ and $PT(skin)_{max}$. If $PT(lesion)_{max} > PT_{th}$, while $PT(skin)_{max} < PT_{th}$, there can be complete destruction of the lesion without permanent damage to the surrounding normal skin, even if the differences in photosensitizer accumulation between the two sites is small.

## B. Mechanisms of Phototoxicity

The mechanism(s) for tumor localization is poorly understood, and the extent to which Photofrin II directly damages tumor cells *in vivo* is also unclear. Instead, experiments in rodent models suggest that the sensitizer may predominantly damage neovascular and vascular elements, with much of the tumor cell death secondary to lack of oxygen and nutrients (24–26). This is not necessarily bad. Vascular damage can be very effective for treating bulk disease (24, 25) and is useful for treating Kaposi's sarcoma and possibly beneficial for warts and psoriasis. However, photosensitizer relying on $^1O_2^*$ production will not be effective after vessel damage blocks tissue perfusion. Treatment failures with Photofrin II also may occur for tumor cells that have alternative oxygen and nutrient supplies (e.g., from adjacent normal tissue vessels that survive the circulatory injury) (24, 25). In humans the relative importance of vascular damage to the observed tumor destruction by Photofrin II is not yet clearly defined.

## C. Skin Optics

### 1. Absorption Bands of Endogenous Chromophores

For visible wavelengths, hemoglobin and melanin are the primary endogenous chromophores in skin. Hemoglobin has significant absorption bands near 425, 544, and 577 nm, so it is necessary to have illumination wavelengths >600 nm to have significant light penetration past the dermal capillary plexi. At wavelengths greater than about 1200 nm, light absorption by water molecules becomes substantial. Therfore, there is a 600–1200 nm "therapeutic window" that permits significant light penetration into tissue. Within this window, longer wavelengths penetrate more deeply because of decreasing absorbance (particularly from melanin) and decreasing light scattering. However, for wavelengths greater than 850–900 nm, the photons may not have sufficient energy to participate in photochemical reactions. Thus the available wavelengths for photodynamic sensitizers are 600–800 nm.

#### 2. *Photosensitizer Absorption*

Photofrin II has a long wavelength absorption peak at 630 nm. This is relatively low in the therapeutic window, and the light has a penetration depth (the depth at which the intensity of the incident light is reduced by 90%) on the order of 3 mm. Thus, superficial lesions can be readily treated with external irradiation. For thicker lesions, it is necesssary to use interstitial irradiation (see below). The topical photosensitizer meso-tetraphenylporphinesulfonate (TPPS) and protoporphyrin IX (PP) also absorb near 630 nm. Several second-generation compounds that are close to undergoing clinical trials absorb around 660 nm. A goal in developing new sensitizers is to find molecules that absorb near 800 nm, providing maximal tissue penetration and permitting the use of high-power diode lasers for irradiation (see below).

### D. Technical Issues

#### 1. *Dosimetry*

For Photofrin II doses above about 1 mg/kg, there is generally an excess of photosensitizer within the lesion, so the amount of photodynamic damage is controlled by limiting the amount of irradiation. To cause uniform photodamage, it is necessary to provide uniform illumination. The light distribution from the cut end of an optical fiber is not uniform, but it can be homogenized by attaching a microlens beam expander to the end of the fiber. As noted above, superficial irradiation is limited to relatively thin lesions because of the decrease in available light with increasing depth into the skin. Uniform superficial illumination may be difficult to achieve when irradiating curved skin surfaces or skin folds. Interstitial radiating fibers that emit light in a cylindrical distribution within the tissue can be used together with superficial irradiation for lesions more than about 3 mm thick.

In some cases, when the dosage of photosensitizer is low and/or when there is facile photobleaching, there may not be an excess amount of photosensitizer. Instead, the tissue photosensitizer level becomes the limiting factor for photodynamic damage. One then can apply essentially "unlimited" light and not worry as much about optical dosimetry or homogeneity of illumination.

#### 2. *Light Sources*

For superficial lesions treated with external irradiation, one can either use an argon-laser-pumped dye laser or a conventional, noncoherent source such as a xenon arc or tungsten–halogen lamp. Laser radiation can be focused into a very small spot, so the light can be readily coupled into optical fibers, which is a convenience, but the total power available is usually less than 3–4 watts. Since typical treatments employ 0.1–0.15 W/cm$^2$, only limited areas of the skin can be treated at one time. While conventional sources cannot be efficiently focused into small volumes or coupled into optical fibers, the total available power can exceed that of a dye laser, so these sources may be superior for treating large areas.

## III. APPLICATIONS

### A. General Principles

#### *1. Photosensitizers*

Most clinical studies have been done with hematoporphyrin derivative or, more recently, Photofrin II. Topical photodynamic therapy using TPPS has been examined in Italy (27), and there has been a recent interesting trial with the topical application of a photosensitizer precursor, δ-aminolevulinic acid, in Canada (28). Several trials of new, "second-generation" photosensitizer are beginning in the United States, Canada, and Europe.

#### *2. Types of Lesions*

The primary indication for photodynamic therapy is the treatment of malignant lesions. However, as discussed below, photodynamic therapy has also been employed on actinic keratoses, psoriasis, and laryngeal warts. Note that in the case of partially treated or recurrent disease, multiple treatment sessions are possible, and photodynamic therapy can be readily followed by conventional surgery.

### B. Clinical Experience

#### *1. Types of Lesions*

In more than 20 studies over the past 10 years, photodynamic therapy has been used for multiple types of cutaneous and subcutaneous malignancies including basal cell carcinoma (4, 7, 13, 14, 22, 28–33), widespread Bowen's disease (13, 30, 34, 35), squamous-cell carcinoma (14, 28–31, 36, 37), metastatic and recurrent breast carcinoma (4, 7, 10, 28, 30, 31), and Kaposi's sarcoma (38). Photodynamic therapy has also been used on melanomas (4, 30, 31), although light-dependent treatment of pigmented lesions or of patients with highly melanized skin is relatively contraindicated. Clearly one would never use photodynamic therapy for a primary melanoma. However it may occasionally be of some benefit for palliation of metastatic lesions. As noted above, there has also been limited experience with photodynamic therapy on patients with nonmalignant lesions, including actinic keratoses (28), psoriasis (39, 40), and laryngeal warts (41).

### 2. *Systemic Photodynamic Therapy for Malignant Lesions*

While the majority of the early trials of photodynamic therapy for cutaneous diseases have had encouraging results, they generally have been pilot or anecdotal studies without optimal documentation of pretreatment conditions or appropriate follow-up. In addition, the choices of HPD/Photofrin II and light doses have been variable. On the basis of current experience, it is likely that in some cases the total photodynamic dose was inadequate (29) or excessive (7). Nevertheless, it is evident from the work that photodynamic therapy can be of significant benefit.

*a. Experience at Roswell Park Cancer Institute.* The most careful studies of the dermatologic use of photodynamic therapy with systemic Photofrin II have been carried out by Wilson et al. at Rosewll Park Cancer Institute (14, 22). Using patients with basal cell carcinomas on the head, face, nose, neck, and torso, the investigators attempted to increase the therapeutic ratio by utilizing a relatively low, 1 mg/kg dosage of Photofrin II together with escalating light dosages ranging from 72 to 288 $J/cm^2$. These clinical trials were the first to attempt to exploit the beneficial effects of the photobleaching described above. The investigators reasoned that with a low dose of Photofrin II, photobleaching of the sensitizer in the epidermis and upper dermis would permit the use of large light dosages (giving effective therapy of deeper portions of the lesion) without significant damage to normal skin (42). This prediction appears to be at least partially confirmed. While there was complete necrosis of the lesions, normal skin within the treatment field generally did not undergo complete necrosis, although erythema and edema were produced (22). Cosmetic results were said to be "excellent," although there was transient hyperpigmentation in about 14% of the patients (22).

In these studies 151 sites were treated in 37 patients. For all but one site, superficial irradiation (using a microlens) was employed; one site received both superficial and interstitial radiation (22). Table 1 gives a summary of the results. As shown in the table, on the basis of an evaluation at 3 months there was an overall complete response rate of 88%, with 12% partial responses. Over a 12–30 month follow-up there was a 10% recurrence rate, with an average time to recurrence of 8.5 months (22). Eleven of the partial responses were retreated, with a 100% complete response and 18% (2 of 11) recurrences (22).

The majority of the treatment failures occurred on the nose, particularly for basal cell carcinoma of the morpheaform type (22). If the nose lesions are excluded, the complete response rate was 92%, with 8% partial responses and 2% recurrences (Table 1). These numbers may actually understate the success rate, since for several extensive or bulky lesions the treatment was deliberately extended over several sessions, producing a partial response by definition. The reasons for the high failure rate on the nose are unclear. Morpheaform-type basal-cell carcinomas may be resistant to photodynamic therapy, but Wilson et

**Table 1** Clinical Responses of Basal Cell Carcinoma Patients to Photodynamic Therapy[a]

| | Total Sites[a] | Complete Responses[b] | Partial Responses[b] | Recurrences[c] |
|---|---|---|---|---|
| All lesions | 151 | 133 (88%) | 18 (12%) | 24 (18%) |
| Sites excluding nose | 96 | 88 (92%) | 8 (8%) | 4 (2%) |
| Nose lesions | 55 | 45 (82%) | 10 (18%) | 20 (40%) |

[a]Thirty-seven patients.
[b]Evaluated 3 months after treatment.
[c]Follow-up period of 23–43 months (average, 29 months). Average time to recurrence was 15.8 ± 10 months.
Source: Refs. 14, 22.

al. speculate that geometric factors, including the curvature and skin folds around the nose, the thicker skin, and high density of sebaceous glands (which may scatter light), may also play a role (22). It is possible that carcinomas on the nose will require routine use of interstitial fibers during irradiation; recent data suggest that this approach significantly improves the outcome (B. D. Wilson, private communication).

*b. Photodynamic Therapy for Kaposi's Sarcoma.* Photodynamic therapy damages blood vessels, so it is logical to expect it to be effective for Kaposi's sarcoma. Results by Schweitzer (38) with oral human immunovirus-associated Kaposi's sarcoma and by Dougherty with classic and HIV-associated Kaposi's sarcoma (private communication) have been encouraging. Compared to ablative laser therapies, photodynamic therapy does not cause aerosolization or volatilization of tissue or viral products. Patient acceptance has been high.

### 3. Photodynamic Therapy for Nonmalignant Lesions

A concern in the application of photodynamic therapy to benign conditions is that the mutagenic and carcinogenic potential of the therapy has not been completely defined. To the extent that the majority of the effects are due to vascular damage, genotoxicity may not be a significant problem, but complete elucidation of mechanisms and effects in humans is still pending.

Photodynamic therapy with a topical precursor to an endogenous photosensitizer has been used with good success on actinic keratoses (28) (see below), and this may be a legitimate alternative to 5-fluorouracil for extensive disease.

One of the most intriguing applications of photodynamic therapy has been in the therapy of psoriasis. Photofrin II should preferentially accumulate in the vicinity of the abnormal microvasculature of the psoriatic lesion, so significant selectivity should be obtained. Irradiation can lead to vascular and cellular damage as well as to production/release of inflammatory mediators (43–45), so multiple therapeutic processes should be possible. In a preliminary study, McCullough and Weinstein treated patients with varying dosages of Photofrin II (0.5–0.75 mg/kg) and with light sources ranging from UVA to blue–green, to red wavelengths. They found that 3 of 12 patients treated once with 630 nm light had long-term remissions extending for more than 7 weeks (40). The treatment affected the psoriatic plaques but not normal skin. At a 0.5 mg/kg dosage of Photofrin II, no eschar formed on the treated lesion and photosensitivity was minimal (J. McCullough, private communication). Although this work needs replication, the approach may have significant potential.

Photodynamic therapy with Photofrin II can effectively destroy wart virus-infected keratinocytes (46), and this technique has been used clinically with encouraging success in patients with recalcitrant laryngeal papillomas (41). Therapeutic alternatives are scarce, so this appears to be an appropriate indication. The role of photodynamic therapy in digital and genital warts is uncertain.

### *4. Photodynamic Therapy with Topical Sensitizers*

Advantages of topical application of the photosensitizer are the avoidance of both systemic photosensitivity and the need for intravenous administration of a drug. Two quite different sensitizers and approaches to therapy have been used in significant clinical trials.

*a. Tetraphenylporphinesulfonate.* Over a 5 year period, Santoro and colleagues have evaluated the effectiveness of topical application of the tetrasodium salt of meso-tetraphenylporphinesulfonate (TPPS) (27). In a series of 292 patients with superficial basal-cell carcinomas that were less than 2 mm thick, three applications of a solution of TPPS in isopropyl alcohol/water/azone (50:48:2) at 24, 6, and 3 hr before treatment followed by 120–150 J/cm$^2$ laser radiation produced a 93.5% complete response rate with about a 20% relapse rate over a subsequent 24 month follow-up (27). About half of the relapses occurred at the margins of the treatment field, which might have been due to an initially inadequate extension of the irradiation field or to a poor choice of the margins of the topical application. In all, 28 patients had recurrences or persistent disease at the treatment margins. Twenty-six of these patients had a second treatment, with a 73% complete response rate. Cosmetic effects were said to be generally satisfactory, although many patients had persistent erythema at the treatment site for some months after therapy (27).

The major disadvantage of the technique is the apparent lack of selectivity of topical TPPS. Santoro et al. noted that it was necessary to use care to confine the painting of the TPPS within the boundaries of the lesion; this was quite tedious with the alcohol/water/azone vehicle (27). In addition, it is necessary to have a good sense for the clinical boundaries of the carcinoma, which may not always be possible.

*b. δ-Aminolevulinic Acid/Protoporphyrin IX.* Kennedy and his colleagues have investigated an alternative, ingenious approach to topical photodynamic therapy: the application of a metabolic *precursor* to an endogenous photosensitizer rather than the sensitizer itself. Kennedy has used topical application of δ-aminolevulinic acid, a precursor of protoporphyrin IX (28). Protoporphyrin IX is an efficient photosensitizer produced as the penultimate step in the biosynthesis of heme. All nonerythroid cells synthesize heme for use in various cytochromes. Normally, endogenous synthesis of δ-aminolevulinic acid is a rate-limiting step and control point in this pathway. If control is transiently bypassed by administration of excess δ-aminolevulinic acid, the cells will make excess protoporphyrin IX that will then be converted into heme. If the rate of conversion of δ-aminolevulinic acid to protoporphyrin IX is faster than that of protoporphyrin IX to heme, there will be a temporary accumulation of the photosensitizer. The relative rates of protoporphyrin IX *vs.* heme production vary with cell type.

Two factors make photodynamic therapy with endogenous protoporphyrin IX formed from topical δ-aminolevulinic acid selective. First, Kennedy et al. found that after topical or systemic administration of δ-aminolevulinic acid in mice, the epidermis and epidermal appendages accumulated protoporphyrin IX but the cells of the dermis, including the vascular endothelium, did not (47). Thus photodynamic therapy with endogenous protoporphyrin IX should selectively damage epidermally derived cells, including malignancies, while sparing the dermis. Photodynamic therapy-induced scarring should be minimal. Second, an additional degree of selectivity comes from the *topical* application of the δ-aminolevulinic acid. Because δ-aminolevulinic acid is a highly polar molecule, its penetration through intact human epidermis is very low. When the stratum corneum is abnormally permeable, as in actinic keratoses and superficial basal and squamous cell carcinomas, a significant amount of δ-aminolevulinic acid can penetrate to the lower epidermis and to the dermis (28). It should also be freely permeable into the base and edges of excisional wounds and into tumor beds (where the stratum corneum has been removed). Thus topographic selectivity can be obtained even if the δ-aminolevulinic acid is applied to both lesional and normal skin.

While protoporphyrin IX is an efficient photosensitizer, it is also very easily photobleached. Thus, there should not be problem with prolonged photoporphyrin IX-induced photosensitivity, because the endogenously produced protoporphyrin IX can decay by both light-independent conversion to heme and light-dependent conversion to "bleached protoporphyrin IX." As discussed above, an advantage of facile photobleaching is that it permits the use of high

## IV. FUTURE DEVELOPMENTS: POTENTIAL RESOLUTION OF SOME CURRENT LIMITATIONS

Many areas of active research are addressing some of the limitations and adverse reactions encountered with current therapy with Photofrin II.

### A. Persistent Photosensitivity

Lower Photofrin II dosages may lead to photosensitizer levels in normal skin that are near or below the threshold for phototoxicity, thus alleviating this problem. In addition, new photosensitizer are being developed with more rapid clearance and possibly higher selectivity.

### B. Limited Penetration of Light into Skin

There will be an increased use of interstitial fibers. More sophisticated placement algorithms are being developed for these fibers. New photosensitizer will have longer-wavelength absorption bands, giving up to two or three times the penetration depth of Photofrin II.

### C. Dosimetry

It is likely that for the same dosage schedule, the amount of photosensitizer that accumulates in a lesion varies from individual to individual and from lesion to lesion in the same patient. Because Photofrin II has a very large therapeutic ratio, there is a tendency to use a light dosage adequate for the lowest range of photosensitizer and to overtreat the other lesions. Instrumentation is currently being developed to measure the amount of photosensitizer in skin by its fluorescence or absorption spectra. These measurements may make it possible to individualize the light dose for each lesion. In addition, increasing evidence suggests that it is more appropriate to calculate the administered dose of sensitizer on the basis of body surface area rather than skin. This approach may lead to more effective therapy for children or small adults who have a smaller ratio of mass to surface area.

### D. Equipment

The argon–pumped-dye laser is bulky, expensive, and difficult to maintain. Laser Therapeutics has developed an alternative system that appears to be more practical and user-friendly. The Laser Therapeutics system employs a solid-state, frequency-doubled Nd: YAG laser (e.g. Laserscope "KTP laser") pumping a simplified, low-maintnenace dye laser. In addition, the next generation of photosensitizer may be at wavelengths where it is practical to use a diode laser to generate the light. Diode lasers are small and reliable, and may cost one-tenth of an argon–dye laser combination.

## REFERENCES


1. Weishaupt KR, Gomer CJ, Dougherty TJ. Identification of singlet oxygen as the cytotoxic agent in photoinactivation of a murine tumor. Cancer Res 1976; 36:2326–2631.
2. Lipson RL, Baldes EJ. The photodynamic properties of particular hematoporphyrin derivatives. Arch Dermatol 1960; 82:508–516.
3. Dougherty TJ, Grindey GB, Fiel R, et al. Photoradiation therapy II. Cure of animal tumors with hematoporphyrin and light. J Natl Cancer Inst 1975; 55:115–121.
4. Dougherty TJ. Photoradiation therapy for cutaneous and subcutaneous malignancies. J Invest Dermatol 1981; 77:122–124.
5. Hayata Y, Kata H, Konaka C, Ono J, Takizawa N. Hematoporphyrin derivative and laser photoradiation in the treatment of lung cancer. Chest 1982; 81:269–273.
6. Tsuchiya A, Obara N, Miwa M, et al. Hematoporphyrin derivative and laser photoradiation in the diagnosis and treatment of bladder cancer. J Urol 1983; 130:79–82.
7. Bandieramonte G, Marchesini R, Melloni E, Andreoli C, di Pietro S, Spinelli P, Fava G, Zunino F, Emanuelli H. Laser phototherapy following HpD administration in superficial neoplastic lesions. Tumori 1984; 70:327–334.
8. Grossweiner LI, Hill JH, Lobraico RV. Photodynamic therapy of head and neck squamous cell carcinoma: optical dosimetry and clinical trial. Photochem Photobiol 1987; 46:911–917.
9. Muller PJ, Wilson BG. Photodynamic therapy of malignant brain tumours: clinical effects, post-operative ICP, and light penetration of the brain. Photochem Photobiol 1987; 46:929–935.
10. Schuh M., Nseyo UO, Potter WR, Dao TL, Dougherty TJ. Photodynamic therapy for palliation of locally recurrent breast carcinoma. J Clin Oncol 1987; 5:1766–1770.
11. Manyak MJ, et al. Photodynamic therapy. J Clin Oncol 1988; 6:380–391
12. McCaughan JS Jr. Photodynamic therapy of malignant tumors. Prog Clin Biol Res 1988; 278:163–169.
13. Buchanan RB, Carruth JA, McKenzie AL, Williams SR. Photodynamic therapy in the treatment of malignant tumours of the skin and head and neck. Eur J Surg Oncol 1989; 15:400–406.
14. Wilson BD, Mang TS, Cooper M, Stoll H. Use of photodynamic therapy for the treatment of extensive basal cell carcinomas. Facial Plast Surg 1989; 6:185–189.
15. Calzavara F, Tomio L, Corti L, Zorat PL, Barone I, Peracchia A, Norberto L, D'Arcais RF, Berti F. Oesophageal cancer treated by photodynamic therapy alone or followed by radiation therapy. J Photochem Photobiol B 1990; 6:167–174.
16. Jocham D, Baumgartner R, Stepp H, Unsold E. Clinical experience with the integral photodynamic therapy of bladder carcinoma. J Photochem Photobiol B 1990; 6:183–187.
17. Karanov S, Kostadinov D, Shopova M, Kurtev P. Photodynamic therapy in lung and gastrointestinal cancers. J. Photochem Photobiol B 1990; 6:175–181.
18. Li JH, Guo ZH, Jin ML, Zhao FY, Cai WM, Gao ML, Shu MY, Zou J. Photodynamic therapy in the treatment of malignant tumours: an analysis of 540 cases. J Photochem Photobiol B 1990; 6:149–155.

19. Nseyo UO, Lundahl SL, Merrill DC. Whole bladder photodynamic therapy: critical review of present-day technology and rationale for development of intravesical laser catheter and monitoring system. Urology 1990; 36:398–402.
20. Patrice T, Foultier MT, Yactayo S, Douet MC, Maloisel F, Le Bodic L. Endoscopic photodynamic therapy with haematoporphyrin derivative in gastroenterology. J Photochem Photobiol B 1990; 6:157–165.
21. Windahl T, Andersson SO, Lofgren L. Photodynamic therapy of localised prostatic cancer (letter). Lancet 1990; 336:1139.
22. Wilson BD, Mang TS, Stoll H, Jones C, Cooper M, Dougherty TJ. Photodynamic therapy for the treatment of basal cell carcinoma. Arch Derm 1992 (in press).
23. Oseroff AR, Ohuoha D, Hasan T, Bommer JC, Yarmush ML: Antibody-targeted photolysis: selective photodestruction of human T-cell leukemia cells using monoclonal antibody-chlorin $e_6$ conjugates. Proc Natl Acad Sci USA 1986; 83:8744–8747.
24. Henderson BW, Doughtery TJ, Malone PB. Studies on the Mechanism of Tumor Destruction by Photoradiation Therapy. New York: Alan R Liss. 1984; 601–612.
25. Star WM, Marijnissen HPA, Hans AE, van den Berg-Blok AE, Versteeg JAC, Franken KAP, Reinhold HS. Destruction of rat mammary tumor and normal tissue microcirculation by hematoporphyrin derivative photoradiation observed in vivo in sandwich observation chambers. Cancer Res 1986; 46:2532–2540.
26. Selman SH, Kreimer Birnbaum M, Klaunig JE, Goldblatt PJ, Keck RW, Britton SL. Blood flow in transplantable bladder tumors treated with hematoporphyrin derivative and light. Cancer Res 1984; 44:1924–1927.
27. Santoro O, Bandieramonte G, Melloni E, Marchesini R, Zunino F, Lepera P, De Palo G. Photodynamic therapy by topical meso-tetraphenylporphinesulfonate tetrasodium salt administration in superficial basal cell carcinomas. Cancer Res 1990; 50:4501–4503.
28. Kennedy JC, Pottier RH, Pross DC. Photodynamic therapy with endogenous protoporphyrin IX: Basic principles and present clinical experience. J Photochem Photobiol B 1990; 6:143–148.
29. Pennington DG, Waner M, Knox A. Photodynamic therapy for multiple skin cancers. Plast Reconstr Surg 1988; 82:1067–1071.
30. McCaughan JS Jr, Guy JT, Hicks W, Laufman L, Nims TA, Walker J. Photodynamic therapy for cutaneous and subcutaneous malignant neoplasms. Arch Surg 1989; 124:211–216.
31. Gilson D, Ash D, Drirer I, Feather JW, Brown S. Therapeutic ratio of photodynamic therapy in the treatment of superficial tumors of skin and subcutaneous tissues in man. Br J Cancer 1988; 58:665–667.
32. Keller GS, Razum NJ, Doiron DR. Photodynamic therapy for nonmelanoma skin cancer. Facial Plast Surg 1989; 6:180–184.
33. Keller GS: Rational anatomical treatment of basal cell carcinoma with photodynamic therapy. In: Mang TS, ed. Optical Fibers in Medicine VI. Bellingham: SPIE, 1991.
34. Robinson PJ, Carruth JAS, Fairris GM. Photodynamic therapy: a better treatment for widespread Bowen's disease. Br J Dermatol 1988; 119:59–61.
35. Jones CM, Cooper M, Mang T, Wilson BD, Stoll HL. Photodynamic therapy in the treatment of unusual Bowen's Disease. J Am Acad Dermatol 1991 (in press).

36. Gross DJ, Waner M, Schosser RH, Dinehart SM. Squamous cell carcinoma of the lower lip involving a large cutaneous surface. Photodynamic therapy as an alternative therapy. Arch Dermatol 1990; 126:1148–1150.
37. Wolf P, Kerl H. Photodynamic therapy in patient with xeroderma pigmentosum. Lancet 1991; 337;1613–1614.
38. Schweitzer VG, Visscher D. Photodynamic therapy for treatment of AIDS-related oral Kaposi's sarcoma. Otolaryngol Head Neck Surg 1990; 102:639–649.
39. McCullough JL, Weinstein GD, Douglas JL, Berns MW. Photosensitizers in dermatology. Photochem Photobiol 1987; 46:77–82.
40. Weinstein GD, McCullough JL, Nelson JS, Berns MW. Low dose Photofrin II photodynamic therapy of psoriasis. J Invest Dermatol 1991; 96:573.
41. Abramson AL, Waner M, Brandsma J. The clinical treatment of laryngeal papillomas with hematoporphyrin therapy. Arch Otolaryngol Head Neck Surg 1988; 114:795–800.
42. Potter WR, Mang TS, Dougherty TJ. The theory of photodynamic dosimetry: consequences of the photodestruction of sensitizer. Photochem Photobiol 1987; 46:97–101.
43. Kamide R, Gigli I, Lim HW: Participation of mast cells and complement in the immediate phase of hematoporphyrin-induced phototoxicity. J Invest Dermatol 1984; 82:485–490.
44. He D, Soter NA, Lim HW. The late phase of hematoporphyrin derivative-induced phototoxicity in mice: release of histamine and histologic changes. Photochem Photobiol 1989; 50:91–95.
45. Nseyo UO, Whalen RK, Duncan MR, Berman B, Lundahl SL. Urinary cytokines following photodynamic therapy for bladder cancer. Urology 1990; 36:167–171.
46. Bernstein EF, Glass JM, DeGraff WG, Schlegel R, Black C, Fisher J, Cook SN, Glatstein E, Russo A, Mitchell JB. In vitro photodynamic treatment of normal and human papilloma virus-transfected keratinocytes with Photofrin II and red light. Arch Dermatol 1991; 127:683–687.
47. Divaris DXG, Kennedy JC, Pottier RH. Phototoxic damage to sebaceous glands and hair follicles of mice after systemic administration of δ-aminolevulinic acid correlates with localized protoporphyrin IX fluorescence. Am J Pathol 1990; 136:891–897.
48. Berlin NI, Neuberger A, Scott JJ. The metabolism of delta aminolevulinic acid. Biochem J 1956; 64:90–100.
49. Berlin NI, Neuberger A, Scott JJ: The metabolism of delta aminolevulinic acid. Biochem J 1956; 64:80–90.
50. Mullooly VM, Abramson AL, Shikowitz MJ. Dihematoporphyrin ether-induced photosensitivity in laryngeal papilloma patients. Lasers Surg Med 1990; 10:349–356.

# Index

# About the Editors

HENRY W. LIM is an Associate Professor of Dermatology at New York University School of Medicine, New York, New York, and Chief of Dermatology Service, at New York Veterans Affairs Medical Center, New York, New York. The author or coauthor of more than 60 book chapters, professional papers, and reviews, he is a Fellow of the American Academy of Dermatology and a member of the Society for Investigative Dermatology, the Dermatology Foundation, the American Dermatological Association, and the American Society for Photobiology, among other organizations. Dr. Lim received the M.D. degree (1975) from the State University of New York, Downstate Medical Center, Brooklyn, New York, and completed his residency training at the Department of Dermatology, New York University School of Medicine.

NICHOLAS A. SOTER is a Professor of Dermatology at New York University School of Medicine, New York, New York, and an attending physician at Tisch Hospital, the University Hospital of New York University, New York, New York. A founding member of the Skin Pharmacology Society and a member of the Society for Investigative Dermatology, the American Academy of Dermatology, the Dermatology Foundation, the American Dermatological Association, and the American Society for Photobiology, among other organizations, he is the author or coauthor of over 170 professional papers, book chapters, and reviews. Dr. Soter received the M.D. degree (1965) from the University of Texas Southwestern Medical School, Dallas, and completed his dermatology residency training at the Baylor University Affiliated Hospitals, Houston, Texas, and Harvard/Massachusetts General Hospital, Boston.